S. Brandt
H. D. Dahmen
Quantum Mechanics on the
Personal Computer

S. Brandt   H.D. Dahmen

Quantum Mechanics
on the
Personal Computer

With a Program Diskette,
69 Figures and 284 Exercises

Springer-Verlag Berlin Heidelberg New York
London Paris Tokyo Hong Kong

Professor Dr. Siegmund Brandt
Professor Dr. Hans Dieter Dahmen
Physics Department, Siegen University, P.O. Box 10 12 40
D-5900 Siegen, Fed. Rep. of Germany

ISBN 3-540-51541-0 Springer-Verlag Berlin Heidelberg New York
ISBN 0-387-51541-0 Springer-Verlag New York Berlin Heidelberg

Printing: Druckhaus Beltz, 6944 Hemsbach/Bergstr.
Binding: J. Schäffer GmbH & Co. KG., 6718 Grünstadt
2154/3150-543210 – Printed on acid-free paper

Preface

Ever since we published our *Picture Book of Quantum Mechanics* we have been asked to make available the programs we wrote to generate the computer graphics that illustrate the book. Supported by a study contract with IBM Germany we were able to improve and generalize the original set of programs considerably. We have called the result INTERQUANTA (the *Interactive Picture Program of Quantum Mechanics*), which we like to abbreviate further by **IQ**.

IQ consists of the program proper and various additional files. These allow classroom demonstrations of prerecorded problems with explanatory text and are an essential help in the organization of coursework. This book contains exercises worked out for a complete course we call *A Computer Laboratory on Quantum Mechanics*.

We have tried out the course with groups of students at Siegen University. Note that the students are not expected to have any knowledge of computer programming. The *Laboratory* is primarily meant to help students in their first encounter with quantum mechanics, but we have also found it useful for students who are already familiar with the basics of the theory. In our experience the *Laboratory* is best offered in parallel with a lecture course such as *Quantum Physics* or *Introductory Quantum Mechanics*.

It is a pleasure to acknowledge the generous help provided by IBM Germany for the preparation of INTERQUANTA. In particular we would like to thank Dr. U. Groh for his competent help with the computer hardware and the systems software.

The program was written mostly in FORTRAN 77, with a few routines in C, developed on an IBM 6150 RT PC computer and later adapted to other systems. The version published with this book runs on the IBM PC and PS/2 or compatible systems; see Appendix B for details. We have also tested versions for the IBM 6150 RT PC, VAX and Atari computers and are willing to make them available for use by universities on request. We should be grateful for comments on the use of the program and suggestions for improvements.

At various stages of the project we were helped considerably by friends and students in Siegen. We would particularly like to thank Martin S. Brandt, Karin Dahmen, Helge Meinhard, Martin Schmidt, Tilo Stroh, Clemens Stupperich and Dieter Wähner for their excellent work.

This book was typeset in our group with the typesetting program TEX by Donald E. Knuth. Our computer graphics coded as POSTSCRIPT files were integrated in the TEX file using a TEX \special command. The complete file was then printed by a POSTSCRIPT driver.

Siegen, Germany
July 1989

Siegmund Brandt
Hans Dieter Dahmen

Contents

1 Introduction

1.1 Interquanta

The language of quantum mechanics is needed to describe nature at the atomic or sub-atomic scale, e.g., the phenomena of atomic, nuclear, or particle physics. But there are many other fields of modern science and engineering in which important phenomena can be explained only by quantum mechanics, for example chemical bonds or the functioning of semiconductor circuits in computers. It is therefore very important for students of physics, chemistry, and electrical engineering to become familiar with the concepts and methods of quantum mechanics.

It is a fact, however, that most students find quantum mechanics difficult and abstract, much more so than classical point mechanics. One easily detects the reason for this by recalling how students learn classical mechanics. Besides learning from lectures they draw on their own experience, on experiments they perform in the laboratory, and on problems they solve on paper. The important concept of a mass point is nothing but that of a very small stone. The experience with throwing stones helps to understand mechanics. Additional experiments are very direct and simple and there is a wealth of problems which are easily solved.

All this is different with quantum mechanics. Although – for all we know – elementary particles are point-like, the concept of the trajectory of a mass point breaks down and has to be replaced by a complex probability amplitude. This function cannot be measured directly; its properties have to be inferred indirectly from experiments involving optical spectra or counting rates, etc. Finally, nearly all nontrivial problems pose severe computational difficulties and require approximative or numerical methods. Thus students can do only a few problems.

Many quantum-mechanical problems can, however, be quite quickly solved numerically by computer. The answer is often very easy to analyze if presented in graphical form. We have written an interactive program taking *alphanumeric input* defining quantum-mechanical problems and yielding *graphical output* to produce a large number of illustrations for an introductory text book on quantum mechanics[1].

Here we present an improved and generalized version of this program which we call INTERQUANTA (**IQ** for short) – the interactive picture program of quantum mechanics. **IQ** can be used in a *computer laboratory on quantum mechanics* and for *classroom demonstrations*.

In the laboratory course students define and solve a quantum-mechanical problem, examine the results, change some parameters in their problem, analyze the new answer

---

[1]S. Brandt and H.D. Dahmen (1985): *The Picture Book of Quantum Mechanics* (Wiley, New York)

given by the computer, etc. By using the program for many different types of problem they acquire experience in quantum mechanics, having performed computer experiments and at the same time having solved problems by numerical methods.

For classroom demonstrations prerecorded input data for selected examples can be used and series of such examples can be shown either fully automatically or by pressing just one key after each example.

1.2 The Structure of this Book

This book consists of a main text and several appendixes. Chapters 2 through 8 of the main text are devoted to the various physics subjects covered by **IQ**. Each chapter begins with a section "Physical Concepts" in which the relevant concepts and formulae are assembled without proofs. Although this book is in no way intended to be a textbook, this section is needed to allow a precise definition of what the program does and also a clear formulation of the exercises. The following sections of each chapter describe in detail the commands the user needs to tackle the physics problems of the chapter. Each chapter is concluded by a collection of exercises. Chapter 9 is devoted to the special functions of mathematical physics relevant to quantum mechanics. In Chap. 2 there is in addition a section "A First Session at the Computer" which in an informal way provides a minimum of general knowledge of **IQ**. The concept of an **IQ** *descriptor* is introduced. This is a set of input data completely defining the physics question and the graphical representation of the answer of a particular problem. Chapter 10 contains hints for the solution of some exercises. It begins with a section on the different possibilities for choosing units for physical quantities. Its content is useful in many exercises for determining numerical values of input parameters and for correctly interpreting the numerical results of computations by **IQ**.

Appendix A is a systematic guide to all **IQ** commands. The Appendixes B through D contain technical information on the installation of the program, on the use of the different graphics devices and related questions.

1.3 The Computer Laboratory

The laboratory is best done on a series of afternoons (at least two hours per session) in a room with several personal computers or graphics terminals or workstations with two students per workstation. Students should have a copy of this book available at each workstation and should also be able to study the book well in advance of sessions with **IQ**. An instructor should be present to answer questions on physics and also, in the beginning, on the use of **IQ**.

The course itself consists in working through (some of) the *exercises* given in the different sections of this book. For most of the exercises an *initial descriptor* is provided with properly chosen graphics and physical parameters. The students are asked to run **IQ** with this descriptor, study the graphical output, and answer questions for which they usually have to change some parameter(s), run **IQ** again, etc. At any stage they can store away their changed descriptors for later use. They can also take hardcopies of all graphical output to perform measurements on or simply to take home. In many exercises it is intended to draw the attention of the student to a particular feature of the plot. This is usually attempted by asking a question that can in most cases be answered by qualitative arguments. Most of these are answered in Chap. 10. Of course, instructors (and students) may define and solve problems not contained in the lists of exercises given.

1.4 The Classroom Demonstrations

IQ provides ready-made classroom demonstrations for the following physics topics:

- free particle motion in one dimension
- bound states in one dimension
- scattering in one dimension
- two-particle systems
- free particle motion in three dimensions
- bound states in three dimensions
- scattering in three dimensions
- functions of mathematical physics

Each demonstration contains many example plots and explanatory text. See Appendix A.5.5.5 for how to run a demonstration and Appendix C.4 for a list of demonstration files. It is easy for an instructor to prepare his or her own demonstration files.

1.5 Literature

Since in the introductory sections "Physical Concepts" we present only a very concise collection of concepts and formulae, the user of **IQ** is urged to study the physics topics in more detail in the textbook literature. Under the heading *'Further Reading'* at the end of our introductory sections we refer the user to the relevant chapters in the following textbooks:

Abramowitz, M., Stegun, I.A. (1965): *Handbook of Mathematical Functions* (Dover Publications, New York)
Alonso, M., Finn, E.J. (1968): *Fundamental University Physics*, Vols. 1–3, (Addison-Wesley, Reading, Mass.)
Kittel, C., Knight, W.D., Ruderman, M.A., Purcell, E.M., Crawford, F.S., Wichmann, E.H., Reif, F. (1965): *Berkeley Physics Course*, Vols. I–IV (McGraw-Hill, New York)
Brandt, S., Dahmen, H.D. (1985): *The Picture Book of Quantum Mechanics* (John Wiley and Sons, New York)
Feynman, R.P., Leighton, R.B., Sands, M. (1965): *The Feynman Lectures on Physics*,

Vols. 1–3 (Addison-Wesley, Reading, Mass.)

Flügge, S. (1971): *Practical Quantum Mechanics*, Vols. 1,2 (Springer-Verlag, Berlin, Heidelberg)

Gasiorowicz, S. (1974): *Quantum Physics* (John Wiley and Sons, New York)

Hecht, E., Zajac, A. (1974): *Optics* (Addison Wesley, New York)

Merzbacher, E. (1970): *Quantum Mechanics (second edition)* (John Wiley and Sons, New York)

Messiah, A. (1970): *Quantum Mechanics*, Vols. 1,2 (North-Holland Publishing Company, Amsterdam)

Schiff, L.I. (1968): *Quantum Mechanics (third edition)* (McGraw-Hill, New York)

2 Free Particle Motion in One Dimension

Contents: Description of a particle as a harmonic wave of sharp momentum and as a wave packet with a Gaussian spectral function. Approximation of a wave packet as a sum of harmonic waves. Analogies in optics: harmonic light waves and light wave packets. Discussion of the uncertainty principle.

2.1 Physical Concepts

2.1.1 Planck's Constant. Schrödinger's Equation for a Free Particle

The fundamental quantity setting the scale of quantum phenomena is *Planck's constant*

$$h = 6.626 \times 10^{-34} \text{Js} \quad , \quad \hbar = h/2\pi \quad .$$

A *free particle* of mass m and velocity v travelling in the x direction with momentum $p = mv$ and kinetic energy $E = p^2/2m$ has a *de Broglie wavelength* $\lambda = h/p$. The harmonic *wave function*

$$\psi_p(x, t) = \frac{1}{(2\pi\hbar)^{1/2}} \exp\left[-\frac{i}{\hbar}(Et - px)\right] \tag{2.1}$$

is called a Schrödinger wave. It has the *phase velocity*

$$v_p = E/p = p/2m \quad .$$

Schrödinger waves are solutions of the *Schrödinger equation for a free particle*

$$i\hbar \frac{\partial}{\partial t}\psi_p(x, t) = -\frac{\hbar^2}{2m}\frac{\partial^2}{\partial x^2}\psi_p(x, t) \quad \text{or} \quad i\hbar\frac{\partial}{\partial t}\psi_p(x, t) = T\,\psi_p(x, t) \tag{2.2}$$

with

$$T = -\frac{\hbar^2}{2m}\frac{\partial^2}{\partial x^2}$$

being the *operator of the kinetic energy*.

2.1.2 The Wave Packet. Group Velocity. Normalization

Since the equation is linear in ψ_p a *superposition*

$$\psi(x,t) = \sum_{n=1}^{N} w_n \psi_{p_n}(x,t) \tag{2.3}$$

of harmonic waves ψ_{p_n} corresponding to different momenta p_n each weighted by a factor w_n also solves the Schrödinger equation. Replacing the sum by an integral we get the wave function of a *wave packet*

$$\psi(x,t) = \int_{-\infty}^{\infty} f(p)\psi_p(x-x_0,t)dp \quad , \tag{2.4}$$

which is determined by the *spectral function* $f(p)$ weighting the different momenta p. In particular, we consider a *Gaussian spectral function*

$$f(p) = \frac{1}{(2\pi)^{1/4}\sqrt{\sigma_p}}\exp\left[-\frac{(p-p_0)^2}{4\sigma_p^2}\right] \tag{2.5}$$

with *mean momentum* p_0 and *momentum width* σ_p.

   Introducing (2.5) into (2.4) we get the wave function of the *Gaussian wave packet*

$$\psi(x,t) = M(x,t)e^{i\phi(x,t)} \tag{2.6}$$

with the *amplitude function*

$$M(x,t) = \frac{1}{(2\pi)^{1/4}\sqrt{\sigma_x}}\exp\left[-\frac{(x-x_0-v_0t)^2}{4\sigma_x^2}\right] \tag{2.7}$$

and the *phase*

$$\phi(x,t) = \frac{1}{\hbar}\left[p_0 + \frac{\sigma_p^2}{\sigma_x^2}\frac{t}{2m}(x-x_0-v_0t)\right](x-x_0-v_0t) + \frac{p_0}{2\hbar}v_0t + \frac{\alpha}{2} \tag{2.8}$$

with *group velocity*

$$v_0 = \frac{p_0}{m} \quad , \tag{2.9}$$

localization in space given by

$$\sigma_x^2 = \frac{\hbar^2}{4\sigma_p^2}\left(1 + \frac{4\sigma_p^4}{\hbar^2}\frac{t^2}{m^2}\right) \quad ,$$

and

$$\tan\alpha = \frac{2}{\hbar}\frac{\sigma_p^2}{m}t \quad .$$

The initial spatial width at $t=0$ is thus $\sigma_{x0} = \hbar/(2\sigma_p)$. In terms of σ_{x0} the time-dependent width becomes

$$\sigma_x^2 = \sigma_{x0}^2\left(1 + \frac{\hbar^2}{4\sigma_{x0}^4}\frac{t^2}{m^2}\right) \quad . \tag{2.10}$$

The absolute square of the wave function

$$\varrho(x,t) =\mid \psi(x,t)\mid^2 = M^2(x,t) = \frac{1}{\sqrt{2\pi}\sigma_x}\exp\left[-\frac{[x-(x_0+v_0t)]^2}{2\sigma_x^2}\right] \qquad (2.11)$$

is the *probability density* for observing the particle at position x and time t. The particular normalization of the Gaussian (2.5) was chosen to ensure the normalization

$$\int_{-\infty}^{\infty}\varrho(x,t)dx = 1 \qquad (2.12)$$

of the probability density. It expresses the fact that there is just one particle in the domain $-\infty < x < \infty$. The widths σ_x and σ_p of the wave packet are connected by *Heisenberg's uncertainty relation*

$$\sigma_x\sigma_p \geq \hbar/2 \quad , \qquad (2.13)$$

the equality holding for a Gaussian wave packet and $t=0$ only.

2.1.3 Analogies in Optics

We now briefly consider also a harmonic electromagnetic wave propagating in vacuum in the x direction. The *electric field strength* is (written as a complex quantity – the physical field strength is its real part)

$$E_k(x,t) = E_0 \exp[-\mathrm{i}(\omega t - kx)] \quad , \qquad (2.14)$$

where the *angular frequency* ω and the *wave number* k are related to the *velocity of light in vacuum* c and the *wavelength* λ by

$$\omega = c\mid k\mid \quad , \quad \lambda = \frac{2\pi c}{\omega} \quad . \qquad (2.15)$$

In analogy to (2.3) and (2.4) we can again form superpositions of harmonic waves as a weighted sum

$$E(x,t) = \sum_{n=1}^{N} w_n E_{k_n}(x,t) \qquad (2.16)$$

or as a wave packet

$$E(x,t) = \int_{-\infty}^{\infty} f(k)E_k(x,t)dk \quad . \qquad (2.17)$$

As spectral function for the wave number k we choose a Gaussian function which is slightly different from the form (2.5)

$$f(k) = \frac{1}{\sqrt{2\pi}\sigma_k}\exp\left[-\frac{(k-k_0)^2}{2\sigma_k^2}\right] \quad . \qquad (2.18)$$

Integration of (2.17) yields

$$E = E_0 \exp\left[-\frac{\sigma_k^2}{2}(ct-x)^2\right]\exp[-\mathrm{i}(\omega_0 t - k_0 x)] \qquad (2.19)$$

and

$$| E |^2 = E_0^2 \exp\left[-\frac{(ct - x)^2}{2\sigma_x^2}\right] \quad , \tag{2.20}$$

with the relation

$$\sigma_x \sigma_k = \tfrac{1}{2} \tag{2.21}$$

between the spatial width σ_x and the width σ_k in wave number of the electromagnetic wave packet. The importance of (2.20) becomes clear through the relation

$$w = \frac{\varepsilon_0}{2} | E |^2 \quad , \tag{2.22}$$

where w is the *energy density* of the electromagnetic field (averaged over a short period of time) and ε_0 is the electric field constant.

Further Reading

Alonso, Finn: Vol. 2, Chaps. 18,19; Vol. 3, Chaps. 1,2
Berkeley Physics Course: Vol. 3, Chaps. 4,6; Vol. 4, Chaps. 5,6,7
Brandt, Dahmen: Chaps. 2,3
Feynman, Leighton, Sands: Vol. 3, Chaps. 1,2
Flügge: Vol. 1, Chap. 2
Gasiorowicz: Chaps. 2,3
Hecht, Zajac: Chaps. 2,7
Merzbacher: Chaps. 2,3
Messiah: Vol. 1, Chaps. 1,2
Schiff: Chaps. 1,2

2.2 A First Session with the Computer

We assume that you are working in a directory in which the program and the other necessary files exist. If they do not, follow the Installation Guide in Appendix B.

2.2.1 Starting IQ

You start the program by typing

    IQ<RET>

(You have to press the RETURN key after each line of input as indicated by the symbol <RET>. We shall in most cases omit writing <RET> where the use of the RETURN key is obvious.) The program displays a few lines of welcome and then answers with

    IQ>

This is the *prompt* you will usually get when the program asks you for input. In the *interactive mode*, which is the usual mode of running, you type a *command* and **IQ** either answers with the prompt `IQ>` or by presenting the result of a computation in graphical form. In this way you have a dialog with **IQ**. In Appendix A all commands are explained in detail. You may study this appendix now but you will probably prefer to look up explanations on individual commands as you need them.

2.2.2 An Automatic Demonstration

Rather than carrying out the *dialog* with **IQ** yourself you may instead just watch the dialog resulting from a list of commands which are prerecorded in a file. Just type

    AM

which is short for *automatic mode*. **IQ** asks you with the prompt

    ENTER 3 CHARACTERS TO COMPLETE INPUT FILE NAME IQ???.INP>

to complete an input file name. Now enter the 3 characters D12 followed by <RET>, lean back, relax, and watch our demonstration. It ends by again displaying the prompt `IQ>`.

2.2.3 A First Dialogue

First you have to choose a *descriptor file* (see Appendix A.1.3) by typing

    CD 012

You may then *list* the titles of the *descriptors* (see Appendix A.1.3.1) in that file by typing

    LT

get the first *descriptor* from disk into memory by typing

    GD 1

and ask **IQ** to *plot* the result of a computation defined by this descriptor by typing

    PL

Your display will now switch to graphics mode and present the contents of Fig. 2.1. Before you can give new input you have to press <RET> to switch the display back to alphanumeric mode.

The descriptor consists of numerical and textual information defining in all detail a problem which is solved by the computer once you give the PL command. You may change that information and with the next PL command the computer will take these changes into account. To *list* the contents of the *descriptor*, type

    LD

The complete list will probably not fit on your screen. You may therefore *list* the two halves of the descriptor separately, the *parameter* part (which you usually want to see) with

        LP

and the *background* part with

        LB

The numerical information in the descriptor comes in sets of 4 numbers, each set preceded by a two-character code, e.g.,

        V0 .000 .000 .000 .000   V1 1.000 -.500 -5.000 .000

To change the first of the 4 numbers in the set called V1 to 2.000 type

        V1(1)4

(for the most general form of command and input see Appendix A.1.2). Asking for a new plot with the PL command you will see that the wave packet shown in Fig. 2.1 now moves twice as fast. In fact the variable we have changed determines the momentum p of the wave packet.

    Now type the two lines

        V0(1) 1
        PL

and you will be shown the real part of the wave function rather than its absolute square. With

        V0(1) 2
        PL

you will get the imaginary part.

    You may also try the command

        NG 0 0 1 1

followed by PL. You will get 4 plots with different pairs of values for p_0 and σ_p. Consult Appendix A.5.2 for details of the NG command which allows you to plot a certain *number of graphs*.

    Having discussed how to change physical parameters we now draw your attention to some of the graphics facilities. First prepare the descriptor to plot a single graph again by typing

        NG 0 0 0 0

Now type

        NL(2) 3
        PL

and you will observe that the wave packet is now drawn for 3 instances of time only. For details of the NL command see Appendix A.3.2.

    In Fig. 2.1 the scale in x extends from -10 to 10. Without changing the size of the plot you may change it to span the interval from -20 to 20 by typing

```
XX(3:4) -20 20
PL
```

You may also change the actual extension of the plot in the x direction by

```
XX(1:2) -10 0
PL
```

For details of the different coordinate systems used by **IQ** and the commands defining them consult Appendix A.2. Note in particular that the three-dimensional coordinate system which is used for physics computations is called the *C3 coordinate* system and denoted by x, y, z. In our example of Fig. 2.1 x corresponds to the position x, y to the time t, and z to the absolute square of the wave function $| \psi(x, t) |^2$.

To *stop* the program you just type the command

```
ST
```

After the message IQ TERMINATED on your display, control is handed back to the operating system of your computer.

2.2.4 A Little Systematics

After this very informal introduction to the dialog with **IQ** we have to get a little more systematic. Start the program again and list a descriptor on your display (or just look at Table A.1). After a line of heading and three lines of text (see Appendix A.4.4) it contains 40 two-letter commands each followed by a set of 4 numbers, which can be divided into four groups: choice part, graphics part, physics part, background part.

The choice part consists of the CH command only, which has the general form

```
CH    n₁    n₂    n₃    n₄
```

The parameter n_1 determines the *plot type* (see Appendix A.3). With the parameters n_2, n_3, n_4 you choose the particular type of physics problem.

The graphics part, comprising the commands NG through BO (see Appendix A.2), defines the graphical appearance of the plot resulting from the PL command.

The physics part, comprising the commands V0 through V9, contains all parameters needed to define in detail the physics problem you have chosen.

The background part (see commands A1 through R2 and texts T1 through F4, Appendix A.4) enables you to steer some features you will usually not want to change which make the plots more understandable, such as scales and arrows.

In the following three sections and in the corresponding sections of later chapters only the choice part and the physics part which you need to produce plots for the physics topics at hand are described. This description is somewhat formal but always follows the same scheme. Once you are accustomed to it you will find it fast to use.

While doing the exercises accompanying each chapter you will always work with a descriptor file corresponding to the physics topics of that chapter. Use the descriptors quoted in the exercises. They are prepared in such a way that you have only to change very few physics-related parameters. Of course, you are free to create complete descriptors and even descriptor files on your own but we suggest that you do that only after you have gained some experience.

2.3 The Time Development of a Gaussian Wave Packet

Aim of this section: Demonstration of the propagation in space and time of a Gaussian wave packet of Schrödinger waves (2.6) and of electromagnetic waves (2.19).

Plot type: 0

C3 coordinates: x: position coordinate x y: time coordinate t
z: $|\Psi|^2$, Re Ψ, or Im Ψ, or $|E|^2$, Re E, or Im E

Input parameters:

    CH 0 12 0 f_{QO}
    V0 f_{RES} V1 q σ_q x_0 f_{pE}
    V2 Δq $\Delta\sigma_q$ $\hbar$ m
    V6 l_{DASH} R

with

$f_{QO} = 0$: quantum-mechanical wave packet

$f_{QO} = 1$: optical wave packet

For $f_{QO} = 0$ the input parameters have the following significance:

$f_{RES} = 0$: function shown is $|\Psi|^2$

$f_{RES} = 1$: function shown is Re Ψ

$f_{RES} = 2$: function shown is Im Ψ

$p_0 = q + h\Delta q$ for $f_{pE} = 0$

$p_0 = \sqrt{2m(q + h\Delta q)}$, for $f_{pE} = 1$, i.e., the input quantities q and Δq are kinetic energies

$\sigma_p = (\sigma_q + v\Delta\sigma_q)p_0$ for $\sigma_q > 0$, i.e., σ_q and $\Delta\sigma_q$ are given in units of p_0

$\sigma_p = |\sigma_q| + v|\Delta\sigma_q|$ for $\sigma_q < 0$, i.e., σ_q and $\Delta\sigma_q$ are given in absolute units

x_0: initial position expectation value

f_{pE}: flag specifying whether input quantities q and Δq are understood as momenta ($f_{pE} = 0$) or kinetic energies ($f_{pE} = 1$)

$\hbar$: numerical value of Planck's constant (default value: 1)

m: mass (default value: 1)

h, v: horizontal and vertical counters in multiple plot

l_{DASH}: dash length (in W3 units) of zero lines (default value: 1/20 of X width of window in W3)

R: radius of circle indicating position of classical particle (default value: $1/100$ of X width of window in W3)

For $f_{QO} = 1$ the meanings of the input parameters are:

$f_{RES} = 0$: function shown is $|E|^2$

$f_{RES} = 1$: function shown is $\mathrm{Re}\, E$

$f_{RES} = 2$: function shown is $\mathrm{Im}\, E$

$k_0 = q + h\Delta q$

$\sigma_k = (\sigma_q + v\Delta\sigma_q)k_0$ for $\sigma_q > 0$, i.e., σ_q and $\Delta\sigma_q$ are given in units of k_0

$\sigma_k = |\sigma_q| + v|\Delta\sigma_q|$ for $\sigma_q < 0$, i.e., σ_q and $\Delta\sigma_q$ are given in absolute units

x_0: initial position expectation value

h, v: horizontal and vertical counters in multiple plot

l_{DASH}: dash length (in W3 units) of zero lines (default value: $1/20$ of X width of window in W3)

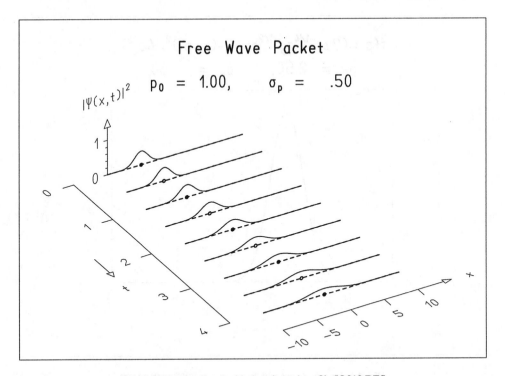

Fig. 2.1. Plot produced with descriptor 1 on file IQ012.DES

Plot recommendation:
Plot a series of lines $y = t = $ const, i.e., use

    NL 0 n_y

Additional plot items:

Item 5: circle symbolizing position of classical particle (drawn only for $f_{RES} = 0$)

Item 7: zero line $z = f(x, t_i)$ for $t = t_i = $ const

Automatically provided texts: TX, T1, T2, TF

Example plot: Fig. 2.1

2.4 The Spectral Function of a Gaussian Wave Packet

Aim of this section: Graphical presentation of the spectral wave function (2.5) and (2.18) for both quantum-mechanical and optical Gaussian wave packets.

Plot type: 2

C3 coordinates: x: momentum p or wave number k
y: spectral function $f(p)$ or $f(k)$

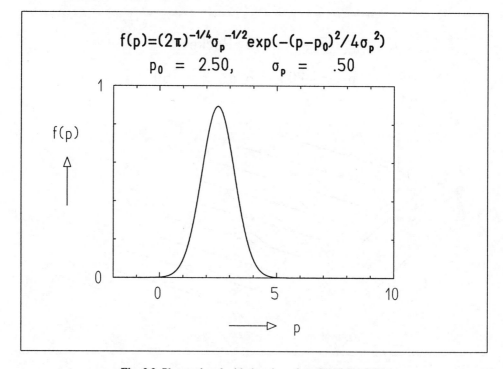

$$f(p) = (2\pi)^{-1/4} \sigma_p^{-1/2} \exp(-(p-p_0)^2/4\sigma_p^2)$$
$$p_0 = 2.50, \qquad \sigma_p = .50$$

Fig. 2.2. Plot produced with descriptor 2 on file IQ012.DES

Input parameters:

CH 2 12 0 f_{QO}

V2 Δq $\Delta\sigma_q$ 0 m

V1 q σ_q 0 f_{pE}

as explained in Sect. 2.3

Automatically provided texts: CA, TX, T1, T2

Example plot: Fig. 2.2

2.5 The Wave Packet as a Sum of Harmonic Waves

Aim of this section: Construction of both quantum-mechanical and optical wave packets as a superposition of harmonic waves of the form (2.3) and (2.16).

The weight factor w_n is chosen to be

$$w_n = \exp\left[-\frac{(p_n - p_0)^2}{4\sigma_p^2}\right]$$

for the quantum-mechanical wave packet and

$$w_n = \exp\left[-\frac{(k_n - k_0)^2}{2\sigma_k^2}\right]$$

for the electromagnetic wave packet, where the parameters in the argument of the exponentials are explained below.

Plot type: 0

C3 coordinates: x: position x y: momentum p_n or wave number k_n
z: Re Ψ_n and Re Ψ, or Im Ψ_n and Im Ψ, or Re E_n and Re E, or Im E_n and Im E

Input parameters:

CH 0 12 1 f_{QO}
VO f_{RES} V1 q σ_q 0 f_{pE}
V2 0 0 $\hbar$ m V3 Δt
V6 l_{DASH} R

with

$f_{QO} = 0$: quantum-mechanical wave packet

$f_{QO} = 1$: optical wave packet

For $f_{QO} = 0$ the input parameters have the following significance:

$f_{RES} = 1$: functions shown are Re Ψ_n and Re Ψ

$f_{RES} = 2$: functions shown are Im Ψ_n and Im Ψ

$p_0 = q$ for $f_{pE} = 0$

$p_0 = \sqrt{2mq}$ for $f_{pE} = 1$, i.e., the input quantity q is a kinetic energy

$\sigma_p = \sigma_q p_0$ for $\sigma_q > 0$, i.e., σ_q is given in units of p_0

$\sigma_p = |\sigma_q|$ for $\sigma_q < 0$, i.e., σ_q is given in absolute units

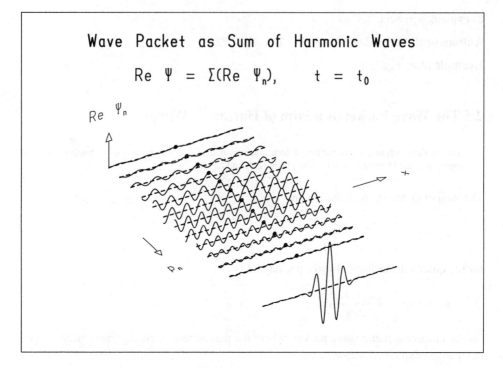

Fig. 2.3. Plot produced with descriptor 3 on file IQ012.DES

f_{pE}: flag specifying whether input quantity q is understood as momentum ($f_{pE} = 0$) or kinetic energy ($f_{pE} = 1$)

$p_n = p_0 - 3\sigma_p + (n-1)\Delta p$: momentum of term number n; $\Delta p = 6\sigma_p/(N-1)$

$\hbar$: numerical value of Planck's constant (default value: 1)

m: mass (default value: 1)

$t = n_t\Delta t$, $n_t = h + v(h_2 - h_1 + 1)$

h, v: horizontal and vertical counters in multiple plot

l_{DASH}: dash length (in W3 units) of zero lines (default value: $1/20$ of X width of window in W3)

R: radius of circle marking a point of fixed phase (default value: $1/100$ of X width of window in W3)

For $f_{QO} = 1$ the meaning of the input parameters is:

$f_{\text{RES}} = 1$: functions shown are $\text{Re}\,E_n$ and $\text{Re}\,E$

$f_{\text{RES}} = 2$: functions shown are $\text{Im}\,E_n$ and $\text{Im}\,E$

$k_0 = q$

$\sigma_k = \sigma_q k_0$ for $\sigma_q > 0$, i.e., σ_q is given in units of k_0

$\sigma_k = |\sigma_q|$ for $\sigma_q < 0$, i.e., σ_q is given in absolute units

$k_n = k_0 - 3\sigma_k + (n-1)\Delta k$: momentum of term number n; $\Delta k = 6\sigma_k/(N-1)$

$t = n_t \Delta t,\ n_t = h + v(h_2 - h_1 + 1)$

h, v: horizontal and vertical counters in multiple plot

l_{DASH}: dash length (in W3 units) of zero lines (default value: $1/20$ of X width of window in W3)

R: radius of circle marking a point of fixed phase (default value: $1/100$ of X width of window in W3)

Plot recommendation:
Plot a series of lines $y = p_i = $ const or $y = k_i = $ const, respectively, i.e., use

> NL 0 , n_y

with $n_y \geq 5$, $N = n_y - 3$ is the number of terms in the sum.

The functions Ψ_n (or E_n) are plotted in the background, the sum Ψ (or E) in the foreground, see example plot.

Additional plot items:

Item 5: circle marking a point of fixed phase (corresponding to $x = 0$ at $t = 0$) on the function Re Ψ_n or Im Ψ_n or Re E_n or Im E_n. This point is shifted to $x = vt$ for $t \neq 0$ with v being the phase velocity

Item 6: zero lines

Automatically provided texts: TX, T1, T2, TF

Example plot: Fig. 2.3

Remarks:
The finite sum $\Psi = \sum \Psi_n$ ($E = \sum E_n$) is an approximation of a Gaussian wave packet if n is large. No attempt has been made to normalize this wave packet. For very small n it becomes evident that Ψ (or E) is periodic in x. For larger n the period is longer than the x interval plotted.

2.6 Exercises

Please note:

(i) you may watch a semiautomatic demonstration of the material of this chapter by typing SA D12

(ii) for the following exercises use descriptor file IQ012.DES

(iii) some of the exercises contain input parameters in physical units. In exercises with dimensionless input data the numerical values of the particle mass and of Planck's constant are meant to be 1 if not stated otherwise in the exercise.

2.3.1 Plot the time development of the absolute square of a Gaussian wave packet using descriptor 4.

2.3.2 Use the flag CH(4) = f_{QO} to plot the corresponding optical wave packet. What is the essential difference between this plot and the one obtained in Exercise 2.3.1?

2.3.3 Return to the quantum-mechanical wave packet, increase its mean momentum p_0 by a factor of 2 and describe the effect on the change of the group velocity.

2.3.4 Perform the corresponding increase in the mean wave number k_0 of the optical wave packet.

2.3.5 Repeat Exercise 2.3.1; plot **(a)** the real part and **(b)** the imaginary part of the wave function. **(c)** Give a reason why after some time the wavelength of the front part of the real and imaginary parts of the wave packet are shorter than close to the rear end.

2.3.6 Repeat Exercise 2.3.2; plot **(a)** the real part and **(b)** the imaginary part of the optical wave packet.

2.3.7 Repeat Exercise 2.3.3 for the real part of the wave function. Explain the change of wavelength observed.

2.3.8 Repeat Exercise 2.3.4 for the real part of the optical wave packet.

2.3.9 (a) Repeat Exercise 2.3.1 but increase the momentum width σ_p by a factor of 2. Explain the change in shape of the wave packet as time passes. **(b)** Repeat the exercise, halving the momentum width. **(c)** Study also the real and imaginary parts.

2.3.10 Adapt Exercise 2.3.9 to the optical wave packet.

For the following exercises study Sect. 10.1 "Units and Orders of Magnitude".

2.3.11 Plot **(a)** the real part, **(b)** the imaginary part, and **(c)** the absolute square of a wave packet of an electron with velocity $1\,\text{ms}^{-1}$ and absolute width $\sigma_p = 0.5 \times 10^{-12}\,\text{eVsm}^{-1}$ moving from an initial position $x_0 = -2\,\text{mm}$ for the instants of time $t_i = 0, 0.5, 1, 1.5, 2, 2.5, 3, 3.5, 4$ ms. Start from descriptor 6 which already contains the correct time intervals. **(d)** Calculate the momentum p_0 of the electron. **(e)** Why is the spreading of the wave packet relatively small in time? **(f)** What is the physical unit along the x axis?

2.3.12 Repeat Exercise 2.3.11 (a–d) with velocity $3\,\text{ms}^{-1}$ for the instants of time $t_i = 0, 0.25, 0.5, 0.75, 1, 1.25, 1.5, 1.75, 2$ ms. Start from descriptor 7.

2.3.13 Plot **(a)** the real part, **(b)** the imaginary part, and **(c)** the absolute square of the wave function of an electron of velocity $2.11\,\text{ms}^{-1}$ and a relative width of $\sigma_p = 0.75p_0$ of the corresponding momentum p_0 for a period $0 \leq t \leq 2\,\text{ms}$ in steps of 0.25 ms. The initial position is $x_0 = -2\,\text{mm}$. Start with descriptor 7. **(d)** Calculate the momentum p_0 of the electron. **(e)** What is the order of the magnitude along the x axis? **(f)** Why do the real and imaginary parts look so different from the earlier exercises and from the picture of the initial descriptor? **(g)** Why are there small wavelengths to either side of the wave packet?

2.4.1 Plot the spectral function corresponding to Exercise 2.3.1 using descriptor 5.

2.4.2 Perform the necessary changes to get the spectral function of the optical wave packet of Exercise 2.3.2.

2.4.3 Plot the spectral functions corresponding to Exercises 2.3.3, 2.3.4, 2.3.9, 2.3.10.

2.4.4 (a) Plot the Gaussian spectral function of an electron of velocity $v_0 = 1\,\text{ms}^{-1}$ and $v_0 = 3\,\text{ms}^{-1}$ for the two widths $\sigma_p = 0.5 \times 10^{-12}\,\text{eV sm}^{-1}$ and $\sigma_p = 10^{-12}\,\text{eV sm}^{-1}$ in a multiple plot with four graphs. Start with descriptor 8. **(b)** Calculate the corresponding momenta. **(c)** What is the physical unit at the abscissa? **(d)** Calculate the corresponding kinetic energies.

2.4.5 (a) Plot the Gaussian spectral function of an electron of kinetic energy $E = 1\,\text{eV}$ and $E = 3\,\text{eV}$ for the two widths $\sigma_p = 0.5 \times 10^{-6}\,\text{eV sm}^{-1}$ and $\sigma_p = 10^{-6}\,\text{eV sm}^{-1}$ in a multiple plot with four graphs. Start with descriptor 9. **(b)** Calculate the corresponding momenta. **(c)** What is the physical unit at the abscissa? **(d)** Calculate the corresponding velocities of the electron. **(e)** To what order is the use of the nonrelativistic formulas still allowed?

2.4.6 Repeat Exercise 2.4.4 for a proton for the two widths $\sigma_p = 0.5 \times 10^{-9}\,\text{eV sm}^{-1}$ and $\sigma_p = 10^{-9}\,\text{eV sm}^{-1}$. Start with descriptor 10.

2.4.7 Repeat Exercise 2.4.5 for a proton for the two kinetic energies $E = 1\,\text{keV}$ and $E = 3\,\text{keV}$ and the two widths $\sigma_p = 0.5 \times 10^{-3}\,\text{eV sm}^{-1}$ and $\sigma_p = 10^{-3}\,\text{eV sm}^{-1}$. Start with descriptor 11.

2.5.1 Plot a wave packet approximated by a finite sum of harmonic waves using descriptor 3.

2.5.2 Study the time development of the harmonic waves and their sum by doing plots for various times (you may do this by using the multiple plot facility). Study the phase velocities of the different harmonic waves and the group velocity of the wave packet.

2.5.3 Repeat Exercises 2.5.1 and 2.5.2 for electromagnetic waves.

2.5.4 Repeat Exercise 2.5.1. Now decrease the number of terms in the sum using the commands NL(2) 13, NL(2) 11, NL(2) 9, etc. Why is the resulting sum periodic in x?

3 Bound States in One Dimension

Contents: Introduction of the time-dependent and stationary Schrödinger equations. Computation of eigenfunctions and eigenvalues in the infinitely deep square-well potential, in the harmonic-oscillator potential and in the general step potential. Motion of a wave packet in the deep square-well potential and in the harmonic-oscillator potential.

3.1 Physical Concepts

3.1.1 Schrödinger's Equation with Potential. Eigenfunctions. Eigenvalues

The motion of a particle under the action of a force given by a *potential $V(x)$* is governed by the *Schrödinger equation*

$$i\hbar \frac{\partial}{\partial t}\psi(x,t) = -\frac{\hbar^2}{2m}\frac{\partial^2}{\partial x^2}\psi(x,t) + V(x)\psi(x,t) \quad . \tag{3.1}$$

With the *Hamiltonian*

$$H = T + V \tag{3.2}$$

it reads

$$i\hbar \frac{\partial}{\partial t}\psi(x,t) = H\psi(x,t) \quad . \tag{3.3}$$

Separation of the variables time t and position x by way of the expression for *stationary wave function*

$$\psi_E(t,x) = e^{-iEt/\hbar}\varphi_E(x) \tag{3.4}$$

leads to the *stationary Schrödinger equation*

$$-\frac{\hbar^2}{2m}\frac{d^2}{dx^2}\varphi_E(x) + V(x)\varphi_E(x) = E\varphi_E(x) \tag{3.5}$$

or equivalently

$$H\varphi_E(x) = E\varphi_E(x) \tag{3.6}$$

for the *eigenfunction $\varphi_E(x)$* of the Hamiltonian belonging to the energy eigenvalue E.

3.1.2 Normalization. Discrete Spectra. Orthonormality

The Hamiltonian H is a *Hermitian* operator for *square-integrable functions* $\varphi(x)$ only. These can be *normalized* to one, i.e.,

$$\int_{-\infty}^{+\infty} \varphi^*(x)\, \varphi(x) dx = 1 \quad . \tag{3.7}$$

For normalizable eigenfunctions the eigenvalues E of H form a set $\{E_1, E_2, \ldots\}$ of discrete real values. This set is the *discrete spectrum* of the Hamiltonian. The corresponding eigenfunctions are called *discrete eigenfunctions* of the Hamiltonian. We will label the eigenfunction belonging to the eigenvalue $E = E_n$ by $\varphi_n(x)$. The eigenfunctions $\varphi_n(x)$, $\varphi_m(x)$ corresponding to different eigenvalues $E_n \neq E_m$ are *orthogonal*:

$$\int \varphi_n^*(x)\, \varphi_m(x) dx = 0 \quad . \tag{3.8}$$

Together with the normalization (3.7) we have the *orthonormality* of the discrete eigenfunctions

$$\int \varphi_n^*(x)\, \varphi_m(x) dx = \delta_{mn} \quad . \tag{3.9}$$

For potentials $V(x)$ bounded below, i.e., $V_0 \leq V(x)$ for all x, the eigenvalues lie in the domain $V_0 \leq E$. For potentials bounded below, tending to infinity at $x \to -\infty$ as well as $x \to +\infty$, all eigenvalues are discrete.

For potentials bounded below tending to a finite limit $V(-\infty)$ or $V(+\infty)$ at either $x \to +\infty$ or $x \to -\infty$ the discrete eigenvalues can occur in the interval $V_0 \leq E_n \leq V_c$ with $V_c = \min(V(+\infty), V(-\infty))$.

3.1.3 The Infinitely Deep Square-Well Potential

$$V(x) = \begin{cases} 0 & , \quad -d/2 \leq x \leq d/2 \\ \infty & , \quad \text{elsewhere} \end{cases} \tag{3.10}$$

d: width of potential.

This potential confines the particle to an interval of length d. The eigenfunctions of the Hamiltonian with this potential are

$$\varphi_n(x) = \sqrt{2/d}\, \cos(n\pi x/d) \quad , \quad n = 1, 3, 5, \ldots$$
$$\varphi_n(x) = \sqrt{2/d}\, \sin(n\pi x/d) \quad , \quad n = 2, 4, 6, \ldots \tag{3.11}$$

belonging to the eigenvalues

$$E_n = \frac{1}{2m} \left(\frac{\hbar n \pi}{d} \right)^2 \quad , \quad n = 1, 2, 3, \ldots \quad . \tag{3.12}$$

The discrete energies E_n are enumerated by the principal quantum number n.

3.1.4 The Harmonic Oscillator

$$V(x) = \frac{m}{2}\omega^2 x^2 \tag{3.13}$$

m: mass of particle
ω: angular frequency

The eigenfunctions of the Hamiltonian of the harmonic oscillator are

$$\varphi_n(x) = (\sqrt{\pi}2^n n! \sigma_0)^{-1/2} H_n\left(\frac{x}{\sigma_0}\right) \exp\left(-\frac{x^2}{2\sigma_0^2}\right) \quad , \quad n = 0, 1, 2, \ldots \tag{3.14}$$

belonging to the eigenvalues

$$E_n = \left(n + \tfrac{1}{2}\right)\hbar\omega \quad . \tag{3.15}$$

n: principal quantum number n of the harmonic oscillator
$H_n(x)$: Hermite polynomial of order n
$\sigma_0 = \sqrt{\hbar/m\omega}$: width of ground-state wave function

3.1.5 The Step Potential

$$V(x) = \begin{cases} V_1 = 0 & , \quad x < x_1 = 0 & \text{region 1} \\ V_2 & , \quad x_1 \le x < x_2 & \text{region 2} \\ \vdots & & \\ V_{N-1} & , \quad x_{N-2} \le x < x_{N-1} & \text{region } N-1 \\ V_N = 0 & , \quad x_{N-1} \le x & \text{region } N \end{cases} \tag{3.16}$$

The potential possesses discrete eigenvalues E_l for $E < 0$ which can again be enumerated by a principal quantum number l.

For an eigenfunction $\varphi_l(x)$ belonging to the eigenvalue E_l the *de Broglie wave number* in region m is

$$\begin{aligned} k_{lm} &= \left|\sqrt{2m(E_l - V_m)}/\hbar\right| & \text{for } E_l > V_m \\ k_{lm} &= \mathrm{i}\,\kappa_{lm} \quad , \quad \kappa_{lm} = \left|\sqrt{2m(V_m - E_l)}/\hbar\right| & \text{for } E_l < V_m \end{aligned} \tag{3.17}$$

The wave function $\varphi_l(x)$ is then given for all the N intervals of constant potential by

$$\varphi_l(x) = \begin{cases} \varphi_{l1}(x) & , \quad \text{region 1} \\ \varphi_{l2}(x) & , \quad \text{region 2} \\ \vdots & \\ \varphi_{lN-1} & , \quad \text{region } N-1 \\ \varphi_{lN} & , \quad \text{region } N \end{cases} \tag{3.18}$$

For $E \neq V_0$ the piece φ_{lm} of the wave function

$$\varphi_{lm}(x) = A_{lm}\mathrm{e}^{\mathrm{i}k_{lm}x} + B_{lm}\mathrm{e}^{-\mathrm{i}k_{lm}x} \quad , \quad x_{m-1} \le x < x_m \quad , \tag{3.19}$$

consists for $E_l > V_m$ of a *right-moving* and a *left-moving harmonic wave* and for $V_m > E_l$ of a *decreasing* and an *increasing exponential function*. For $E_l = V_m$ the piece φ_{lm} is a straight line

$$\varphi_{lm}(x) = A_{lm} + B_{lm}x \quad , \quad x_{m-1} \leq x < x_m \quad . \tag{3.20}$$

Because of the normalizability (3.7), the bound-state wave function $\varphi_l(x)$ must decrease exponentially in the regions

$$m = 1: \quad \varphi_{l1} = B_{l1}e^{\kappa_{l1}x} \quad -\infty < x < x_1 = 0$$
$$m = N: \quad \varphi_{lN} = A_{lN}e^{-\kappa_{lN}x} \quad x_{N-1} \leq x < \infty \quad , \tag{3.21}$$

i.e., $A_{l1} = 0$, $B_{ln} = 0$.

The requirement of exponential behavior stipulates $E < 0$. The coefficients A_{lm}, B_{lm} are determined from the requirement of the wave function being *continuous* and *continuously differentiable* at the values x_m, $m = 1, \ldots, N - 1$. For $E \neq V_m, V_{m+1}$, these *continuity conditions* read

$$A_{lm}e^{ik_{lm}x_m} + B_{lm}e^{ik_{lm}x_m} = A_{lm+1}e^{ik_{lm+1}x_m} + B_{lm+1}e^{-ik_{lm+1}x_m}$$

$$k_{lm}(A_{lm}e^{ik_{lm}x_m} - B_{lm}e^{-ik_{lm}x_m}) = k_{lm+1}(A_{lm+1}e^{ik_{lm+1}x_m} - B_{lm+1}e^{-ik_{lm+1}x_m}) \quad .\tag{3.22}$$

If $E_l = V_m$ or $E_l = V_{m+1}$ the left-hand or right-hand side has to be replaced by using (3.20). The system represents a set of $2(N - 1)$ linear homogeneous equations for $2(N - 1)$ unknown coefficients A_{lm}, B_{lm}. It has a non-trivial solution only if its determinant $D(E)$ vanishes. This requirement leads to a transcendental equation for the eigenvalues E_l present in the wave numbers k_{lm}. In general, its solution can be obtained numerically only and is calculated by the computer by finding the zeros of the function

$$D = D(E) \tag{3.23}$$

which coincide with the values E_l at which the determinant vanishes. Once the eigenvalue E_l is determined as a single zero of the transcendental equation the system of linear equations can be solved yielding the coefficients A_{lm}, B_{lm} as functions of one of them. This undetermined coefficient is then fixed by the normalization condition (3.7). The number of the eigenstates $\varphi_l(x)$ of step potentials is finite; thus they do not form a complete set. In Chap. 4 we present the continuum eigenfunctions which supplement the $\varphi_l(x)$ to a complete set of functions.

3.1.6 Time-Dependent Solutions

Since the time-dependent Schrödinger equation (3.1) is linear, the time-dependent harmonic waves, (3.4),

$$\psi_n(x, t) = e^{-iEt/\hbar}\varphi_n(x) \tag{3.24}$$

can be superimposed with time-independent spectral coefficients w_n yielding the solution

$$\psi(x, t) = \sum_n w_n e^{-iEt/\hbar}\varphi_n(x) \tag{3.25}$$

of the Schrödinger equation. Since the eigenfunctions $\varphi_n(x)$, with n belonging to the discrete spectrum, confine the particle to a bounded region in space, the solutions $\psi(t, x)$ do so for all times.

3.1.7 Harmonic Particle Motion. Coherent States. Squeezed States

For the time $t = 0$ we choose an initial Gaussian wave packet with its maximum placed at the position x_0:

$$\psi(x, 0) = \frac{1}{(2\pi)^{1/4}\sigma_{x0}} \exp\left\{-\frac{(x - x_0)^2}{4\sigma_{x0}^2}\right\} \quad ; \tag{3.26}$$

x_0 : location of position of maximum

$\sigma_0 = \sqrt{\hbar/m\omega}$: ground-state width

σ_{x0}: initial width of wave packet

We decompose $\psi(x, 0)$ into a sum over the complete set of real eigenfunctions $\varphi_n(x)$ of the harmonic oscillator

$$\psi(x, 0) = \sum_{n=0}^{\infty} w_n \, \varphi(x) \quad . \tag{3.27}$$

The orthonormality condition (3.7) is used to determine the coefficients

$$w_n = \int_{-\infty}^{+\infty} \varphi_n(x) \, \psi(x, 0) dx \quad . \tag{3.28}$$

The time-dependent solution $\psi(x, t)$ is then given with these coefficients by (3.25). This expansion can be summed up. For brevity we present only its absolute square explicitly:

$$|\psi(x, t)|^2 = \frac{1}{\sqrt{2\pi}\sigma(t)} \exp\left\{-\frac{(x - x_0\cos\omega t)}{2\sigma^2(t)}\right\} \quad . \tag{3.29}$$

It represents a Gaussian wave packet moving with a mean position

$$x = x_0\cos\omega t \tag{3.30}$$

oscillating harmonically in time and with an in-general time-dependent width

$$\sigma(t) = \frac{\sigma_0}{2\sqrt{2}\sigma_{r0}}(4\sigma_{r0}^4 + 1 + (4\sigma_{r0}^4 - 1)\cos(2\omega t))^{1/2} \quad ; \tag{3.31}$$

$\sigma_{r0} = \sigma_{x0}/\sigma_0$: relative initial width of wave packet

The time-dependent width itself oscillates harmonically with double angular frequency 2ω about an average width

$$\bar{\sigma} = \frac{\sigma_0}{2\sqrt{2}\sigma_{r0}}(4\sigma_{r0}^4 + 1)^{1/2} \quad . \tag{3.32}$$

The *coherent state* is distinguished by a time-independent width $\sigma = \sigma_0/\sqrt{2}$. It is of central importance in quantum optics and quantum electronics, e.g., in lasers and quantum oscillations in electrical circuits. States with oscillating widths are called *squeezed states*.

3.1.8 Particle Motion in a Deep Square Well

For time $t = 0$ we choose an initial wave packet with a bell shape

$$\psi(x, 0) = \sum_{n=N_1}^{N_2} w_n \, \varphi_n(x) \quad ; \tag{3.33}$$

$\varphi_n(x)$: eigenfunctions of infinitely deep square well
w_n: spectral weights

The spectral weights w_n are taken as the values of a Gaussian spectral distribution in momentum space at the discrete values $k_n = n\pi/d$ of the wave numbers allowed in the infinitely deep square well. The Gaussian is centered at p_0. Its complex phase factor puts at $t = 0$ the initial position expectation value of the wave packet to $x = x_0$. The result is a "Gaussian" wave packet inside the infinitely deep square well. The explicit formulae for the spectral weights used are

$$w_n = (2\pi)^{1/4} \sqrt{\sigma_r} [e^{-\sigma_r^2 (p_0 d/\hbar + n\pi)^2} e^{in\pi x_0/d} + e^{-\sigma_r^2 (p_0 d/\hbar - n\pi)^2} e^{-in\pi x_0/d}] \quad ,$$
$$n = 1, 3, 5, \ldots$$

$$w_n = -i(2\pi)^{1/4} \sqrt{\sigma_r} [e^{-\sigma_r^2 (p_0 d/\hbar + n\pi)^2} e^{in\pi x_0/d} - e^{-\sigma_r^2 (p_0 d/\hbar - n\pi)^2} e^{-in\pi x_0/d}] \quad ,$$
$$n = 2, 4, 6, \ldots \quad . \tag{3.34}$$

x_0: expectation value of initial position
p_0: expectation value of initial momentum
d: width of potential
σ_{x0}: initial width of wave packet
$\sigma_r = \sigma_{x0}/d$: relative initial width
N_1, N_2: lower and upper limits of summation

For reasonable localization of the wave packet within the deep well the relative initial width σ_r must be small compared to one.

The moving wave packet is obtained from the time-dependent solution (3.25) with the spectral weights (3.34). In the harmonic oscillator the expectation values of position and of momentum of the wave packet coincide with its classical position and momentum.

For a wave packet in the infinitely deep square well the motion of the classical particle is after some time drastically different from the motion of the position expectation value of the wave packet. The reason for this phenomenon is the broadening of the wave packet. As soon as its width substantially exceeds the width of the well, the probability density of the particle fills the whole well and its position expectation value just rests at the center of the well. Thus the original amplitude of the oscillating expectation value within some inner range of the well decreases to zero and the particle rests at the center of the well.

However, the broadening of the wave packet in the infinitely deep square well cannot go on forever as in the case of the free motion of a particle. In the infinitely deep square well all time-dependent processes are periodic in time with the period T_1. This can be calculated with the following arguments. The energy of the ground state of an infinitely deep square well is

$$E_1 = \frac{\hbar^2}{2m} \frac{\pi^2}{d^2} \quad .$$

The corresponding angular frequency $\omega_1 = E_1/\hbar$ determines a period T_1 for all time-dependent processes in this system:

$$T_1 = 2\pi/\omega_1 = 4md^2/(\pi\hbar) = 8md^2/h \quad .$$

This means, in particular, that after the time T_1 a wave packet moving in the well assumes its initial shape, i.e., the shape it had at $t = 0$. The classical particle of momentum p_0 and mass m bouncing back and forth between the two walls of the well has the bouncing period

$$T_c = 2d/v_0 = 2md/p_0 \quad .$$

At the time T_1, the wave packet has regained its initial width so that its position expectation values show again the bouncing behavior of the initial narrow wave packet. However, its location coincides with the classical particle only if the quantum-mechanical and classical periods T_1 and T_c are compatible. Actually, after half the quantum-mechanical period the wave packet already assumes its original width, however with the opposite of the initial momentum. Since the classical particle position and the expectation value of the wave packet have to coincide at times $0, T_1, 2T_1, \ldots$, the initial momentum p_0 must be chosen so that

$$T_1 = MT_c \quad , \quad M = 1, 2, \ldots \quad ,$$

i.e.,

$$p_0 = M\frac{\pi}{2d}\hbar = M\frac{h}{4d} \quad .$$

Further Reading

Alonso, Finn: Vol. 3, Chaps. 2,6
Berkeley Physics Course: Vol. 4, Chaps. 7,8
Brandt, Dahmen: Chaps. 4,6
Feynman, Leighton, Sands: Vol. 3, Chaps. 13,14,16
Flügge: Vol. 1, Chap. 2A
Gasiorowicz: Chaps. 3,4
Merzbacher: Chaps. 3,4,5,6
Messiah: Vol. 1, Chaps. 2,3
Schiff: Chaps. 2,3,4

3.2 Eigenstates in the Infinitely Deep Square-Well Potential and in the Harmonic-Oscillator Potential

Aim of this section: Computation and presentation of the eigenfunctions (3.11) and eigenvalue spectrum (3.12) for the deep square-well potential (3.10) and of the eigenfunctions (3.14) and eigenvalues (3.15) for the harmonic-oscillator potential (3.13).

Plot type: 2

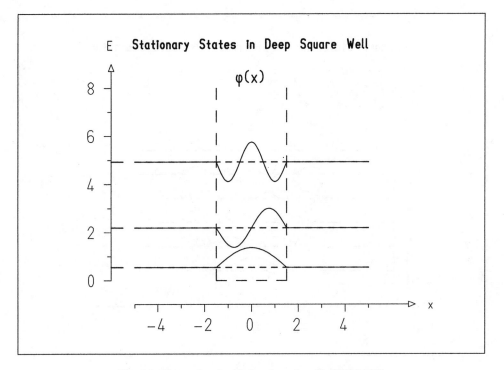

Fig. 3.1. Plot produced with descriptor 1 on file IQ010.DES

C3 coordinates: x: position x y: $\varphi_n(x)$ or $\varphi_n^2(x)$ and E_n and $V(x)$

Input parameters:

CH 2 10 f_{PROBLEM}
V0 f_{RES} V1 0 E_{max}
V2 0 0 $\hbar$ m V3 p_{POT}
V6 l_{DASH} s V7 f_{POT} f_{EF} f_{TS} f_{EV}

with

$f_{\text{PROBLEM}} = 2$: for deep square well

$f_{\text{PROBLEM}} = 3$: for harmonic oscillator

$f_{\text{RES}} = 0$: function shown is $\varphi_n^2(x)$

$f_{\text{RES}} = 1$: function shown is $\varphi_n(x)$

E_{max}: only eigenvalues $E_i < E_{\text{max}}$ are considered (default value: upper boundary of y coordinate of C3 window which coincides with upper boundary for y scale which is used as the scale of the term scheme.)

$\hbar$: Planck's constant (default value: 1)

m: mass of particle (default value: 1)

$p_{\text{POT}} = d$: i.e., width of deep square-well potential (default value: 1) for $f_{\text{PROBLEM}} = 2$

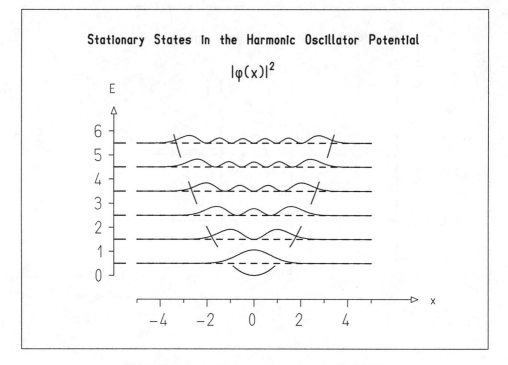

Fig. 3.2. Plot produced with descriptor 2 on file IQ010.DES

$p_{POT} = \omega$: i.e., angular frequency (default value: 1) for $f_{PROBLEM} = 3$

l_{DASH}: dash length for potential in W3 coordinates (default value: $1/10$ of X width of window in W3). Dash length for eigenvalues is $l_{DASH}/2$

s: scale factor for plotting $\varphi_n(x)$ or $\varphi_n^2(x)$ (default value: 1). (Since $\varphi_n(x)$ is shown with the eigenvalue E_n serving as zero line, actually the function $f(x) = E_n + s\varphi_n(x)$ is plotted)

$f_{POT} = 0$: normally

$f_{POT} = 1$: if plot of potential $V(x)$ is *not* wanted

$f_{EF} = 0$: normally

$f_{EF} = 1$: if plot of eigenfunction $\varphi_n(x)$ or of $\varphi_n^2(x)$ is *not* wanted

$f_{TS} = 0$: normally

$f_{TS} = 1$: if term scheme on left side of plot is *not* wanted

$f_{EV} = 0$: normally

$f_{EV} = 1$: if eigenvalues *not* wanted

Additional plot items:

Item 6: potential

Item 7: eigenvalues serving as zero lines and in term scheme

Automatically provided texts: TX, T1, T2

Example plots: Fig. 3.1 and Fig. 3.2

Remarks:
The y scale (controlled by the commands Y1 and Y2) is used as the scale for the term scheme. The lines in the term scheme extend in X from the scale to half the distance to the lower boundary in X of the W3 window.

3.3 Eigenstates in the Step Potential

Aim of this section: Computation and presentation of eigenfunctions (3.18) and the corresponding eigenvalues of bound states in a step potential (3.16).

Two cases are treated separately:

a) the potential values in the regions $2, 3, \ldots, N-1$ are chosen completely *arbitrarily*

b) the potential is *quasiperiodic*, i.e., $V_2 = V_4 = \ldots V_{N-1}$, $V_3 = V_5 = V_{N-2}$, N odd

Plot type: 2

C3 coordinates: x: position x y: $\varphi_n(x)$ or $\varphi_n^2(x)$ and E_n and $V(x)$

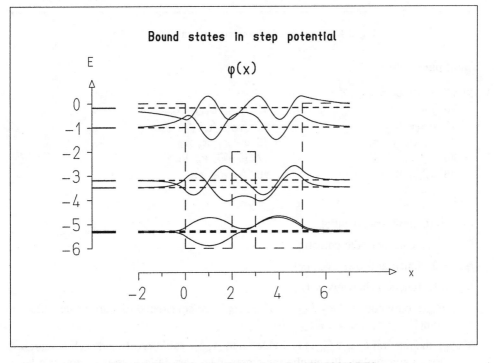

Fig. 3.3. Plot produced with descriptor 3 on file IQ010.DES

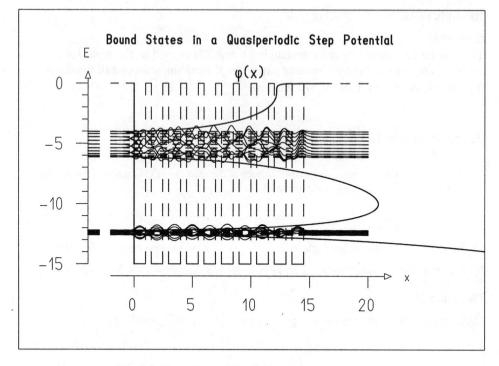

Fig. 3.4. Plot produced with descriptor 4 on file IQ010.DES

Input parameters:

CH 2 10 0 f_{CASE}
V0 f_{RES} V1 E_{min} E_{max} N_{search} ε
V2 p_{STEP} 0 $\hbar$ m V3 f_{POT} f_{EF} f_{TS} f_{EV}
V4 x_a x_b x_c x_d V5 x_e x_f x_g x_h
V6 V_a V_b V_c V_d V7 V_e V_f V_g V_h
V8 l_{DASH} 0 s V9 f_D s_D x_D

with

f_{CASE} = 0: arbitrary potential

f_{CASE} = 1: quasiperiodic potential

f_{RES} = 0: function shown is $\varphi_n^2(x)$

f_{RES} = 1: function shown is $\varphi_n(x)$

E_{min}, E_{max}: only eigenvalues $E_{min} < E < E_{max}$ are considered (default values: $E_{min} = \min(V_i, i = 1, \ldots, N)$, $E_{max} = 0$)

N_{search}: the interval (E_{min}, E_{max}) is divided into N_{search} subintervals and each is searched for an eigenvalue. If neighboring eigenvalues vary little from one another a large value has to be chosen for N_{search} (default value: 100)

ε: an accuracy parameter for the solution of the eigenvalue problem. The smaller the value of ε is chosen the higher is the accuracy (default value: 1)

p_{STEP}: $N = p_{STEP} + 2, p_{STEP} \leq 8$ for $f_{CASE} = 0$
$\qquad N = 2p_{STEP} + 1, p_{STEP} \leq 10$ for $f_{CASE} = 1$

$\hbar$: Planck's constant (default value: 1)

m: mass of particle (default value: 1)

$x_a, x_b, \ldots$: $x_1 = 0, x_2 = x_a, x_3 = x_b, \ldots$ for $f_{CASE} = 0$
$\qquad x_1 = 0, x_2 = x_a, x_3 = x_2 + x_b, x_4 = x_3 + x_a, x_5 = x_4 + x_b, \ldots$ for $f_{CASE} = 1$

$V_a, V_b, \ldots$: $V_1 = 0, V_2 = V_a, V_3 = V_b, \ldots, V_N = 0$, for $f_{CASE} = 0$,
$\qquad V_1 = 0, V_2 = V_a, V_3 = V_b, V_4 = V_a, V_5 = V_b, \ldots, V_{N-1} = V_a, V_N = 0$ for $f_{CASE} = 1$

l_{DASH}: dash length for potential in W3 coordinates (default value: $1/10$ of X width of window in W3). Dash length for eigenvalues is $l_{DASH}/2$

s: scale factor for plotting $\varphi_n(x)$ or $\varphi_n^2(x)$ (default value: 1). (Since $\varphi_n(x)$ is shown with the eigenvalue E_n serving as zero line, actually the function $f(x) = E_n + s\varphi_n(x)$ is plotted)

$f_{POT} = 0$: normally

$f_{POT} = 1$: if plot of potential $V(x)$ is *not* wanted

$f_{EF} = 0$: normally

$f_{EF} = 1$: if plot of eigenfunction $\varphi_n(x)$ or of $\varphi_n^2(x)$ is *not* wanted

$f_{TS} = 0$: normally

$f_{TS} = 1$: if term scheme on left side of plot is *not* wanted

$f_{EV} = 0$: normally

$f_{EV} = 1$: if eigenvalues are *not* wanted

$f_D \leq 0$: function $D = D(E)$ not shown

$f_D > 0$: the function $D = D(E)$, (3.23), which has zeros for the eigenvalue $E = E_n$ is shown. The function value is plotted along the x axis, the independent variable along the y axis. To allow convenient scaling, actually $f_D \sinh^{-1}[s_D D(E)]$ is plotted, where s_D is a scale factor. Usually $f_D = 1, s_D = 0$ will suffice

s_D: scale factor (default value $s_D = 1$) for plotting the function $D = D(E)$, see above

x_D: x position (in C3 coordinates) of zero line for function $D = D(E)$

Additional plot items:

Item 5: function $D = D(E)$

Item 6: potential

Item 7: eigenvalues serving as zero lines and in term scheme

Automatically provided texts: TX, T1, T2

Example plots: Fig. 3.3 and Fig 3.4

Remarks:
The y scale (controlled by the commands Y1 and Y2) is used as scale for the term scheme. The lines in the term scheme extend in X from the scale to half the distance to the lower boundary in X of the W3 window.

3.4 Harmonic Particle Motion

Aim of this section: Demonstration of the motion of coherent states and squeezed states in the harmonic-oscillator potential. Presentation of absolute square (3.29) of the wave function and also of its real and imaginary part.

Plot type: 0

C3 coordinates: x: position coordinate x y: time coordinate t
z: $|\Psi|^2$, Re Ψ, or Im Ψ

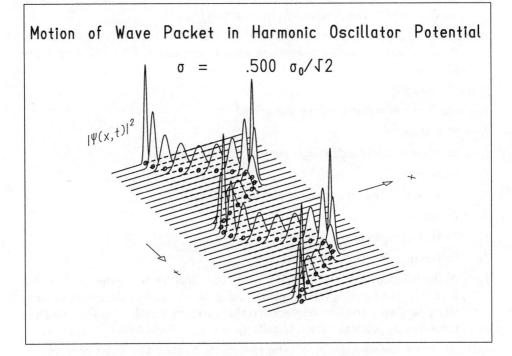

Fig. 3.5. Plot produced with descriptor 5 on file IQ010.DES

Input parameters:

```
CH 0 10 0
V0  fRES                        V1  fσ  x0  ω
V2  0 0 ℏ m                     V3  R
                                V5  lDASH
```

with

$f_{\mathrm{RES}} = 0$: function shown is $|\Psi(x, t)|^2$

$f_{\mathrm{RES}} = 1$: function shown is Re $\Psi(x, t)$

$f_{\mathrm{RES}} = 2$: function shown is Im $\Psi(x, t)$

f_σ: initial width is $\sigma_{x0} = f_\sigma \sigma_0/\sqrt{2}$ (default value: $f_\sigma = 1$)

x_0: initial position expectation value

ω: angular frequency

$\hbar$: Planck's constant (default value: 1)

m: mass of particle (default value: 1)

R: radius (in W3 units) of circle symbolizing classical particle (default value: $R = l_{\mathrm{DASH}}/8$)

l_{DASH}: dash length (in W3 units) of zero line (default value: $1/10$ of x width of window in W3)

Plot recommendation:
Plot a series of lines $y = t = $ const, i.e., use

    NL 0 n_y

Additional plot items:

Item 5: circle indicating position of classical particle (only for $f_{\mathrm{RES}} = 0$)

Item 7: zero line spanning the region in x between the turning points of the classical particle $-|x_0| < x < |x_0|$

Automatically provided texts: TX, T1, T2, TF

Example plot: Fig. 3.5

3.5 Particle Motion in the Infinitely Deep Square-Well Potential

Aim of this section: Study of the motion of a wave packet (3.33) in the deep square-well potential.

The sum (3.33) extends over integer values n ranging from

$$N_1 = N_0 - 5\sigma_0$$

to

$$N_2 = N_0 + 5\sigma_0$$

with N_0 as the nearest integer in the neighborhood of the number $|p_0|d/(\hbar\pi)$. The number σ_0 is the nearest integer of $1/\sigma_r$. If N_1 turns out to be less than one, N_1 is set to one.

Plot type: 0

C3 coordinates: x: position coordinate x y: time coordinate t
z: $|\Psi|^2$, Re Ψ, or Im Ψ

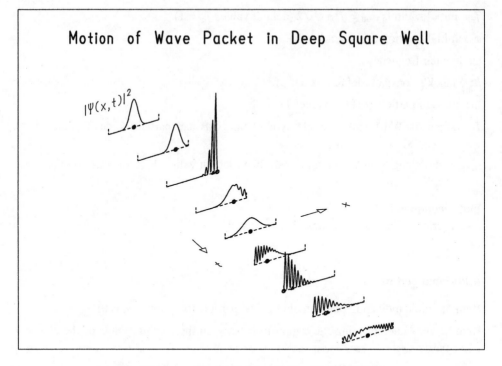

Fig. 3.6. Plot produced with descriptor 6 on file IQ010.DES

Input parameters:

```
CH 0 10 1
V0 fRES                          V1 σx0 x0 p0
V2 d 0 ℏ m                       V3 R
V4 WPOT                          V5 lDASH
```

with

$f_{RES} = 0$: function shown is $|\Psi(x,t)|^2$

$f_{RES} = 1$: function shown is $\mathrm{Re}\,\Psi(x,t)$

$f_{RES} = 2$: function shown is $\mathrm{Im}\,\Psi(x,t)$

d: width of potential well

σ_{x0}: initial width of wave packet (default value: 1)

x_0: initial position expectation value

p_0: initial momentum expectation value

$\hbar$: Planck's constant (default value: 1)

m: mass (default value: 1)

R: radius (in W3 units) of circle symbolizing classical particle (default value: $R = l_{DASH}/8$)

W_{POT}: height (in W3 units) of vertical lines indicating the position of the walls of the potential well (default value: 1)

l_{DASH}: dash length (in W3 units) of line indicating the potential (default value: 1/10 of X width of window in W3)

Plot recommendation:
Plot a series of lines $y = t = $ const, i.e., use

```
NL 0 n_y
```

Additional plot items:

Item 5: circle indicating position of classical particle (only for $f_{RES} = 0$)

Item 6: line indicating potential

Item 7: triangle indicating position expectation value

Automatically provided texts: T1, T2, TF

Example plot: Fig. 3.6

3.6 Exercises

Please note:

(i) you may watch a semiautomatic demonstration of the material of this chapter by typing
SA D10

(ii) for the following exercises use descriptor file IQ010.DES

(iii) some of the exercises contain input parameters in physical units. In exercises with dimensionless input data the numerical values of the particle mass and of Planck's constant are meant to be 1 if not stated otherwise in the exercise.

3.2.1 Plot **(a)** the eigenfunctions, **(b)** the probability densities of an infinitely deep square-well potential of width $d = 3$ for the energy range $0 \le E_i \le E_{max}$, $E_{max} = 30$. Start from descriptor 7.

3.2.2 What are the expectation values of position and momentum of the eigenfunctions of problem 3.2.1?

3.2.3 Rewrite the eigenfunctions $\varphi_n(x)$ of (3.11) in terms of the functions $\varphi_n^{(\pm)}(x) = (1/\sqrt{d})e^{\pm in\pi x/d}$, $n = 1, 2, 3, \ldots$, possessing nonvanishing momentum expectation value $p_n^{(\pm)} = \pm \hbar \pi/d$. **(a)** Why are the wave functions $\varphi_n^{(\pm)}(x)$ not among the eigenfunctions of the infinitely deep potential? **(b)** What is the classical interpretation of the eigenfunctions $\varphi_n(x)$ in terms of the $\varphi_n^{(\pm)}(x)$?

3.2.4 Calculate the eigenfunctions and eigenvalues in the energy range $0 \le E_i \le E_{max} = 2\,\text{eV}$ for an electron in the infinitely deep square-well potential of width **(a)** 3 nm, **(b)** 5 nm. Start from descriptor 8. **(c)** Calculate a rough estimate of the ground state energy using Heisenberg's uncertainty relation.

3.2.5 Repeat Exercise 3.2.4 for the proton for the energy range of the eigenfunctions $0 \leq E_i \leq E_{max}$, $E_{max} = 1$ meV. Start from descriptor 9.

3.2.6 (a) Plot the wave functions and the energy levels E_n in the range $0 \leq E_i \leq E_{max}$, $E_{max} = 20$ eV for an electron (mass m_e) in an infinitely deep potential of width $d = 3a$. Here, $a = \hbar c/(\alpha m_e c^2) = 0.05292$ nm is the Bohr radius of the innermost orbit of the hydrogen atom. Read off the energy E_1 of the lowest eigenfunction φ_1. Start with descriptor 10. **(b)** Calculate the potential energy of an electron in Coulomb potential $V(r) = -\alpha \hbar c/r$ at the Bohr radius $r = a$. **(c)** Calculate the sum of the kinetic energy E_1 of the lowest eigenfunction φ_1 as determined in (a) and the potential energy as calculated in (b). Compare it to the binding energy of the electron in the hydrogen ground state.

3.2.7 Plot the eigenfunctions of the harmonic oscillator for an electron with the angular frequencies **(a)** $\omega = 10^{15}$ s^{-1}, **(b)** $\omega = 1.5 \times 10^{15}$ s^{-1}, **(c)** $\omega = 2 \times 10^{15}$ s^{-1} and, **(d–f)** the corresponding probability densities. Start with descriptor 11. **(g)** What is the physical unit of the energy scale? **(h)** Calculate the spring constants $D = m_e \omega^2$. **(i)** What is the physical unit of the x scale? **(j)** Why is the probability density highest close to the wall of the potential?

3.2.8 Plot **(a,b)** the eigenfunctions, **(c,d)** the probability density of the proton in the harmonic oscillator potential $V = Dx^2/2$ with the constants $D = 0.1$ eV m^{-2} and $D = 1$ keV m^{-2} respectively. Start from descriptor 12. **(e)** Calculate the angular frequency ω of the oscillator. **(f)** Calculate the width of the ground state of the harmonic oscillator $\sigma_0 = \sqrt{\hbar m^{-1} \omega^{-1}}$.

3.3.1 Plot **(a)** the eigenfunctions, **(b)** the probability distributions for a square-well potential of width $d = 4$ and depth $V_0 = -6$. Start from descriptor 14. **(c)** Read the energy eigenvalues E_i off the screen and calculate the differences $\Delta_i = E_i - V_0$.

3.3.2 Repeat Exercise 3.3.1 for the widths **(a,b,c)** $d = 2$, **(d,e,f)** $d = 6$. Start from descriptor 14.

3.3.3 (a) Calculate the lowest-energy eigenvalues E_i' in the infinitely deep square-well for the widths $d = 2, 4, 6$. **(b)** Compare the E_i' to the differences Δ_i for the square-well potentials of Exercises 3.3.1, 3.3.2 of corresponding widths. **(c)** Explain why the separation of the eigenvalues E_i is smaller than that of the E_i'.

3.3.4 Plot **(a)** the eigenfunctions, **(b)** the probability densities of a double-well potential

$$V(x) = \begin{cases} 0 & , \quad x < 0 \\ -4 & , \quad 0 \leq x < 1.625 \\ -1.33 & , \quad 1.625 \leq x < 1.875 \\ -4 & , \quad 1.875 \leq x < 3.5 \\ 0 & , \quad 3.5 \leq x \end{cases} \quad .$$

Start with descriptor 15. **(c)** Why is the third eigenfunction a horizontal straight line in the central region? **(d)** Why have the two lowest eigenfunctions energies close to each other? **(e)** Why are they symmetric and antisymmetric?

3.3.5 Plot **(a)** the eigenfunctions, **(b)** the probability densities of a double-well potential

$$V(x) = \begin{cases} 0 & , & x < 0 \\ -4 & , & 0 \le x < 1 \\ -1.33 & , & 1 \le x < 2 \\ -4 & , & 2 \le x < 3.5 \\ 0 & , & 3.5 \le x \end{cases}.$$

Start with descriptor 15. **(c)** Why do the wave functions exhibit no symmetric pattern?

3.3.6 Plot **(a)** the eigenfunctions, **(b)** the probability densities of an asymmetric double-well potential

$$V(x) = \begin{cases} 0 & , & x < 0 \\ -5.72 & , & 0 \le x < 1 \\ -1 & , & 1 \le x < 1.5 \\ -4 & , & 1.5 \le x < 3.5 \\ 0 & , & 3.5 \le x \end{cases}.$$

Start with descriptor 15. **(c)** Why is the ground-state wave function given by a straight line in the second well?

3.3.7 Plot the eigenfunctions in a set of asymmetric potential wells given by the potentials

$$V(x) = \begin{cases} 0 & , & x < 0 \\ -4 & , & 0 \le x < 1 \\ -1.33 & , & 1 \le x < 2 \\ -4 & , & 2 \le x < d \\ 0 & , & d \le x \end{cases},$$

where the right edge d is to be set to **(a)** $d = 4$, **(b1)** $d = 4.7$, **(b2)** $d = 4.8$, **(b3)** $d = 4.9$, **(b4)** $d = 5$, **(c)** $d = 6$. Start from descriptor 15. In order to facilitate a direct comparison of the plots (b1–b4) show them in a combined plot; see Appendix A.5.3. An example of a mother descriptor 18 which in V0(1), V0(2), V0(3), V0(4) quotes the descriptors 7,8,9,10. Try it out by PL 18. Now modify descriptor 15 according to question (b1) and store away the modified descriptor with ND. Repeat this procedure for (b2), (b3), (b4). With LT you will get the list of titles of all descriptors and at its end the four descriptors just stored away and their numbers. Enter these four numbers in V0(1), V0(2), V0(3), V0(4) of descriptor 18. **(d)** By typing PL plot the graphs corresponding to (b1), (b2), (b3), (b4) in one combined plot. **(e)** Explain the behavior of the second and third eigenstate in terms of admixtures of eigenstates in the two single wells.

3.3.8 Repeat Exercise 3.3.7 for the probability densities.

3.3.9 We consider a quasiperiodic potential which consists of r equal potential wells of width 1 and depth -15 with a separation of 0.5. Plot the wave functions in an increasing number of equal wells **(a)** $r = 1$, **(b)** $r = 2$, **(c)** $r = 3$, **(d)** $r = 4$, **(e)** $r = 5$, **(f)** $r = 6$, **(g)** $r = 7$, **(h)** $r = 8$, **(i)** $r = 9$, **(j)** $r = 10$. Start from descriptor 13. **(k)** Give a qualitative reason for the occurrence of two bands of states in the quasiperiodic potential.

3.3.10 Repeat Exercise 3.3.9 for a quasiperiodic potential with 10 wells of width 0.2 and depth -50 and separation of 0.15 between the wells for $\hbar = 0.658$ and $m = 5.685$. **(a)** Plot the term scheme. Start from descriptor 13. **(b)** Plot the wave functions of the lowest band. **(c)** Plot the wave functions of the second-lowest band. **(d)** Plot the wave functions of the highest band. **(e)** Explain the symmetry structure of the wave functions in a band. Start from the form of the wave functions in the wide square-well potential that is obtained by taking all the walls out.

3.3.11 (a–d) Repeat Exercise 3.3.10 (a–d) for a quasiperiodic potential of a depth of -40 and with the value -5 for the potential in the regions of the intermediate walls separating the narrow wells. Start from descriptor 13. **(e)** Show the two highest states of the lowest band in a separate plot. Switch off the plotting of the potential with V3(1) $f_{POT} = 1$. **(f)** Why are the two highest states separated from the lower eight by a somewhat larger energy gap?

3.4.1 Plot **(a)** the real part, **(b)** the imaginary part, **(c)** the absolute square of a wave packet initially at rest moving in a harmonic-oscillator potential. For $\hbar$ and particle mass use the default values. The oscillator frequency is $\omega = \pi$. The initial data of the wave packet are the initial location $x_0 = -3$ and the relative initial width $f_\sigma = 0.5$. Start from descriptor 16. **(d)** What is the period of the time evolution of the wave function? **(e)** What is the period of the time evolution of the absolute square? **(f)** What is the most general requirement for the periodicity of a wave function describing a physical process periodic in time?

3.4.2 Plot the absolute square of the wave packets initially at rest at $x_0 = -6$ moving in a harmonic-oscillator potential of angular frequency $\omega = \pi$ for the three relative widths **(a)** $f_\sigma = 0.5$, **(b)** $f_\sigma = 2$, **(c)** $f_\sigma = 1$ for two periods of the oscillation. Start from descriptor 17. **(d)** Explain the oscillation of the widths observed in (a) and (b) in terms of a classical particle with inaccurately known initial values of location and momentum.

3.4.3 Plot **(a)** the real part, **(b)** the imaginary part, **(c)** the absolute square of a wave packet initially at rest in the central position of the harmonic-oscillator potential. Choose the relative initial width as $f_\sigma = 1.75$. Start with descriptor 17. **(d)** Why does the wave packet periodically change its width?

3.4.4 (a,b,c) Repeat Exercise 3.4.3 (a,b,c) for the relative width $f_\sigma = 1$. **(d)** Why does the wave packet not change its width over time?

3.5.1 Plot the motion of a "Gaussian" wave packet with initial values $x_0 = 0$, $p_0 = 5$, $\sigma_{x0} = 0.75$, in an infinitely deep square-well potential of width $d = 10$ for the time intervals given in time steps $\Delta t = 0.5$. Choose **(a)** $0 \le t \le 10$, **(b)** $10 \le t \le 20$, **(c)** $20 \le t \le 30$. Start from descriptor 19. **(d)** Why does the wave packet disperse in time? **(e)** Why does the wave exhibit a wiggly shape when close to the wall?

3.5.2 Plot the motion of a "Gaussian" wave packet with initial values $x_0 = 0$, $p_0 = 5$, $\sigma_{x0} = 0.5$, in an infinitely deep square-well potential of width $d = 10$ for the time intervals **(a)** $0 \le t \le 10$, **(b)** $10 \le t \le 20$, **(c)** $20 \le t \le 30$, **(d)** $30 \le t \le 40$, **(e)** $40 \le t \le 50$, **(f)** $50 \le t \le 60$, **(g)** $60 \le t \le 70$, **(h)** $70 \le t \le 80$, **(i)** $80 \le t \le 90$, **(j)** $90 \le t \le 100$, **(k)** $100 \le t \le 110$, **(l)** $110 \le t \le 120$, **(m)** $120 \le t \le 130$ in time steps $\Delta t = 1$. Start from descriptor 19. **(n)** Calculate the time T_1 in which the initial wave packet is re-established. **(o)** Look at the wave packet at time $T_1/2$. **(p)** Why do the classical position of the particle and the position expectation value of the wave packet coincide only at the beginning of the motion. **(q)** Why are there long time intervals in which the position expectation value is almost at rest?

3.5.3 Plot the motion of a "Gaussian" wave packet in an infinitely deep square well for the initial values $x_0 = 2$, $p_0 = 5.184$, $\sigma_{x0} = 0.5$, for a particle of mass 1 ($\hbar = 1$) in time steps of $\Delta t = 0.5$ for the intervals **(a)** $0 \le t \le 10$, **(b)** $58.66 \le t \le 68.66$,

(c) $122.32 \le t \le 132.32$. Start from descriptor 19. (d) Why does the expectation value of the wave packet at $t = 63.66$ coincide with the position of the classical particle?

3.5.4 Plot the motion of a "Gaussian" wave packet in the infinitely deep square well with initial conditions $x_0 = 2$, $p_0 = 1.09956$, $\sigma_{x0} = 0.6$. (a) Start with a plot of the motion of the position expectation value of the wave packet during the time interval $0 \le t \le 127.32$. Subdivide this interval into 40 time steps. To make the wave function disappear decrease the accuracy y to AC(1)= 26. For the projection angles choose PJ(1)= 60 and PJ(2)= −90. Start from descriptor 19. One observes a time interval during which the expectation value of the position of the wave packet is almost at rest. (b) What is the expectation value of the energy of the wave packet during this interval? (c) What is the approximate expectation value of the momentum of the wave packet during this time interval? Plot the probability density of the Gaussian wave packet (set AC(1) back to default) for (d) $0 \le t \le 20$, (e) $20 \le t \le 40$, (f) $40 \le t \le 60$, (g) $60 \le t \le 80$, (h) $80 \le t \le 100$, (i) $100 \le t \le 120$, (j) $120 \le t \le 140$.

3.5.5 (a–g) Repeat Exercise 3.5.4 (a, d, e,. . . ,j) with $x_0 = 0$, $p_0 = 10.9956$, $\sigma_{x0} = 0.6$, for a particle of mass $m = 10$. Start from descriptor 19. (h) Why does the position expectation value of the wave packet in this case follow the motion of the classical particle much longer than in Exercise 3.5.4?

4 Scattering in One Dimension

Contents: Continuum eigenfunctions and continuous spectra. Boundary conditions and stationary solutions of the Schrödinger equation for step potentials. Continuum normalization. Motion of wave packets in step potentials. Transmission and reflection coefficients. Unitarity and Argand diagram. Tunnel effect. Resonances.

4.1 Physical Concepts

4.1.1 Stationary Scattering States. Continuum Eigenstates and Eigenvalues. Continuous Spectra

For a potential with at least one finite limit

$$V(+\infty) = \lim_{x \to \infty} V(x) \quad \text{or} \quad V(-\infty) = \lim_{x \to -\infty} V(x) \tag{4.1}$$

there are normalized eigenstates φ_n only for energies $E < V_c = \min(V(+\infty), V(-\infty))$. In addition to these discrete eigenvalues with normalizable eigenfunctions the Schrödinger equation with a potential satisfying (4.1) also possesses eigenvalues with eigenfunctions which are *not normalizable*. Their fall-off for large values of $|x|$ is not sufficiently fast for the integral of the absolute square $|\varphi|^2$ extended over the whole x axis to have a finite value. Therefore these eigenfunctions are not normalizable and do not represent actual physical states. The eigenvalues E belonging to the non-normalizable eigenfunctions are no longer discrete points but form continuous sets of values, e.g., intervals or a half axis of energy values. The set of continuous eigenvalues is called the *continuous spectrum*, the corresponding non-normalizable eigenfunctions are called *continuum eigenfunctions* $\varphi(E, x)$. They are solutions of the stationary Schrödinger equation (3.6)

$$H\varphi(E, x) = (T + V)\varphi(E, x) = E\varphi(E, x) \quad . \tag{4.2}$$

If a nonsingular potential fulfills the relation $V(x) < V_c$ only for a finite number of regions of finite lengths on the x axis, the continuous spectrum is bounded by

$$V_c \leq E \quad . \tag{4.3}$$

Normalizable solutions of the time-dependent Schrödinger equation can be formed as linear superpositions of these continuum eigenfunctions.

4.1.2 Time-Dependent Solutions of the Schrödinger Equation

Since the continuum eigenfunctions $\varphi(E, x)$ are not normalizable they extend over the x axis to $+\infty$ or $-\infty$, depending on the values of $V(+\infty)$ and $V(-\infty)$. Thus the continuum eigenfunctions can be used to form moving wave packets far away from the region where the potential actually exerts a force on the particle, i.e., in regions of constant or almost constant potential:

$$\psi(x, 0) = \int_{V_c}^{\infty} w(E)\, \varphi(E, x)dE \quad . \tag{4.4}$$

The time-dependent solution of the Schrödinger equation having $\psi(x, 0)$ as initial state at $t = 0$ then takes the form

$$\psi(x, t) = \int_{V_c}^{\infty} w(E)e^{-iEt/\hbar}\varphi(E, x)dE \quad . \tag{4.5}$$

4.1.3 Right-Moving and Left-Moving Stationary Waves of a Free Particle

Equation (2.1) describes the harmonic wave associated with a particle of mass m and momentum p. Since $E = p^2/2m$ is quadratic in p, (2.1) represents two solutions $p = \pm|p|$, with $|p| = \left|\sqrt{2mE}\right|$, for each energy value E. Thus the wave functions (2.1) can also be interpreted as belonging to one of two sets

$$
\begin{aligned}
\psi_+(E, x, t) &= \frac{1}{(2\pi\hbar)^{1/2}} \exp\left\{-\frac{i}{\hbar}(Et - |p|x)\right\} \\
\psi_-(E, x, t) &= \frac{1}{(2\pi\hbar)^{1/2}} \exp\left\{-\frac{i}{\hbar}(Et + |p|x)\right\} \quad .
\end{aligned}
\tag{4.6}
$$

With a spectral function being different from zero for positive values of p only, the superposition (2.4) formed with $\psi_+(E, x, t)$ represents a *right-moving wave packet*, i.e., a wave packet propagating from smaller x values to larger ones. For the same spectral function the wave packet formed with $\psi_-(E, x, t)$ is *left-moving*, i.e., propagating from larger to smaller x values. Actually, the two harmonic waves themselves propagate to the right and to the left, respectively. In analogy to (3.4) they can be factorized:

$$\psi_\pm(E, x, t) = e^{-iEt/\hbar}\varphi_\pm(E, x) \quad , \quad \varphi_\pm(E, x) = (2\pi\hbar)^{-1/2}e^{\pm i|p|x/\hbar} \quad ; \tag{4.7}$$

that is into a solely time-dependent exponential and eigenfunctions $\varphi_+(E, x)$, $\varphi_-(E, x)$ of the kinetic energy T, i.e., the Hamiltonian $H = T$ of a free particle,

$$H\varphi_\pm(E, x) = E\varphi_\pm(E, x) \quad . \tag{4.8}$$

The time-dependent solutions $\psi_+(x, t)$, superpositions of the stationary waves $\varphi_+(E, x)$,

$$\psi_+(x, t) = \int w(p)e^{-iEt/\hbar}\varphi_+(E(p), x)dp \quad , \tag{4.9}$$

only represent right-moving wave packets. Analogously, replacing φ_+ by φ_- leads to left-moving wave packets. For $t = 0$ and real $w(p)$ these wave packets are centered around $x = 0$. If we want to place a right-moving wave packet at $t = t_0$ around $x = x_0$ we have to substitute t with $t - t_0$ and the Gaussian spectral function $f(p)$, (2.5), with

$$w(p) = f(p)e^{-ipx_0/\hbar} \quad . \tag{4.10}$$

This allows us to construct the wave packets incident on a step potential.

The eigenfunctions belonging to different energy eigenvalues E, E' are orthogonal:

$$\int_{-\infty}^{+\infty} \varphi_{\pm}^{*}(E', x)\varphi_{\pm}(E, x)dx = 0 \quad , \tag{4.11}$$

as are those for equal energy eigenvalues but different subscript signs, e.g.,

$$\int_{-\infty}^{\infty} \varphi_{+}^{*}(E, x)\varphi_{-}(E, x)dx = 0 \quad . \tag{4.12}$$

The stationary wave functions $\varphi_{\pm}(E, x)$ are two continuum eigenfunctions to the same energy eigenvalue E, which is therefore called *two-fold degenerate*.

4.1.4 Orthogonality and Continuum Normalization of Stationary Waves of a Free Particle. Completeness

Since the integral over the absolute squares $|\varphi_+|^2$ or $|\varphi_-|^2$ does not exist, a normalization to unity is not possible. The normalization of discrete eigenfunctions is replaced by the *continuum normalization*

$$\int_{-\infty}^{\infty} \varphi_{s'}^{*}(E(p'), x)\varphi_s(E(p), x)dx = \delta_{s's}\delta(p' - p) \quad , \quad s' = \pm \quad , \quad s = \pm \quad . \tag{4.13}$$

This ensures that the normalization of the wave packet is equal to one if the spectral function $w(p)$ is correctly normalized to one:

$$\int_{-\infty}^{\infty} w^{*}(p)w(p)dp = 1 \quad . \tag{4.14}$$

The set of functions $\varphi_s(E, x)$ is *complete*: any absolute-square-integrable function $\varphi(x)$ can be represented by an integral (Fourier's theorem):

$$\varphi(x) = \sum_{s=\pm} \int_{0}^{\infty} w_s(p) \, \varphi_s(E(p), x)dp \quad . \tag{4.15}$$

4.1.5 Boundary Conditions for Stationary Scattering Solutions in Step Potentials

In the step potential, (3.16), the wave numbers k_j , (3.17), determine the solutions (3.18) for given E. The stationary solutions

$$\varphi_j(E, x) = \varphi_{j+}(E, x) + \varphi_{j-}(E, x) \tag{4.16}$$

are superpositions of two exponentials of opposite exponents in the region j: $x_{j-1} \leq x < x_j$:

$$\varphi_{j+}(E, x) = A_j' e^{ik_j x} \quad , \quad \varphi_{j-}(x) = B_j' e^{-ik_j x} \quad . \tag{4.17}$$

The scattering of a right-moving wave packet incident from $-\infty$ is possible for $E \geq V_0 = 0$. We have to distinguish two cases: $E \geq V_N$ and $E < V_N$.

i) For $E \geq V_N$, $k_N = \left|\sqrt{2m(E - V_N)}/\hbar\right|$, an outgoing wave

$$\varphi_N(E, x) = A_N' e^{ik_N x} \tag{4.18}$$

propagates inside the region $x \geq x_{N-1}$, i.e., to the right of the step potential.

ii) For $E < V_N$, $k_N = i\kappa_N$, $\kappa_N = \left|\sqrt{2m(V_N - E)}/\hbar\right|$, there is only an exponentially decreasing wave function

$$\varphi_N(E, x) = A_N' e^{-\kappa_N x} \tag{4.19}$$

in the region $x \geq x_{N-1}$.

The scattering of a left-moving wave packet incident from $+\infty$ is possible for $E \geq V_N$. Again, we have to distinguish two cases: $E \geq V_1 = 0$ and $E < V_1 = 0$.

i) For $E \geq V_1 = 0$, $k_1 = \left|\sqrt{2mE}/\hbar\right|$, there exists only an outgoing wave in the region $-\infty < x < x_1 = 0$:

$$\varphi_1(E, x) = B_1' e^{-ik_1 x} \quad . \tag{4.20}$$

ii) For $E < V_1 = 0$, $k_1 = i\kappa_1$, $\kappa_1 = \left|\sqrt{2m|E|}\right|$, there exists only a wave function

$$\varphi_1(E, x) = B_1' e^{\kappa_1 x} \tag{4.21}$$

decreasing exponentially towards $-\infty$ in the region $-\infty < x < x_1 = 0$.

In the following discussion we shall restrict ourselves to right-moving incoming waves. For this scattering situation the boundary condition is given by (4.18) or (4.19) depending either on the relation $E \geq V_N$ or $E < V_N$.

4.1.6 Stationary Scattering Solutions in Step Potentials

The *stationary solutions* of the Schrödinger equation for a right-moving incoming wave incident on a step potential (3.16) with N regions is of the form

$$\begin{aligned}
\varphi_1(E, x) &= A_1' e^{ik_1 x} + B_1' e^{-ik_1 x} \quad &\text{region 1} \\
\varphi_2(E, x) &= A_2' e^{ik_2 x} + B_2' e^{-ik_2 x} \quad &\text{region 2} \\
&\vdots \quad \vdots \\
\varphi_N(E, x) &= A_N' e^{ik_N x} \quad &\text{region } N
\end{aligned} \tag{4.22}$$

The $(2N - 1)$ coefficients A_j', B_j' are again determined from the requirement of the wave function being continuous and continuously differentiable at the values $x_m, m = 1, \ldots, N - 1$. This leads once more to the conditions (3.34) for $E \neq V_m, V_{m+1}$. For $E = V_m$ or $E = V_{m+1}$, (3.20) has to be used. Again this yields $2(N - 1)$ linear algebraic equations, now, however, for $(2N - 1)$ coefficients A_j', B_j'. Thus, for every value $E \geq V_0 = 0$, a number of $2(N - 1)$ coefficients can be determined as functions of one of them. We single out the coefficient A_1' as the independent one. Its size determines the amplitude of the right-moving wave coming in from $-\infty$. Thus it regulates the strength of the incoming current. It will either be fixed in (4.25) below by a normalization, or simply be set to one.

Since for any real value $E \geq V_0 = 0$ of the energy we find a stationary solution in the step potential, all values $E \geq V_0$ form the continuous spectrum of the Hamiltonian. All corresponding stationary solutions $\varphi(E, x)$ are continuum eigenfunctions with right-moving outgoing waves. There is a further set of eigenfunctions of $E \geq V_N$ for scattering processes where the incoming particles move in from $+\infty$ which we do not further discuss.

4.1.7 Constituent Waves

The pieces $\varphi_j(E, x)$ in region j ($x_{j-1} \leq x < x_j$) of the stationary wave function $\varphi(E, x)$ consist of a *right-moving* and a *left-moving constituent wave*

$$\varphi_{j+}(E, x) = A'_j e^{ik_j x} \quad \text{and} \quad \varphi_{j-}(E, x) = B'_j e^{-ik_j x} \quad , \tag{4.23}$$

if $E > V_j$. For $E < V_j$ the wave number becomes imaginary: $k_j = i\kappa_j$, κ_j real. In this case,

$$\varphi_{j+}(E, x) = A'_j e^{-\kappa_j x} \quad \text{and} \quad \varphi_{j-}(E, x) = B'_j e^{\kappa_j x}$$

represent a decreasing and an increasing exponential, respectively;

x : position variable
$k_j = i\kappa_j$, $\kappa_j = \left| \sqrt{2m(V_j - E)}/\hbar \right|$: wave vector for $E < V_j$
A'_j, B'_j : complex amplitudes

4.1.8 Normalization of Continuum Eigenstates

As all eigenvectors of Hermitian operators the continuum eigenfunctions belonging to different eigenvalues E, E', are orthogonal,

$$\int_{-\infty}^{+\infty} \varphi^*(E', x)\varphi(E, x)dx = 0 \quad . \tag{4.24}$$

The normalization condition for continuum eigenfunctions for $E = E'$ is in analogy to (4.13) given by ($E = p^2/2m$, $E' = p'^2/2m$)

$$\int_{-\infty}^{+\infty} \varphi^*(E', x)\varphi(E, x)dx = \delta(p' - p) \quad . \tag{4.25}$$

Again this ensures that the normalization of a right-moving wave packet is unity if the spectral function $w(p)$ is normalized as in (4.14). It is the normalization (4.25) that fixes the independent coefficient A_1 in the stationary scattering solution (4.22).

4.1.9 Harmonic Waves in a Step Potential

The time-dependent harmonic waves

$$\psi(E, x, t) = e^{-iEt/\hbar}\varphi(E, x) \quad , \tag{4.26}$$

$\varphi(E, x)$: right-moving incident wave continuum eigenfunction (4.22)
$E = p^2/2m$: energy eigenvalue

x: position
t: time,

are solutions of the time-dependent Schrödinger equation. In the regions j with $E > V_j$ they are harmonic waves. Also the time-dependent stationary waves can be decomposed into time-dependent right-moving and left-moving constituent waves:

$$\begin{aligned}
\psi_{j+}(E, x, t) &= e^{-iEt/\hbar}\varphi_{j+}(E, x) \\
\psi_{j-}(E, x, t) &= e^{-iEt/\hbar}\varphi_{j-}(E, x)
\end{aligned} \tag{4.27}$$

4.1.10 Time-Dependent Scattering Solutions in a Step Potential

If we want to describe a particle coming in from the left by a right-moving Gaussian wave packet of spatial width σ_x, as in classical mechanics, we have to set at the initial time $t = 0$ its position to x_0 and its average momentum to p_0. This is accomplished with the time-dependent superposition

$$\psi(x, t) = \int w(p)e^{-iEt/\hbar}\varphi(E(p), x)dp \tag{4.28}$$

of the continuum eigenfunctions $\varphi(E, x)$ with the Gaussian spectral function

$$w(p) = \frac{1}{(2\pi)^{1/4}\sqrt{\sigma_p}} \exp\frac{(p - p_0)^2}{4\sigma_p^2} e^{-ipx_0/\hbar} \quad , \tag{4.29}$$

$E = p^2/2m$: energy
p: momentum
x: position
t: time
p_0: momentum expectation value of incident wave packet
x_0: position expectation value of incident wave packet
$\sigma_p = \hbar/2\sigma_{x0}$ momentum width of incident wave packet
σ_{x0}: spatial width of initial wave packet
$\varphi(E, x)$: right-moving stationary scattering wave

The constituent waves $\psi_{j+}(E, x, t)$, $\psi_{j-}(E, x, t)$ of the wave packet can be formed with the stationary constituent waves

$$\psi_{j\pm}(x, t) = \int w(p)e^{-iEt/\hbar}\varphi_{j\pm}(E(p), x)dp \quad , \tag{4.30}$$

which are right-moving or left-moving.

4.1.11 Transmission and Reflection. Unitarity. The Argand Diagram

For $E \geq V_N$, i.e., $k_N = \left|\sqrt{2m(E - V_N)}/\hbar\right|$, the solution (4.22) is interpreted in the following way:

i) $A_1'e^{ik_1x}$ is the right-moving harmonic wave coming in from $-\infty$

ii) $A_N'e^{ik_Nx}$ is the transmitted wave. It is right moving, going out to $+\infty$

iii) $B_1'e^{-ik_1x}$ is the reflected wave. It is left moving, going out to $-\infty$

For $E < V_N$, i.e., $k_N = i\kappa_N$, $\kappa_N = \left|\sqrt{2m(V_N - E)}/\hbar\right|$, the solution (4.22) contains the term

$$A'_N e^{ik_N x} = A'_N e^{-\kappa_N x} \quad , \tag{4.31}$$

which represents an exponentially decreasing function in the region N. It approaches zero for $x \rightarrow +\infty$. Thus there is no transmission for $E < V_N$. The incoming wave $A'_1 e^{ik_1 x}$ is *totally reflected* to produce the left-moving reflected wave $B'_1 e^{-ik_1 x}$, which goes out to $-\infty$.

For $E > V_N$ the complex functions $A_N = \sqrt{k_1/k_N} A'_N(E)$ and $B_1 = B'_1(E)$ are called the *transmission* and *reflection coefficients*, respectively. Their normalization is fixed by setting the independent coefficient $A'_1 = 1$. They depend on the energy E of the incoming wave and fulfill the *unitarity relation*

$$|A_N|^2 + |B_1|^2 = |A_1|^2 = 1 \quad . \tag{4.32}$$

This relation states that, for varying energy E, the complex quantities $A_N(E)$ and $B_1(E)$ move inside a circle of radius 1 around the origin in the complex plane. This representation of the coefficients A_N, B_1 in the complex plane is known as the *Argand diagram*. The coefficients A_N and B_1 are also called the *scattering-matrix elements* or *S-matrix elements* of transmission and reflection, respectively. Accordingly, (4.32) is called a unitarity relation of the S matrix. A detailed discussion of the physical interpretation of the prominent features of the Argand diagram, e.g., in relation to resonances, is given in Sect. 10.2.

For $E < V_N$ the complex function $B_1(E)$ is the reflection coefficient. There is no transmission of a wave that goes out to infinity. For the normalization $A'_1 = 1$ the reflection coefficient fulfills for $E < V_N$ the unitarity relation

$$|B_1|^2 = |A_1|^2 = 1 \quad . \tag{4.33}$$

Thus, for $E < V_N$, the complex reflection coefficient $B_1(E)$ moves for varying energy E on the unit circle in the complex plane. Related quantities are the transition-matrix elements or *T-matrix elements* T_T, T_R of transmission and reflection:

$$T_T(E) = (A_N(E) - 1)/2i \quad \text{and} \quad T_R(E) = B_1(E)/2i \quad . \tag{4.34}$$

The T-matrix elements fulfill the T-matrix unitarity relation

$$\operatorname{Im} T_T = |T_T|^2 + |T_R|^2 \quad . \tag{4.35}$$

This states that the element $T_T(E)$ moves for varying energy E inside a circle of radius $1/2$ with its center at the point $i/2$ in the complex plane. The element $T_R(E)$ varies inside the circle of radius $1/2$ around the origin of the complex plane.

4.1.12 The Tunnel Effect

We consider a simple potential with three regions

$$V(x) = \begin{cases} 0 & , \quad x < x_1 = 0 \quad \text{region 1} \\ V_0 & , \quad 0 \le x < d \quad \text{region 2} \\ 0 & , \quad d \le x \quad \text{region 3} \end{cases} \quad . \tag{4.36}$$

x: position coordinate
d: width of potential
$V_0 > 0$: potential height

For energies $0 < E < V_0$ a classical particle will be reflected. Quantum mechanics allows a nonvanishing transmission coefficient A_3:

$$|A_3|^2 = \frac{4E(V_0 - E)}{4E(V_0 - E) + V_0^2 \sinh^2 \kappa d} \quad , \quad \kappa = \left| \sqrt{2m(V_0 - E)}/\hbar \right| \quad . \tag{4.37}$$

Thus there is a nonvanishing probability of the particle being transmitted from region 1 into the classically forbidden region 3, if $E < V_0$. This phenomenon is called the *tunnel effect*.

For general potentials, the tunnel effect means that penetration through a repulsive wall is possible if the incident energy is larger than the potential on the other side of the wall.

4.1.13 Resonances

In step potentials with N regions and $V_1 = 0$, transmission is possible for positive energies if $E > V_N$. The transmission coefficient A_N varies with the energy E of the incident particle.

The maxima of the absolute square $|A_N|^2$ of the transmission coefficient are called *transmission resonances*. The energies at which these maxima occur are the *resonance energies*. Because of the unitarity relation (4.32) the absolute square $|B_N|^2$ of the reflection coefficient exhibits a minimum at the resonance energy of transmission. Therefore, in a plot of the energy dependence of the absolute square of the wave function, transmission resonances can be recognized at energies where the interference pattern of the incoming and reflected wave in the region 1 is least prominent or absent.

4.1.14 Phase Shifts upon Reflection at a Steep Rise or Deep Fall of the Potential

We study the reflection and transmission in two adjacent regions l and $l + 1$ with large differences in the values V_l, V_{l+1} of the potentials:

$$V(x) = \begin{cases} V_l & , \quad x_{l-1} \leq x < x_l \quad \text{region } l \\ V_{l+1} & , \quad x_l \leq x < x_{l+1} \quad \text{region } l + 1 \end{cases} \quad . \tag{4.38}$$

A particle with the kinetic energy E is incident on the potential step at $x = x_l$. The wave function in the regions l and $l + 1$ is given by

$$\begin{aligned} \varphi_l(E, x) &= A_l' e^{ik_l x} + B_l' e^{-ik_l x} \\ \varphi_{l+1}(E, x) &= A_{l+1}' e^{ik_{l+1} x} + B_{l+1}' e^{-ik_{l+1} x} \end{aligned} \quad . \tag{4.39}$$

The continuity conditions to be satisfied at $x = x_l$ are

$$\begin{aligned} A_l' e^{ik_l x_l} + B_l' e^{-ik_l x_l} &= A_{l+1}' e^{ik_{l+1} x_l} + B_{l+1}' e^{-ik_{l+1} x_l k} \\ A_l' e^{ik_l x_l} - B_l' e^{-ik_l x_l} &= \frac{k_{l+1}}{k_l} (A_{l+1}' e^{ik_{l+1} x_l} - B_{l+1}' e^{-ik_{l+1} x_l}) \end{aligned} \quad . \tag{4.40}$$

This leads to the solutions

$$
A_l' e^{ik_l x_l} = \frac{1}{2}\left(1 + \frac{k_{l+1}}{k_l}\right) A_{l+1}' e^{ik_{l+1}x_l} + \frac{1}{2}\left(1 - \frac{k_{l+1}}{k_l}\right) B_{l+1}' e^{-ik_{l+1}x_l}
$$

$$
B_l' e^{-ik_l x_l} = \frac{1}{2}\left(1 - \frac{k_{l+1}}{k_l}\right) A_{l+1}' e^{ik_{l+1}x_l} + \frac{1}{2}\left(1 + \frac{k_{l+1}}{k_l}\right) B_{l+1}' e^{-ik_{l+1}x_l} \ . \tag{4.41}
$$

i) Reflection and transmission at a sudden increase in potential energy ($V_l \ll V_{l+1}$).
For a particle with kinetic energy $E \geq V_{l+1}$ closely above the potential value V_{l+1} in the region ($l + 1$) the quotient of the wave numbers satisfies $k_{l+1}/k_l \ll 1$. In this case (4.41) yields for kinetic energies E slightly larger than V_{l+1}

$$
B_l' \approx A_l' \qquad \text{for} \qquad E \geq V_{l+1} \ . \tag{4.42}
$$

We conclude that the reflected wave, i.e., the left-moving constituent wave in region l

$$
\varphi_{l-}(E, x) = B_l' e^{ik_l x} \approx A_l' e^{ik_l x} \ , \tag{4.43}
$$

does not show a *phase shift* compared to the incident wave, i.e., to the left-moving constituent wave in this region

$$
\varphi_{l+}(E, x) = A_l' e^{ik_l x} \ . \tag{4.44}
$$

The analogous situation in optics is the reflection of light on a "thinner medium", which does not exhibit a phase shift either. The analogy rests on the relation of the wave numbers k_l and k_{l+1} in the two adjacent regions. In optics and in quantum mechanics reflection on a thinner medium requires $k_l > k_{l+1}$. Actually, to obtain a vanishing phase shift in quantum mechanics the relation has to be stronger, i.e., $k_l \gg k_{l+1}$.

ii) Reflection and transmission at a sudden decrease in potential energy ($V_l \gg V_{l+1}$).
For a particle of kinetic energy $E \geq V_l$ close above the potential value V_l in the region l the quotient of wave numbers satisfies $k_{l+1}/k_l \gg 1$. For kinetic energies E slightly larger than V_l, (4.41) then leads to the relation

$$
B_l' \approx -A_l' \qquad \text{for} \qquad E \geq V_l \tag{4.45}
$$

which is tantamount to a *phase shift* of π between the reflected wave, i.e., the left-moving constituent wave in region l

$$
\varphi_{l-}(E, x) = B_l' e^{-ik_l x} \approx -A_l' e^{-ik_l x} = A_l' e^{-i(k_l x + \pi)} \ , \tag{4.46}
$$

and the incident or right-moving constituent wave in this region

$$
\varphi_{l+}(E, x) = A_l' e^{ik_l x} \ . \tag{4.47}
$$

This corresponds to the reflection of light on a denser medium (Sect. 4.7 "Analogies in Optics"). Both the quantum-mechanical and the optical situation are characterized by $k_l < k_{l+1}$. In quantum mechanics the phase shift upon reflection on a "denser medium" approaches the value π for the limiting case $k_l \ll k_{l+1}$ only.

iii) Reflection at a high potential step
A particle with a kinetic energy E satisfying $V_l < E \ll V_{l+1}$ is only reflected at $x = x_l$, there is vanishing transmission, i.e., $A_{l+1} = 0$. In region ($l + 1$) the wave number is imaginary $k_{l+1} = i\kappa_{l+1}$, and furthermore $\kappa_{l+1}/k_l \gg 1$. This leads to the relation

$$B'_l \approx -A'_l \qquad \text{for} \qquad V_l < E \ll V_{l+1} \quad . \tag{4.48}$$

As under ii), we conclude that the reflected wave in region l suffers a phase shift of π. This situation is analogous to the reflection at a fixed end.

4.1.15 Transmission Resonances upon Reflection at "Thinner and Denser Media"

We investigate a simple repulsive potential of three regions

$$V(x) = \begin{cases} 0 & x < 0 & \text{region 1} \\ V_0 > 0 & 0 \le x < x_1 = d & \text{region 2} \\ 0 & x_1 \le x & \text{region 3} \end{cases} \tag{4.49}$$

A particle of kinetic energy E slightly larger than V_0 is incident on this potential from the left. Reflection occurs at $x = 0$ and $x = d$. At $x = 0$ the reflection occurs as in optics on a "thinner medium"; thus the reflected wave in region 1 suffers no phase shift. At $x = d$ reflection occurs on a "denser medium" and thus with a phase shift of π for the reflected wave in region 2. The left-moving constituent wave $\varphi_{1-}(E, x)$ in region 1 can be thought of as consisting of two parts interfering with each other: the one reflected at $x = 0$ on a thinner medium and the other reflected at $x = d$ on a denser medium and transmitted into region 1 at $x = 0$. The phase difference of the two parts consists of the phase shift π upon reflection on the denser medium at $x = d$ and the phase shift due to the longer path $k_2(2d)$ of the wave in region 2. The *total phase shift* amounts to

$$\delta = 2k_2 d + \pi \quad . \tag{4.50}$$

For destructive interference of the two parts making up the reflected, i.e., left-moving, constituent wave $\varphi_{1-}(E, x)$ in region 1 this phase difference has to be equal to an odd multiple of π. Thus a transmission resonance for the potential (4.49) under the condition $E - V_0 \ll V_0$ occurs if

$$2k_2 d + \pi = (2l + 1)\pi \quad \text{or} \quad k_2 = l\pi/d \quad \text{for} \quad l = 1, 2, 3, \ldots \quad . \tag{4.51}$$

For the corresponding wavelength we find

$$\lambda_2 = 2\pi/k_2 = 2d/l \quad , \qquad l = 1, 2, 3, \ldots \quad , \tag{4.52}$$

i.e., whenever the wavelength in region 2 is an integer fraction of twice the width of the step potential, transmission is at a maximum. The largest wavelength for which this happens is just twice the width of the potential region. It should be remembered however that the validity of the simple formula (4.51) hinges on the condition $k_2 \ll k_1$ at resonance energy E_l, i.e.,

$$l\hbar\frac{\pi}{d} \ll \sqrt{2mE_l} \quad . \tag{4.53}$$

Under this condition the resonance energies of the kinetic energy of the incident particles are

$$E_l = V_0 + l^2 \frac{1}{2m}\left(\frac{\hbar\pi}{d}\right)^2 \quad . \tag{4.54}$$

The spacing of the resonances increases like l^2 for not too large values of the integer l.

4.1.16 The Quantum-Well Device and the Quantum-Effect Device

Two of the very recent developments in circuit elements based on the tunnel effect are the quantum-well device and, yet more recently, the quantum-effect device. For an introductory article we refer the reader to R.T. Bates "Quantum-Effect Device: Tomorrow's Transistor?" in Scientific American Vol. 258, No. 3, p. 78 (March 1988).

A *quantum-well device* (QWD) with one-dimensional confinement is an arrangement of five layers of material, Fig. 4.1a. The two outer layers are n-doped gallium arsenide, GaAs. The two slices to the left and right of the middle layer are made of aluminium gallium arsenide, AlGaAs. The middle slice is gallium arsenide GaAs. The band structure of AlGaAs is such that no classical electron current flowing in the outer n-doped GaAs can pass it. The middle layer acts like a potential well between the two AlGaAs layers, which act like two barriers. Thus the one-dimensional potential representing the quantum-well device possesses five regions with $V_1 = 0$, $V_2 > 0$, $V_3 < V_2$, $V_4 = V_2$, $V_5 = 0$, Fig. 4.1b.

The electrons in the first region, usually called the emitter, can be transmitted into the fifth region, the collector, only if their initial energy in region 1 matches a resonance energy in the well. In this case the tunnel effect through the barrier (region 2) into the well (region 3) and from here through the second barrier (region 4) into region 5 leads to a sizable transmission coefficient. The adaptation of the resonance energy in the well can be facilitated by connecting the material in regions 1 and 5 to a battery. By varying the voltage between emitter and collector, Fig. 4.1c, the potential can be changed and thus the resonance energy. This effect can be used to steer the current through the quantum-well device.

Another possible way to influence the current is to connect a third electrical contact (base) to the middle layer (region 3) of the quantum-well device. This contact can be used to change the potential V_3 in the well for a fixed voltage between emitter and collector. A circuit element of this kind is called a *quantum-effect device*.

Further Reading

Alonso, Finn: Vol. 3, Chap. 2
Berkeley Physics Course: Vol. 4, Chaps. 7,8
Brandt, Dahmen: Chaps. 4,5
Feynman, Leighton, Sands: Vol. 3, Chaps. 9,16
Flügge: Vol. 1, Chap. 2A
Gasiorowicz: Chap. 5
Merzbacher: Chaps. 3,4,5,6
Messiah: Vol. 1, Chaps. 2,3
Schiff: Chaps. 2,3,4

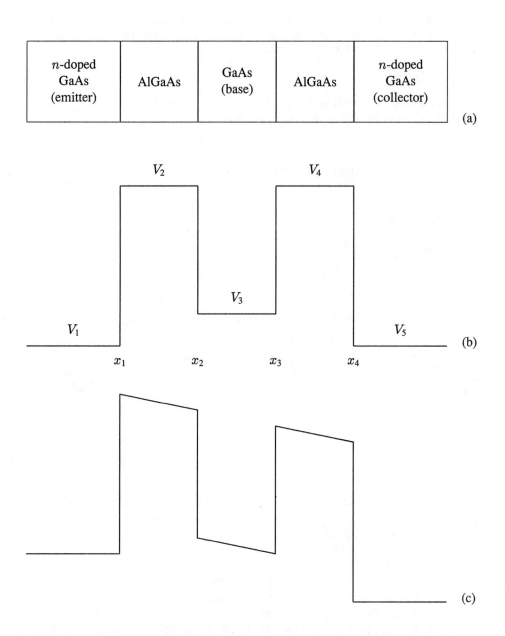

Fig. 4.1. Quantum-well device (a) layers of different materials (b) potential at zero voltage (c) potential with voltage between emitter and collector

4.2 Stationary Scattering States in the Step Potential

Aim of this section: Computation and demonstration of the stationary solution (4.22) of the Schrödinger equation for a right-moving incoming wave in a step potential (3.16) as a function of position x and energy E or momentum p.

Plot type: 0

C3 coordinates: x: position x y: energy E or momentum $p = \sqrt{2mE}$
z: $|\varphi(E, x)|^2$, Re $\varphi(E, x)$, or Im $\varphi(E, x)$ and $V(x)$ and E

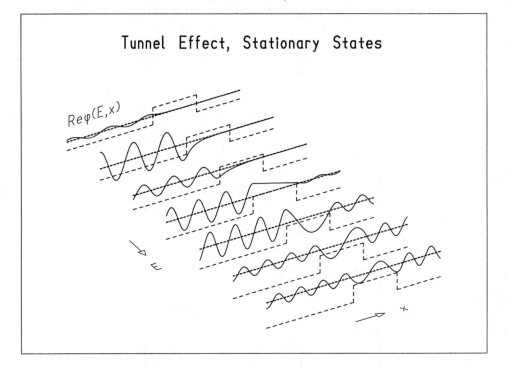

Fig. 4.2. Plot produced with descriptor 1 on file IQ011.DES

Input parameters:

```
CH 0 11 1 0
V0 fRES                        V1 0 0 x0 fpE
V2 N 0 ℏ m                     V3 fCONSTIT
V4 x2 x3 x4 x5                 V5 x6 x7 x8 x9
V6 V2 V3 V4 V5                 V7 V6 V7 V8 V9
V8 lDASH 0 s
```

with

$f_{\text{RES}} = 0$: function shown is $|\varphi(E, x)|^2$ or $|\varphi_{j\pm}(E, x)|^2$

$f_{\text{RES}} = 1$: function shown is Re $\varphi(E, x)$ or Re $\varphi_{j\pm}(E, x)$

$f_{\mathrm{RES}} = 2$: function shown is $\operatorname{Im} \varphi(E, x)$ or $\operatorname{Im} \varphi_{j\pm}(E, x)$

x_0: phase fixing position

$f_{pE} = 0$: momentum $p = \sqrt{2mE}$ is allocated to C3 coordinate y

$f_{pE} = 1$: energy E is allocated to C3 coordinate y

N: number of regions, $2 \le N \le 10$. (For input values < 2 (> 10) the minimum (maximum) permissible value is used.)

$\hbar$: Planck's constant (default value: 1)

m: mass of particle (default value: 1)

$f_{\mathrm{CONSTIT}} = 0$: full solution $\varphi(E, x)$ is computed

$f_{\mathrm{CONSTIT}} = +j$: constituent solution $\varphi_{j+}(E, x)$ is computed

$f_{\mathrm{CONSTIT}} = -j$: constituent solution $\varphi_{j-}(E, x)$ is computed

$x_2, x_3, \ldots, x_{N-1}$: boundaries between individual regions

$V_2, V_3, \ldots, V_N$: potential in individual regions. (For $N = 10$ the potential V_{10} is set to zero)

l_{DASH}: dash length (in W3 units) of lines indicating the potential $V(x)$ (default value: $1/10$ of X width of window in W3). Dash length for zero line is $l_{\mathrm{DASH}}/2$

s: scale factor for plotting $|\varphi(E, x)|^2$ (or $\operatorname{Re} \varphi(E, x)$ or $\operatorname{Im} \varphi(E, x)$) (default value: 1). (Since $|\varphi(E, x)|^2$ is shown with the energy E serving as zero line, actually the function $f(x) = E + s|\varphi(E, x)|^2$ is plotted.)

Plot recommendation:
Plot a series of lines $y = p = \mathrm{const}$ or $y = E = \mathrm{const}$, respectively, i.e., use

```
NL 0 n_y
```

Additional plot items:

Item 6: line indicating potential $V(x)$

Item 7: line indicating the energy $z = E = \mathrm{const}$ and serving as zero line for the function $|\varphi(E, x)|^2$ (or $\operatorname{Re} \varphi(E, x)$ or $\operatorname{Im} \varphi(E, x)$)

Automatically provided texts: T1, T2, TF

Example plot: Fig. 4.2

Remarks:
Note that the constituent solutions $\varphi_{j+}, \varphi_{j-}$ apply only to region j although they are drawn in the complete x range determined by the C3 window, unless you restrict the range to region j. Note also that only the sum φ_j is a solution of the stationary Schrödinger equation.

4.3 Scattering of a Harmonic Wave by the Step Potential

Aim of this section: Computation and demonstration of a time-dependent harmonic wave (4.26) coming in from the left and scattered by a step potential and of the right-moving and left-moving constituent waves (4.27).

Plot type: 0

C3 coordinates: x: position x y: time t
z: $|\Psi(E, x, t)|^2$, $\mathrm{Re}\,\Psi(E, x, t)$, or $\mathrm{Im}\,\Psi(E, x, t)$, and $V(x)$ and E

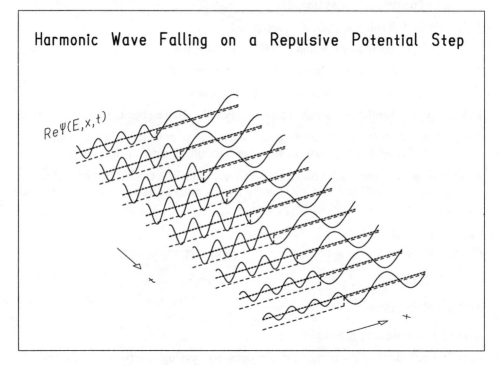

Fig. 4.3. Plot produced with descriptor 2 on file IQ011.DES

Input parameters:

```
CH 0 11 0 0
V0  fRES                        V1  q 0 x0 fpE
V2  N 0 ℏ m                     V3  fCONSTIT
V4  x2 x3 x4 x5                 V5  x6 x7 x8 x9
V6  V2 V3 V4 V5                 V7  V6 V7 V8 V9
V8  lDASH 0 s
```

with

$f_{\mathrm{RES}} = 0$: function shown is $|\Psi(E, x, t)|^2$ or $|\Psi_{j\pm}(E, x, t)|^2$

$f_{\mathrm{RES}} = 1$: function shown is $\mathrm{Re}\,\Psi(E, x, t)$ or $\mathrm{Re}\,\Psi_{j\pm}(E, x, t)$

$f_{RES} = 2$: function shown is Im $\Psi(E, x, t)$ or Im $\Psi_{j\pm}(E, x, t)$

q: $E = q^2/2m$ (i.e., input parameter q is interpreted as momentum) for $f_{pE} = 0$

$\qquad E = q$ (i.e., input parameter q is interpreted as energy) for $f_{pE} = 1$

$f_{pE} = 0$: input parameter q is interpreted as momentum

$f_{pE} = 1$: input parameter q is interpreted as energy

All other variables: As in Sect. 4.2

Plot recommendation:
Plot a series of lines $y = t = $ const, i.e., use

$\qquad$ NL 0 n_y

Additional plot items:

Item 6: line indicating potential $V(x)$

Item 7: line indicating the energy $z = E = $ const and serving as zero line for the function $|\Psi(E, x, t)|^2$ (or Re $\Psi(E, x, t)$ or Im $\Psi(E, x, t)$)

Automatically provided texts: T1, T2, TF

Example plot: Fig. 4.3

Remarks:
Note that the constituent solutions $\Psi_{j\pm}(E, x, t)$ have physical significance only in region j although they are drawn in the complete x range determined by the C3 window, unless you restrict the range to region j. Note also that only the sum $\Psi_j(E, x, t)$ is a solution of the Schrödinger equation.

4.4 Scattering of a Wave Packet by the Step Potential

Aim of this section: Study of the time development (4.28) of a Gaussian wave packet scattered by a step potential. Study of the constituent waves (4.30).

The integration over p has to be performed numerically and is thus approximated by a sum. The larger the number of terms in that sum the better the approximation but also the longer the time needed for computation. You specify the number of terms

$$N_{sum} = 2N_{int} + 1$$

through the input parameter N_{int} (see below). Note that for $N_{int} = 0$ you will just get the harmonic wave of the preceding section. As in Sect. 2.5 the solution obtained numerically as a sum with a finite number of terms will be periodic in x. That is, only in a limited x region will you get a good approximation to the true solution. The patterns for $\Psi(x)$ (or $\Psi_{j+}(x)$ and $\Psi_{j-}(x)$) will repeat themselves periodically along the x direction, the period Δx becoming larger as N_{int} increases. You will have to make sure that the x coordinate of your C3 window is small compared to Δx.

Plot type: 0

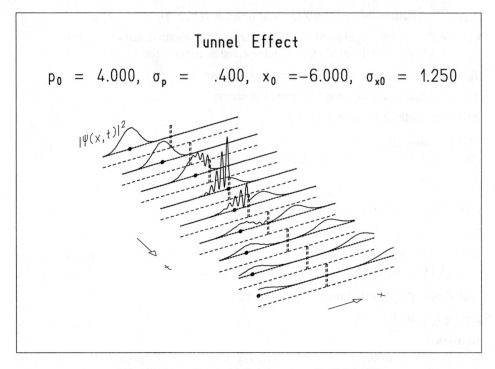

Fig. 4.4. Plot produced with descriptor 3 on file IQ011.DES

C3 coordinates: x: position x y: time t
z: $|\Psi(x,t)|^2$, $\mathrm{Re}\,\Psi(x,t)$, or $\mathrm{Im}\,\Psi(x,t)$, and $V(x)$ and E

Input parameters:

> CH 0 11 0 0
> V0 f_{RES} V1 q σ_q x_0 f_{pE}
> V2 N N_{int} $\hbar$ m V3 f_{CONSTIT}
> V4 x_2 x_3 x_4 x_5 V5 x_6 x_7 x_8 x_9
> V6 V_2 V_3 V_4 V_5 V7 V_6 V_7 V_8 V_9
> V8 l_{DASH} R s

with

$f_{\mathrm{RES}} = 0$: function shown is $|\Psi(x,t)|^2$ or $|\Psi_{j\pm}(x,t)|^2$

$f_{\mathrm{RES}} = 1$: function shown is $\mathrm{Re}\,\Psi(x,t)$ or $\mathrm{Re}\,\Psi_{j\pm}(x,t)$

$f_{\mathrm{RES}} = 2$: function shown is $\mathrm{Im}\,\Psi(x,t)$ or $\mathrm{Im}\,\Psi_{j\pm}(x,t)$

q: $p_0 = q$ for $f_{pE} = 0$ (i.e., input parameter q is interpreted as a momentum expectation value)

q: $p_0 = \sqrt{2mq}$ for $f_{pE} = 1$ (i.e., input parameter q is interpreted as the kinetic energy corresponding to a momentum expectation value)

σ_q: $\sigma_p = \sigma_q p_0$ for $\sigma_q > 0$ (i.e., input parameter is given in units of p_0)
$\qquad \sigma_p = |\sigma_q|$ for $\sigma_q < 0$ (i.e., input parameter is given in absolute units)

x_0: initial position expectation value of wave packet

$f_{pE} = 0$: input parameter q is interpreted as momentum

$f_{pE} = 1$: input parameter q is interpreted as energy

N_{int}: $N_{sum} = 2N_{int} + 1$ is the number of terms in a sum approximating the momentum integration, see above. If $N_{int} < 0$ then **IQ** sets $N_{int} = 0$. If $N_{int} > 40$ then **IQ** sets $N_{int} = 40$

R: radius (in W3 units) of a circle symbolizing the position of a classical particle on the zero line (default value: $l_{DASH}/4$)

All other variables: As in Sect. 4.2

Plot recommendation:
Plot a series of lines $y = t = $ const, i.e., use

    NL 0 n_y

Additional plot items:

Item 5: circle symbolizing position of classical particle (drawn only for $f_{RES} = 0$)

Item 6: line indicating potential $V(x)$

Item 7: line indicating the energy $E_0 = p_0^2/2m$ and serving as zero line for the function $|\Psi(x,t)|^2$ (or Re $\Psi(x,t)$ or Im $\Psi(x,t)$)

Automatically provided texts: TX, T1, T2, TF

Example plot: Fig. 4.4

Remarks:
Note that the constituent wave packets $\Psi_{j\pm}(x,t)$ have physical significance only in region j although they are drawn in the complete x range determined by the C3 window, unless you restrict the range to region j. Note that only the sum $\Psi_j(x,t)$ is solution of the Schrödinger equation.

4.5 Transmission and Reflection. The Argand Diagram

Aim of this section: Presentation of the complex transmission coefficient $A_N(E)$ and the complex reflection coefficient $B_1(E)$ and of the complex T-matrix elements of transmission $T_T(E)$ and of reflection $T_R(E)$, see (4.34).

If $C(E)$ is one of these quantities, we want to illustrate its energy dependence by 4 different graphs:

- The Argand diagram Re $\{C(E)\}$ vs. Im $\{C(E)\}$
- The real part Re $\{C(E)\}$ as a function of E
- The imaginary part Im $\{C(E)\}$ as a function of E
- The absolute square $|C(E)|^2$ as a function of E

Plot type: 2

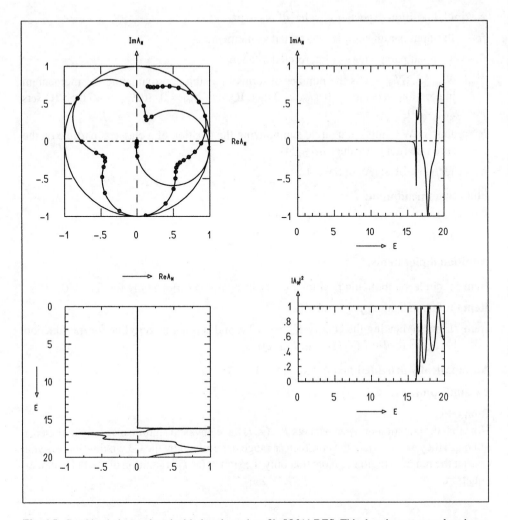

Fig. 4.5. Combined plot produced with descriptor 4 on file IQ011.DES. This descriptor quotes descriptors 5, 6, 7 and 8 to generate the individual plots situated in the top left, top right, bottom left and bottom right corners, respectively

C3 coordinates: x: E or Re $\{C\}$ y: E or Im $\{C\}$ or $|C|^2$

Input parameters:

CH 2 11 f_C
NL 0 f_{VAR}
V0 f_{RES}
V2 N 0 $\hbar$ m
V4 x_2 x_3 x_4 x_5 V5 x_6 x_7 x_8 x_9
V6 V_2 V_3 V_4 V_5 V7 V_6 V_7 V_8 V_9

with

$f_C = 0:\ C(E) = T_\mathrm{T}(E)$

$f_C = 1:\ C(E) = T_\mathrm{R}(E)$

$f_C = 2:\ C(E) = A_N(E)$

$f_C = 3:\ C(E) = B_1(E)$

$f_\mathrm{VAR} = 0:\ x = E, y = f(C(E))$ as given by f_RES

$f_\mathrm{VAR} = 1:\ y = E, x = f(C(E))$ as given by f_RES

$f_\mathrm{VAR} = 2:$ Argand diagram ($x = \mathrm{Re}\,\{C(E)\}$, $y = \mathrm{Im}\,\{C(E)\}$), $\mathrm{RP}(1) \le E \le \mathrm{RP}(2)$, see Appendix A.3.4.1

$f_\mathrm{RES} = 0:\ f(C(E)) = |C(E)|^2$

$f_\mathrm{RES} = 1:\ f(C(E)) = \mathrm{Re}\,\{C(E)\}$

$f_\mathrm{RES} = 2:\ f(C(E)) = \mathrm{Im}\,\{C(E)\}$

All other variables: As in Sect. 4.2

Additional plot items:

Item 5: polymarkers which you can place at intervals $\Delta E = \mathrm{const}$ on the Argand diagram, see Appendix A.3.4

Item 6: unitarity circle (corresponding to $|B_1(E)|^2 = 1$ or $|B_1(E)|^2 = 0$)

Automatically provided texts: T1, T2

Example plot: Fig. 4.5

Remarks:
It is customary to draw an Argand diagram ($\mathrm{Im}\,\{C(E)\}$ vs. $\mathrm{Re}\,\{C(E)\}$) and graphs $\mathrm{Im}\,\{C(E)\}$ and $\mathrm{Re}\,\{C(E)\}$ in such a way that the graphs appear to be projections to the right and below the Argand diagram, respectively. You can do that by using a *combined plot* descriptor which in turn quotes several individual descriptors (see Appendix A.5.3) as in the example plot, Fig. 4.5.

4.6 Exercises

Please note:

(i) you may watch a semiautomatic demonstration of the material of this chapter by typing SA D11

(ii) for the following exercises use descriptor file IQ011.DES

(iii) if not stated otherwise in the exercises the numerical values of the mass of the particle and of Planck's constant are put to 1.

4.2.1 Plot **(a)** the real part, **(b)** the imaginary part, **(c)** the absolute square of the scattering wave function for the potential

$$V(x) = \begin{cases} 0 & , \quad x < 0 \\ 2 & , \quad 0 \le x < 2 \\ 0 & , \quad 2 \le x \end{cases} .$$

Start from descriptor 17. **(d)** Why is the wave function a linear function in the region of the potential for one of the energies? **(e)** Explain the trend of the transmission coefficient for increasing energies in the plot.

4.2.2 (a,b,c) Repeat Exercise 4.2.1 (a,b,c) for the potential

$$V(x) = \begin{cases} 0 & , \quad x < 0 \\ 2 & , \quad 0 \le x < 0.5 \\ 0 & , \quad 0.5 \le x \end{cases} .$$

(d) Why does the tunnel probability increase in comparison with Exercise 4.2.1? **(e)** Why is the absolute square constant in the region beyond the potential? **(f)** Why does the absolute square show a wave pattern in the region to the left of the potential well?

4.2.3 Plot **(a)** the real part, **(b)** the imaginary part, **(c)** the absolute square of the scattering wave function for the potential

$$V(x) = \begin{cases} 0 & , \quad x < 0 \\ 2 & , \quad 0 \le x < 0.5 \\ 0 & , \quad 0.5 \le x < 1.5 \\ 2 & , \quad 1.5 \le x < 2 \\ 0 & , \quad 2 \le x \end{cases} .$$

Start from descriptor 17. **(d)** Why is the amplitude of the wave pattern of the absolute square to the left of the double well very small for one of the energies? At which energy does this phenomenon occur?

4.2.4 Study **(a)** the real part, **(b)** the imaginary part, **(c)** the absolute square of the wave function in the potential of Exercise 4.2.3 in the neighborhood of the particular energy determined in 4.2.3 (d). Choose in particular the energy interval $1.08 \le E \le 1.14$. Start from descriptor 18. **(d)** Why does the resonance phenomenon occur at the value 1.12 for the energy? **(e)** Is there another resonance at lower energies?

4.2.5 Study the resonances of a repulsive double-well potential:

$$V(x) = \begin{cases} 0 & , \quad x < 0 \\ 2.5 & , \quad 0 \le x < 0.5 \\ 0 & , \quad 0.5 \le x < 2.5 \\ 2.5 & , \quad 2.5 \le x < 3 \\ 0 & , \quad 3 \le x \end{cases} .$$

(a) Plot the absolute square of the wave function in the energy region $0.4 \le E \le 2.4$ in ten intervals. In which energy regions do you see indications for the occurrence of resonances? Determine their energies roughly by changing the limits of the energy interval. Start from descriptor 18. **(b)** Plot the absolute square of the wave function in the neighborhood of the first resonance as an interval of 0.04 in steps of 0.008 energy units. Place the interval in such a way that the resonating wave function comes out best. **(c)** Plot the real part of the wave function in the interval of (b). **(d)** Repeat (b) for the neighborhood of the second resonance. Choose an interval of 0.5 energy units in steps of 0.1. **(e)** Plot the real part of the wave function in the interval of (d). **(f)** What distinguishes the two resonances from each other? How is the difference in the two wave functions correlated to their difference in energy?

4.2.6 Study the resonances of an asymmetric triple-well potential:

$$V(x) = \begin{cases} 0 & , & x < 0 \\ 7 & , & 0 \le x < 0.5 \\ 0 & , & 0.5 \le x < 2.5 \\ 7 & , & 2.5 \le x < 3 \\ 0 & , & 3 \le x < 4.5 \\ 7 & , & 4.5 \le x < 5 \\ 0 & , & 5 \le x \end{cases}.$$

Plot the absolute square of the wave function in the energy regions **(a)** $0.01 \le E \le 1.01$, **(b)** $1.01 \le E \le 2.01$, **(c)** $2.01 \le E \le 3.01$, **(d)** $3.01 \le E \le 4.01$, **(e)** $4.01 \le E \le 5.01$, **(f)** $5.01 \le E \le 6.01$, **(g)** $6.01 \le E \le 7,01$, in steps of $\Delta E = 0.1$. Find the energies of the resonances in these regions. Start from descriptor 18. For **(h–l)** start from descriptors 19, 20, ..., 23, respectively. **(h)** Plot the first resonance in the left-hand well as one of the center plots in a set of 10 plots with an energy resolution 0.01. **(i)** Plot the second resonance in the left-hand well as the center plot in a set of 10 plots with an energy resolution 0.01. **(j)** Plot the third resonance in the left-hand well as the center plot in a set of 10 plots with an energy resolution 0.05. **(k)** Plot the first resonance in the right-hand well in a set of 10 plots with an energy resolution of 0.02. **(l)** Plot the second resonance in the right-hand well in a set of 10 plots with an energy resolution of 0.05.

4.2.7 Plot the transmission probability $|A_N|^2$ for the potential of Exercise 4.2.6. Start from descriptor 24. Determine the energies of the maxima of $|A_N|^2$ and compare them with the resonance energies of Exercise 4.2.6.

4.2.8 Study the resonance behavior in a double-well potential corresponding to the left-hand potential of Exercise 4.2.6:

$$V(x) = \begin{cases} 0 & , & x < 0 \\ 7 & , & 0 \le x < 0.5 \\ 0 & , & 0.5 \le x < 2.5 \\ 7 & , & 2.5 \le x < 3 \\ 0 & , & 3 \le x \end{cases}.$$

Start from descriptor 18. Plot the absolute square of the wave function in the energy regions **(a)** $0.01 \le E \le 1.01$, **(b)** $1.01 \le E \le 2.01$, **(c)** $2.01 \le E \le 3.01$, **(d)** $3.01 \le E \le 4.01$, **(e)** $4.01 \le E \le 5.01$, **(f)** $5.01 \le E \le 6.01$, **(g)** $6.01 \le E \le 7.01$, in steps of $\Delta E = 0.1$. Find the energies of the resonances in these regions. Start from descriptor 18. **(h)** Plot the first resonance as one of the center plots in a set of 10 plots with an energy resolution 0.01. **(i)** Plot the second resonance as one of the center plots in a set of 10 plots with an energy resolution 0.01. **(j)** Plot the third resonance as one of the center plots in a set of of 10 plots with an energy resolution 0.05.

4.2.9 Plot the transmission probability $|A_N|^2$ for the potential of Exercise 4.2.8. Start from descriptor 24. Determine the energies of the maxima of $|A_N|^2$ and compare them with the resonance energies of Exercise 4.2.8.

4.2.10 (a–i) Repeat Exercise 4.2.8 (a–i) for the double-well potential corresponding to the right-hand potential of Exercise 4.2.6:

$$V(x) = \begin{cases} 0 & , \ x < 2.5 \\ 7 & , \ 2.5 \leq x < 3 \\ 0 & , \ 3 \leq x < 4.5 \\ 7 & , \ 4.5 \leq x < 5 \\ 0 & , \ 5 \leq x \end{cases}.$$

Start from descriptor 18. Choose as energy resolution in (a–g) $\Delta E = 0.1$, (h) $\Delta E = 0.02$, (i) $\Delta E = 0.05$.

4.2.11 Plot the transmission probability $|A_N|^2$ for the potential of Exercise 4.2.10. Start from descriptor 24. Determine the energies of the maxima of $|A_N|^2$ and compare them to the resonance energies of Exercise 4.2.10.

4.2.12 We consider a potential with five steps (a model of the quantum-well device)

$$V(x) = \begin{cases} V_1 & , \quad x < 0 \quad \text{region 1} \\ V_2 & , \quad 0 \leq x < 0.5 \quad \text{region 2} \\ V_3 & , \quad 0.5 \leq x < 1 \quad \text{region 3} \\ V_4 & , \quad 1 \leq x < 1.5 \quad \text{region 4} \\ V_5 & , \quad 1.5 \leq x \quad \text{region 5} \end{cases}.$$

For the potential values $V_1 = 0$, $V_2 = 7$, $V_3 = -0.5$, $V_4 = 6.5$, $V_5 = -1$, the energy E_{res} for a transmission resonance is $E_{res} = 3.67$. Plot the square of the wave function for **(a)** "zero voltage" $V_1 = 0$, $V_2 = 7$, $V_3 = 0$, $V_4 = 7$, $V_5 = 0$, **(b)** a voltage below resonance $V_1 = 0$, $V_2 = 7$, $V_3 = -0.25$, $V_4 = 6.75$, $V_5 = -0.5$, **(c)** resonance voltage $V_1 = 0$, $V_2 = 7$, $V_3 = -0.5$, $V_4 = 6.5$, $V_5 = -1$, **(d)** a voltage above resonance $V_1 = 0$, $V_2 = 7$, $V_3 = -0.75$, $V_4 = 6.25$, $V_5 = -1.5$. Start from descriptor 17. **(e)** For comparison put the descriptors (a–d) into one combined plot. Start from the mother descriptor 25. The voltage U at the quantum well is given by the difference $U = V_1 - V_5$. The variation of the voltage is $\Delta U = \Delta V_1 - \Delta V_5 = -\Delta V_5$ since the potential V_1 is kept fixed in the cases (a–d). The electric current I through the quantum-well device is proportional to the absolute square $|A_5|^2$ of the coefficient A_5: $I = \alpha |A_5|^2$. Thus the variation ΔI of the electric current is proportional of the variation $\Delta |A_5|^2$ of the absolute square of the coefficient A_5: $\Delta I = \alpha \Delta |A_5|^2$. For a given variation $\Delta U = -\Delta V_5$ of the voltage the quotient $R = -\Delta U / \Delta I = -\Delta V_3 / \Delta I$ of ΔU and the corresponding variation ΔI of the electric current is called the differential resistance. It is given by $R = -(1/\alpha)\Delta U / \Delta |A_5|^2$. What is the trend of the differential restistance that can be learned for $\Delta U = 0.5$ from the comparison if the situations (ab), (bc), (cd)?

4.2.13 Use the potential of Exercise 4.1.12 with the potential parameters at resonance $V_1 = 0$, $V_2 = 60$, $V_3 = -2$, $V_4 = 58$, $V_5 = -4$. There is a lowest transmission resonance in this step potential at $E_1 = 8.464$. Plot the square of the wave function in the neighborhood of the lowest resonance for **(a)** a voltage below resonance $V_1 = 0$, $V_2 = 60$, $V_3 = -1.95$, $V_4 = 58.05$, $V_5 = -3.9$, **(b)** a voltage close below resonance $V_1 = 0$, $V_2 = 60$, $V_3 = -1.993$, $V_4 = 58.007$, $V_5 = -3.986$, **(c)** resonance voltage $V_1 = 0$, $V_2 = 60$, $V_3 = -2$, $V_4 = 58$, $V_5 = -4$, **(d)** a voltage above resonance $V_1 = 0$, $V_2 = 60$, $V_3 = -2.007$, $V_4 = 57.993$, $V_5 = -4.014$. Start from descriptor 17. **(e)** For comparison put the graphs (a–d) into one combined plot. Start from the mother descriptor 25. **(f)** Why is the variation of the differential resistance much faster than in Exercise 4.2.12?

4.2.14 Repeat Exercise 4.1.13 for a much higher voltage between emitter and collector so that the step potential has at $E_{res} = 8.464$ the parameters $V_1 = 0$, $V_2 = 60$, $V_3 = -33.25$,

$V_4 = 26.75$, $V_5 = -6.65$. Plot the square of the wave function in the neighborhood of
(a) the voltage below resonance $V_1 = 0$, $V_2 = 60$, $V_3 = -33.1$, $V_4 = 26.9$, $V_5 = -66.2$,
(b) the voltage below resonance $V_1 = 0$, $V_2 = 60$, $V_3 = -33.2$, $V_4 = 26.8$, $V_5 = -66.4$,
(c) the resonance voltage $V_1 = 0$, $V_2 = 60$, $V_3 = -33.25$, $V_4 = 26.75$, $V_5 = -66.5$, (d) the
voltage above resonance $V_1 = 0$, $V_2 = 60$, $V_3 = -33.3$, $V_4 = 26.7$, $V_5 = -66.6$. Start
from descriptor 17. (e) For comparison put the plots (a–d) into one combined plot. Start
from the mother descriptor 25. (f) Why is the variation of the differential resistance for
this second resonance slower than for the lowest resonance as studied in Exercise 4.2.13?

4.2.15 This exercise models a quantum-effect device. Make use of the potential of five
regions of Exercise 4.2.12. Plot the square of the wave function for different voltages at the
base (region 3), i.e., for the different potential values V_3 (a) below resonance $V_3 = -1.9$,
(b) slightly below resonance $V_3 = -1.98$, (c) at resonance $V_3 = -2$, (d) above resonance
$V_3 = -2.03$, whereas the potential values in the other regions remain unchanged at $V_1 = 0$,
$V_2 = 40$, $V_4 = 38$, $V_5 = -4$. Start from descriptor 17. (e) Put the four plots (a–d) into one
combined plot. Start from the mother descriptor 25. (f) For the case of a quantum-effect
device the variation of the potential V_3 leads to a change $\Delta U = -\Delta V_3$ in the voltage
$U = V_1 - V_3$. Thus the differential resistance (see Exercise 4.2.12 (f)) is accordingly
defined as $R = \Delta U/\Delta I = -\Delta V_3/\Delta I$. What is the trend of the differential resistance that
can be learned from the comparison of the situations (ab), (bc), (cd)?

4.2.16 Make use of the potential of five regions as in Exercise 4.2.12. The potential in
four of the five regions is kept fixed to the values $V_1 = 0$, $V_2 = 10$, $V_4 = 9.5$, $V_5 = -1$. In
region 3 the potential values are changed. Plot the square of the wave function (a) below
resonance $V_3 = -0.2$, (b) at resonance $V_3 = -0.5$, (c) slightly above resonance $V_3 = -0.8$,
(d) above resonance $V_3 = -1$. Start from descriptor 17. (e) Put the four plots (a–d) into
one combined plot. Start from mother descriptor 25.

4.2.17 Use a potential of seven regions

$$V(x) = \begin{cases} V_1 \;, & x < 0 & \text{region 1} \\ V_2 \;, & 0 \leq x < 0.5 & \text{region 2} \\ V_3 \;, & 0.5 \leq x < 1 & \text{region 3} \\ V_4 \;, & 1 \leq x < 1.5 & \text{region 4} \\ V_5 \;, & 1.5 \leq x < 2 & \text{region 5} \\ V_6 \;, & 2 \leq x < 2.5 & \text{region 6} \\ V_7 \;, & 2.5 \leq x & \text{region 7} \end{cases}$$

The values $V_1 = 0$, $V_2 = 10$, $V_4 = 9.5$, $V_6 = 9$, $V_7 = -1$, are kept fixed. The values in
the regions 3 and 5 are varied: (a) below resonance $V_3 = V_5 = -0.35$, (b) at resonance
$V_3 = V_5 = -0.5$, (c) slightly above resonance $V_3 = V_5 = -0.65$, (d) above resonance
$V_3 = V_5 = -0.75$. Start from descriptor 17. (e) Put the four plots (a–d) into one combined
plot. Start from mother descriptor 25. (f) Why is the variation of the differential resistance
in this case much faster than in Exercise 4.2.16?

4.3.1 Plot (a) the real part, (b) the imaginary part, (c) the absolute square of a harmonic
wave of momentum $p_0 = 2.2$ for the time range $0 \leq t \leq 6$ in the potential
$$V(x) = \begin{cases} 0 \;, & x < 0 \\ 32 \;, & 0 \leq x \end{cases} \;.$$
Start from descriptor 26. (d) Why does the wave function look similar to the reflection of
a wave on a fixed end?

4.3.2 (a,b,c) Repeat Exercise 4.3.1 (a,b,c) for the potential
$$V(x) = \begin{cases} 0 & , \quad x < 0 \\ -3000 & , \quad 0 \le x \end{cases}.$$
(d) Why does the wave function look similar to the reflection of a light wave on a denser medium?

4.3.3 (a,b,c) Repeat Exercise 4.3.1 (a,b,c) for a potential
$$V(x) = \begin{cases} 0 & , \quad x < 0 \\ 32 & , \quad 0 \le x \end{cases}$$
for a harmonic wave of momentum $p_0 = 8$ for the time range $0 \le t \le 6$. Start from descriptor 26. **(d)** Why does the reflection pattern now look like one at a thinner medium?

4.3.4 Plot **(a)** the real part, **(b)** the imaginary part, **(c)** the absolute square of a harmonic wave of momentum $p_0 = 1.2$ for the time range $0 \le t \le 6$ in the potential of Exercise 4.2.2. Start from descriptor 26. **(d)** Why do the plots 4.2.2 (c) at $p_0 = 1.2$ and 4.3.4 (c) look alike?

4.3.5 Plot **(a)** the real part, **(b)** the imaginary part, **(c)** the absolute square of a harmonic wave of momentum $p_0 = 1.2$ for the time range $0 \le t \le 6$ in the potential of Exercise 4.2.3. Start from descriptor 26. **(d)** Why is the amplitude of the real part of wave function to the right of the potential barrier time independent? **(e)** Why is the amplitude of the wave function to the left of the potential barrier time dependent?

4.4.1 Plot the motion of the wave packet incident on a step potential
$$V(x) = \begin{cases} 0 & , \quad x < 0 \\ 6 & , \quad x \ge 0 \end{cases}$$
for the width $\sigma_p = 0.3$ and the energies **(a)** $E = 2$, **(b)** $E = 6.5$, **(c)** $E = 8$. Start from descriptor 27. **(d)** Describe the trend of the reflection probability. **(e)** Why is the transmitted wave packet in (b) faster than the classical particle?

4.4.2 Plot the motion of a wave packet incident on a down-step potential
$$V(x) = \begin{cases} 0 & , \quad x < 0 \\ -6 & , \quad x \ge 0 \end{cases}$$
for the width $\sigma_p = 0.3$ and the energies **(a)** $E = 1$, **(b)** $E = 2$, **(c)** $E = 4$. The initial position expectation value of the wave packet is $x_0 = -6$. Start from descriptor 27.

4.4.3 (a–c) Repeat Exercise 4.4.2 for the potential
$$V(x) = \begin{cases} 0 & , \quad x < 0 \\ -12 & , \quad x \ge 0 \end{cases}.$$
(d) Why does the reflection probability increase with the height of the step (see Exercise 4.4.2)?

4.4.4 Plot the transmission probability $|A_N|^2$ for the potential
$$V(x) = \begin{cases} 0 & , \quad x < 0 \\ 16 & , \quad 0 \le x < 4 \\ 0 & , \quad 4 \le x \end{cases}$$
for the energy range $15 \le E \le 22$. Determine the energies of the four transmission resonances ($|A_N| = 1$) in this energy range. Start from descriptor 28.

4.4.5 (a,b) Plot the scattering of the wave packet of width $\sigma_p = 0.05$ at the repulsive square-well potential of Exercise 4.4.4 for the lowest two resonance energies determined in Exercise 4.4.4. Choose 8 intervals in the time range $0 \le t \le 12$. Start from descriptor 27. **(c)** Explain the occurrence of the resonance phenomena. Give a qualitative argument for the energies at which they occur. **(d)** For the last four consecutive time instants determine the ratios of the maxima of the wave functions within the range of the barrier. Explain why these ratios are approximately equal.

4.4.6 (a) Plot the scattering of the wave packet of width $\sigma_p = 0.05$ at the repulsive square-well potential of Exercise 4.4.4 for the third resonance energy determined in Exercise 4.4.4. Choose 8 intervals in the time range $0 \le t \le 12$. **(b)** Study the time range $8 \le t \le 12$ in 8 intervals. Start from descriptor 27.

4.4.7 (a) Repeat Exercise 4.4.6 (a) for the fourth resonance energy determined in Exercise 4.4.4. Choose 8 intervals in the time range $0 \le t \le 8$. **(b)** Study the time range $6 \le t \le 8$ in 8 intervals. Start from descriptor 27. **(c)** Why is the exponential decay faster with higher resonance energy? Compare with the result of Exercises 4.4.2 and 4.4.3 and look at the Argand diagram of Exercise 4.4.4.

4.4.8 In the energy range $0 \le E \le 40$ study the energy dependence of the complex transmission amplitude **(a)** $|A_N|^2$ (descriptor 29), **(b)** Re A_N (descriptor 30), **(c)** Im A_N (descriptor 31), **(d)** Re A_N and Im A_N in an Argand plot (descriptor 32) for the potential

$$V(x) = \begin{cases} 0 & , \quad x < 0 \\ -16 & , \quad 0 \le x < 4 \\ 0 & , \quad 4 \le x \end{cases} .$$

(e) Put the above plots into one multiple plot with the positioning **(d)** upper-left field, **(c)** upper-right field, **(b)** lower-left field, **(a)** lower-right field. Start from descriptor 33. **(f)** Relate the prominent features of the absolute square, the real and imaginary parts to the ones of the Argand plot.

4.4.9 Study the energy dependence of the quantities **(a)** $|T_T|^2$, **(b)** Re T_T, **(c)** Im T_T, **(d)** Im T_T vs. Re T_T in an Argand plot for the potential of Exercise 4.4.8. Use the same descriptors as in Exercise 4.4.8. **(e)** Put the above plots into one multiple plot with the same positioning as in Exercise 4.4.8 (e). Start from descriptor 33. **(f)** Relate the behavior of this quantity to the complex transmission amplitude A_N.

4.4.10 Repeat Exercise 4.4.9 for the complex reflection amplitude B_1.

4.4.11 Repeat Exercise 4.4.9 for the quantity T_R.

4.4.12 Repeat Exercise 4.4.8 with the energy range $0 \le E \le 1$ for the potential

$$V(x) = \begin{cases} 0 & , \quad x < 0 \\ -5 & , \quad 0 \le x < 1 \\ 0 & , \quad 1 \le x \end{cases} .$$

Determine the energy of the lowest resonance.

4.4.13 Plot the bound states in the potential of Exercise 4.4.12 and determine their energy eigenvalues. Start from descriptor 34.

4.4.14 Plot the time development of a wave packet under the action of the potential of Exercise 4.4.12. The wave packet comes in from the initial position $x_0 = -6$ at $t = 0$ with the energy $E = 0.1$ and the absolute width $\sigma_p = 0.05$. The time ranges are **(a)** $0 \le t \le 40$, **(b)** $40 \le t \le 80$, **(c)** $80 \le t \le 120$. They should be subdivided into 8 intervals each. Start from descriptor 35. **(d)** Using also the result of Exercise 4.4.13 interpret the behavior of the wave function inside the potential region.

4.4.15 (a–e) Repeat Exercise 4.4.8 (a–e) for the potential

$$V(x) = \begin{cases} 0 & , & x < 0 \\ 16 & , & 0 \le x < 0.2 \\ 0 & , & 0.2 \le x < 2.2 \\ 16 & , & 2.2 \le x < 2.4 \\ 0 & , & 2.4 \le x \end{cases} .$$

(f) Determine the four lowest resonance energies.

4.4.16 Plot the time development of the wave packet under the action of the potential of Exercise 4.4.15 starting at $t = 0$ at $x_0 = -6$ with the lowest resonance energy and the relative width $\sigma_p/p_0 = 0.3$. **(a)** Plot nine instants in the time range $0 \le t \le 16$. Start from descriptor 36. **(b)** Plot 11 instants in the time range $8 \le t \le 24$. Change the range in x to $-3 \le x \le 3$. Start from descriptor 37.

4.4.17 Repeat Exercise 4.4.16 for the second resonance energy for the relative width $\sigma = 0.15$. **(a)** $0 \le t \le 6$, **(b)** $5 \le t \le 9$.

4.4.18 Repeat Exercise 4.4.16 for the third resonance energy for the relative width $\sigma_p/p_0 = 0.1$. **(a)** $0 \le t \le 4$, **(b)** $2 \le t \le 4$.

4.4.19 Repeat Exercise 4.4.16 for the fourth resonance energy for the relative width $\sigma_p/p_0 = 0.05$. **(a)** $0 \le t \le 3$, **(b)** $2.5 \le t \le 3$.

4.4.20 Study the behavior of a wave packet in a five-step potential modelling a quantum-effect device

$$V(x) = \begin{cases} V_1 & , & x < 0 & \text{region 1} \\ V_2 & , & 0 \le x < 0.5 & \text{region 2} \\ V_3 & , & 0.5 \le x < 1 & \text{region 3} \\ V_4 & , & 1 \le x < 1.5 & \text{region 4} \\ V_5 & , & 1.5 \le x < 2 & \text{region 5} \end{cases} .$$

The potential values $V_1 = 0$, $V_2 = 10$, $V_4 = 10.5$, $V_5 = 1$, remain unchanged. The value in region 3 varies **(a)** below resonance $V_3 = 1.5$, **(b)** slightly below resonance $V_3 = 1$, **(c)** at resonance $V_3 = 0.5$, **(d)** above resonance $V_3 = -0.5$. For the above cases (a–d) plot the absolute square of the wave function of a wave packet with the initial data at $t = 0$ $p_0 = 5.410$, $\sigma_p = 0.01p_0$, $x_0 = -6$ moving in the interval $-50 \le x \le 50$ during the time interval $0 \le t \le 20$ in time steps of 1. Start from descriptor 27. **(e)** Put the four plots (a–d) into a combined plot. Start from mother descriptor 33.

4.5.1 In the range $0.1 \le E \le 40$ study the energy dependence of the complex transmission amplitude **(a)** $|A_N|^2$ (descriptor 29) and read off the resonance energies, **(b)** Re A_N (descriptor 30), **(c)** Im A_N (descriptor 31), **(d)** Im A_N vs. Re A_N in an Argand plot (descriptor 32) for the potential

$$V(x) = \begin{cases} 0 & , & x < 0 \\ 16 & , & 0 \le x < 1 \\ 0 & , & 1 \le x \end{cases} .$$

(e) Put the above plots into one multiple plot with the positioning (d) upper-left field, (c) upper-right field, (b) lower-left field, (a) lower-right field. Start from descriptor 33. **(f)** Relate the prominent features of the absolute square, the real and imaginary parts to the ones of the Argand plot. **(g)** Calculate the resonance energies according to (4.54) and compare them to the values read off $|A_N|^2$ in (a).

4.5.2 In the range $0.1 \leq E \leq 40$ study the energy dependence of the quantities **(a)** $|T_T|^2$, **(b)** $\operatorname{Re} T_T$, **(c)** $\operatorname{Im} T_T$, **(d)** $\operatorname{Im} T_T$ vs. $\operatorname{Re} T_T$ in an Argand plot for the potential of Exercise 4.5.1. Use the same descriptors as in Exercise 4.5.1. **(e)** Put the above plots into one multiple plot with the same positioning as in Exercise 4.5.1 (e). Start from descriptor 33. **(f)** Relate the behavior of this quantity to the complex transmission amplitude A_N.

4.5.3 Repeat Exercise 4.5.2 for the complex reflection amplitude B_1.

4.5.4 Repeat Exercise 4.5.2 for the quantity T_R.

4.5.5 In the range $0.1 \leq E \leq 40$ study the energy dependence of the complex transmission amplitude **(a)** $|A_N|^2$ (descriptor 29) and read off the resonance energies, **(b)** $\operatorname{Re} A_N$ (descriptor 30), **(c)** $\operatorname{Im} A_N$ (descriptor 31), **(d)** $\operatorname{Im} A_N$ vs. $\operatorname{Re} A_N$ in an Argand plot (descriptor 32) for the potential

$$V(x) = \begin{cases} 0 & , \quad x < 0 \\ 16 & , \quad 0 \leq x < 2 \\ 0 & , \quad 2 \leq x \end{cases} .$$

(e) Put the above plots into one multiple plot with the positioning (d) upper-left field, (c) upper-right field, (b) lower-left field, (a) lower-right field. Start from descriptor 33. **(f)** Relate the prominent features of the absolute square, the real and imaginary parts to the ones of the Argand plot. **(g)** Calculate the resonance energies according to (4.54) and compare them to those the read off $|A_N|^2$ in (a).

4.5.6 In the range $0.1 \leq E \leq 40$ study the energy dependence of the quantities **(a)** $|T_T|^2$, **(b)** $\operatorname{Re} T_T$, **(c)** $\operatorname{Im} T_T$, **(d)** $\operatorname{Im} T_T$ vs. $\operatorname{Re} T_T$ in an Argand plot for the potential of Exercise 4.5.5. Use the same descriptors as in Exercise 4.5.5. **(e)** Put the above plots into one multiple plot with the same positioning as in Exercise 4.5.5 (e). Start from descriptor 33. **(f)** Relate the behavior of this quantity to the complex transmission amplitude A_N.

4.5.7 Repeat Exercise 4.5.6 for the complex reflection amplitude B_1.

4.5.8 Repeat Exercise 4.5.6 for the quantity T_R.

4.5.9 Explain qualitatively the differences in the behavior of the quantities A_N, T_T, B_1, T_R for the two potentials used in Exercises 4.5.1 – 4.5.4 and Exercises 4.5.5 – 4.5.8, respectively. In particular, argue why the distances of the energies at which $|A_N|^2 = 1$ vary with the potential the way it is observed in Exercises 4.5.1 and 4.5.5.

4.5.10 In the range $0.1 \leq E \leq 40$ study the energy dependence of the complex transmission amplitude **(a)** $|A_N|^2$ (descriptor 29) and read off the resonance energies, **(b)** $\operatorname{Re} A_N$ (descriptor 30), **(c)** $\operatorname{Im} A_N$ (descriptor 31), **(d)** $\operatorname{Im} A_N$ vs. $\operatorname{Re} A_N$ in an Argand plot (descriptor 32) for the potential

$$V(x) = \begin{cases} 0 & , \quad x < 0 \\ 16 & , \quad 0 \leq x < 4 \\ 0 & , \quad 4 \leq x \end{cases} .$$

(e) Put the above plots into one multiple plot with the positioning (d) upper-left field, (c) upper-right field, (b) lower-left field, (a) lower-right field. Start from descriptor 33. **(f)** Relate the prominent features of the absolute square, the real and imaginary parts to the ones of the Argand plot. **(g)** Calculate the resonance energies according to (4.54) and compare them to the values read off $|A_N|^2$ in (a).

4.5.11 In the range $0.1 \leq E \leq 40$ study the energy dependence of the quantities **(a)** $|T_T|^2$, **(b)** $\operatorname{Re} T_T$, **(c)** $\operatorname{Im} T_T$, **(d)** $\operatorname{Im} T_T$ vs. $\operatorname{Re} T_T$ in an Argand plot for the potential of Exercise 4.5.10. Use the same descriptors as in Exercise 4.5.10. **(e)** Put the above plots into one multiple plot with the same positioning as in Exercise 4.5.10 (e). Start from descriptor 33. **(f)** Relate the behavior of this quantity to the complex transmission amplitude A_N.

4.5.12 Repeat Exercise 4.5.11 for the complex reflection amplitude B_1.

4.5.13 Repeat Exercise 4.5.11 for the quantity T_R.

4.5.14 In the ranges **(a)** $0 \leq E \leq 60$, **(b)** $0 \leq E \leq 15$ plot the energy dependence of the complex transmission amplitude $|A_N|^2$ for a quasiperiodic potential of 9 regions with four square wells of width 1 and depth -44 and three separating walls of width 0.2 and depth 0. **(c)** Why do the lowest transmission resonances form a band of four?

4.7 Analogies in Optics

For a right-moving plane wave of light vertically incident on glass or other dielectrics we study reflection and refraction. We choose the x axis normal to the plane surface of the glass. The dielectric may consist of layers $1, 2, \ldots, N$ of different refractive indices $n'_1, \ldots, n'_N$. This divides the x axis into N regions:

$$n(x) = \begin{cases} n_1 = 1 & , \quad x < x_1 = 0 & \text{region } 1 \\ n_2 & , \quad x_1 \leq x < x_2 & \text{region } 2 \\ \vdots & & \\ n_{N-1} & , \quad x_{N-2} \leq x < x_{N-1} & \text{region } N-1 \\ n_N & , \quad x_{N-1} \leq x & \text{region } N \end{cases} \qquad (4.55)$$

For simplicity we have set $n_1 = 1$. Hereby, all the different n_i are relative refractive indices $n_\ell = n'_\ell/n'_1$, with n'_ℓ ($\ell = 1, \ldots, N$) being the absolute refractive index. For all further considerations in this system we may suppress the coordinates y and z so that we deal with a one-dimensional problem. The complex electric field strength (2.14) can be factorized into time-dependent and purely x-dependent factors

$$E_c(\omega, x, t) = e^{-i\omega t} E_s(\omega, x) \quad , \quad E_s(\omega, x) = A e^{ikx} \quad , \quad k = \omega/c \qquad (4.56)$$

The time-independent factor $E_s(\omega, x)$ is called the *stationary electric field strength*. For $k > 0$, the real part of $E_c(\omega, x, t)$

$$\operatorname{Re} E_c(\omega, x, t) = |A| \cos(\omega t - kx - \alpha) \qquad (4.57)$$

represents a right-moving harmonic wave. Here we have decomposed the complex amplitude A into modulus $|A|$ and phase α:

$$A = |A| e^{i\alpha} \quad . \qquad (4.58)$$

Therefore, for $k > 0$ we call

$$A e^{ikx} = E_{s+}(\omega, x) \qquad (4.59)$$

a "right-moving" stationary electric field strength. By the same token, for $k > 0$,

$$A e^{-ikx} = E_{s-}(\omega, x) \qquad (4.60)$$

is called a "left-moving" stationary field strength.

If a right-moving incoming monochromatic light wave of angular frequency ω and wave number $k = \omega/c$ falls onto the arrangement of dielectrics (4.55) we have an outgoing wave in region N,

$$E_N(\omega, x, t) = A'_N e^{ik_N x} \quad , \quad k_N = n_N k_1 \quad , \quad \text{region } N \quad , \tag{4.61}$$

only. For all other regions there is a reflected left-moving wave in addition to the right-moving one. Thus, the stationary electric field in any region ℓ, $1 \le \ell \le N - 1$, is a superposition:

$$E_l(\omega, x) = A'_\ell e^{ik_\ell x} + B'_\ell e^{-ik_\ell x} \quad , \quad k_\ell = n_\ell k_1 \quad , \quad \text{region } \ell \quad . \tag{4.62}$$

The complex electric field strength is given by

$$E_c(\omega, x, t) = e^{-i\omega t} E_s(\omega, x) \tag{4.63}$$

with

$$E_s(\omega, x) = \begin{cases} E_1(\omega, x) & , \quad x < x_1 = 0 \\ E_2(\omega, x) & , \quad 0 \le x < x_2 \\ \vdots \\ E_N(\omega, x) & , \quad x_{N-1} \le x \end{cases} \tag{4.64}$$

$\omega = c_\ell k_\ell$: angular frequency ($\ell = 1, \ldots, N$)

$k_\ell = n_\ell k_1 = n'_\ell k$: wave number in region ℓ

$c_\ell = c/n'_\ell = c_1/n_\ell$: speed of light in region ℓ

c : speed of light in vacuum

n'_ℓ : absolute refractive index

$n_\ell = n'_\ell/n'_1$: relative refractive index

The expression (4.63) solves Maxwell's equations if the coefficients A'_ℓ and B'_ℓ in (4.62) are determined such that the function $E_s(\omega, x)$ is continuous and continuously differentiable at the end points of the regions $1, \ldots, N - 1$, i.e.,

$$\begin{aligned} E_\ell(\omega, x_\ell) &= E_{\ell+1}(\omega, x_\ell) \\ \frac{dE_\ell}{dx}(\omega, x_\ell) &= \frac{dE_{\ell+1}}{dx}(\omega, x_\ell) \end{aligned} \quad , \quad \ell = 1, \ldots, N - 1 \quad . \tag{4.65}$$

This yields a system of equations

$$\begin{aligned} A'_\ell e^{ik_\ell x_\ell} + B'_\ell e^{-ik_\ell x_\ell} &= A'_{\ell+1} e^{ik_{\ell+1} x_\ell} + B'_{\ell+1} e^{-ik_{\ell+1} x_\ell} \\ k_\ell (A'_\ell e^{ik_\ell x_\ell} - B'_\ell e^{-ik_\ell x_\ell}) &= k_{\ell+1} (A'_{\ell+1} e^{ik_{\ell+1} x_\ell} - B'_{\ell+1} e^{-ik_{\ell+1} x_\ell}) \end{aligned} \tag{4.66}$$

for $\ell = 1, \ldots, N - 1$. The condition (4.61) in region N is implemented by setting $B_N = 0$. The $(2N - 1)$ coefficients A'_ℓ, B'_ℓ are determined by the system (4.66) of $(2N - 2)$ equations. Choosing again A'_1 as the independent variable, determining the incoming flux of light, (4.66) constitutes a system of $(2N - 2)$ inhomogeneous linear equations, the term with A'_1 being the inhomogeneity. Its solution yields the coefficients $A'_2, \ldots, A'_N$ and $B'_1, \ldots, B'_{N-1}$ as functions of the wave number $k = k_1$ of the incident wave.

The energy density of the electromagnetic field of light in vacuum averaged over one period $T = 2\pi/\omega$ is given by (2.22). In glass with refractive index n_ℓ it is

$$w_\ell = n_\ell^2 \frac{\varepsilon_0}{2} E_\ell^* E_\ell \quad . \tag{4.67}$$

The *average density of the energy flux* in the wave in glass is

$$S_\ell = w_\ell c_\ell = n_\ell c \frac{\varepsilon_0}{2} E_\ell^* E_\ell \quad , \tag{4.68}$$

where $c_\ell = c/n_\ell$ is the speed of light in glass. Because of the discontinuity of n when passing from one material to the other, neither of the two quantities is continuous. Therefore we plot in addition to Re E_ℓ and Im E_c the absolute square $E_c^* E_c$.

For stationary waves the current densities S of the electromagnetic energy of right-moving and left-moving currents are proportional to the squares of the transmission and reflection coefficients:

$$A_\ell = \sqrt{n_\ell} A_\ell' \quad , \quad B_\ell = \sqrt{n_\ell} B_\ell' \quad . \tag{4.69}$$

Current conservation of the electromagnetic energy simply states that the sum of the transmitted and reflected currents is equal to the incoming current. As a conventional normalization we shall set $A_1 = 1$. Then the conservation of the current of electromagnetic energy leads to the *unitarity relation*

$$|A_N|^2 + |B_1|^2 = |A_1|^2 = 1 \quad , \tag{4.70}$$

which is of the same form as (4.32) for the transmission and reflection coefficients in one-dimensional quantum mechanics.

A *transmission resonance* occurs for those values of the wave number k for which $|A_N|$ is at a maximum and $|B_1|$ therefore at a minimum. For three regions with refractive indices n_1, n_2, n_3 the resonant wave numbers can be determined by simple arguments. In this arrangement there are two surfaces at $x_1 (= 0)$ and $x_2 (= d)$ where reflection occurs. The coefficient $|B_1|$ is at a minimum if the two waves reflected at x_1 and x_2 interfere destructively in the region $x < x_1 = 0$ to the left of $x_1 = 0$. Since there is a phase shift of π for reflection on a denser medium we have to distinguish two cases:

i)

$$1 = n_1 < n_2 < n_3 \quad \text{or} \quad 1 = n_1 > n_2 > n_3 \tag{4.71}$$

In these cases the relative phase shift of the two waves reflected at x_1 and x_2 is simply given by $\delta = 2k_2 d$, $k_2 = n_2 k$. For maximally destructive interference the phase shift δ has to be equal to odd multiples of π.

$$2k_2 d = (2m+1)\pi \quad \text{or} \quad k_2 = (2m+1)\pi/2d, \quad m = 0, 1, 2, \ldots \quad . \tag{4.72}$$

For the wavelength λ_2 in region 2 we find in terms of the thickness of the material

$$\lambda_2 = 4d/(2m+1) \quad . \tag{4.73}$$

The resonant wavelengths in region 2 are odd fractions of $4d$. The longest resonant wavelength ($m = 0$) is then four times the thickness of the middle layer of material. This is the well-known $d = \lambda/4$ condition that ensures a minimization of reflection for light of this wavelength in an optical system with three different refractive indices. It is used to produce antireflex lenses etc. by coating the surface of the glass with a material transparent for visible light and a thickness of $d = \lambda/4$ for an average wavelength of the visible spectrum. In order to achieve the absence of reflection for this wavelength – and therefore little reflection for neighboring wavelengths – the refractive indices have to be chosen suitably according to

$$n_2 = \sqrt{n_1 n_3} \quad . \tag{4.74}$$

For coated lenses in air n_1 is the refractive index of air, n_2 that of the coating and n_3 that of the glass.

ii)

$$1 = n_1 < n_2 > n_3 \quad \text{or} \quad 1 = n_1 > n_2 < n_3 \quad . \tag{4.75}$$

For these cases in addition to the relative phase shift $\delta = 2k_2 d$ caused by the difference $2d$ in the length of the light path there is the phase shift of π from the reflection at the denser medium. For maximally destructive interference we find the condition

$$2k_2 d + \pi = (2m + 1)\pi \quad \text{or} \quad k_2 = m\pi/d \quad , \quad m = 1, 2, \ldots \quad . \tag{4.76}$$

For the wavelength λ_2 in region 2 we find

$$\lambda_2 = 2d/m \quad . \tag{4.77}$$

The longest resonant wavelength is 2d. All others are integer fractions of $2d$.

The stationary waves can be superimposed to form a wave packet of finite energy content in analogy to (2.17). Using the harmonic electric field strength (4.63), with the stationary field strength (4.64), we find that

$$E_c(x, t) = E_0 \int f(k)e^{-ikx_0} E_c(\omega, x, t)dk \tag{4.78}$$

is a Gaussian wave packet centered at $t = 0$ around the initial position $x = x_0$, if we choose the Gaussian spectral function

$$f(k) = \frac{1}{\sqrt{2\pi}\sigma_k} \exp\left[-\frac{(k - k_0)^2}{2\sigma_k^2}\right] \quad . \tag{4.79}$$

Further Reading

Alonso, Finn: Vol. 3, Chaps. 19,20
Berkeley Physics Course: Vol. 3, Chaps. 4,5,6
Brandt, Dahmen: Chap. 2
Feynman, Leighton, Sands: Vol. 2, Chaps. 32,33
Hecht, Zajac: Chaps. 4,7,8,9

4.8 Reflection and Refraction of Stationary Electromagnetic Waves

Aim of this section: Computation and demonstration of the stationary electric field E_s (4.64) for a right-moving incoming wave in a system (4.55) as a function of position x and wave number k of dielectrics.

Plot type: 0

C3 coordinates: x: position x y: wave number k z: $|E_s|^2$, Re E_s, or Im E_s

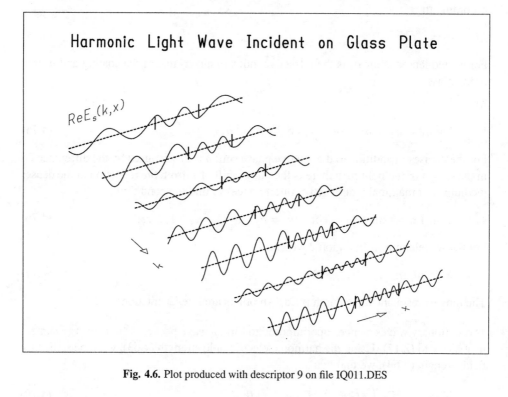

Fig. 4.6. Plot produced with descriptor 9 on file IQ011.DES

Input parameters:

```
CH 0 11 1 1
VO fRES              V1 0 0 x0
V2 N 0               V3 fCONSTIT
V4 x2 x3 x4 x5       V5 x6 x7 x8 x9
V6 n2 n3 n4 n5       V7 n6 n7 n8 n9
V8 lDASH 0 H
```

with

$f_{RES} = 0$: function shown is $|E_s(k, x)|^2$ or $|E_{sj\pm}(k, x)|^2$

$f_{RES} = 1$: function shown is Re $E_s(k, x)$ or Re $E_{sj\pm}(k, x)$

$f_{RES} = 2$: function shown is Im $E_s(k, x)$ or Im $E_{sj\pm}(k, x)$

x_0: phase-fixing position

N: number of regions, $2 \leq N \leq 10$. (For input values < 2 (> 10) the minimum (maximum) permissible value is used.)

$f_{\text{CONSTIT}} = 0$: full solution $E_s(k, x)$ is computed

$f_{\text{CONSTIT}} = +j$: constituent solution $E_{sj+}(k, x)$ is computed

$f_{\text{CONSTIT}} = -j$: constituent solution $E_{sj-}(k, x)$ is computed

$x_2, x_3, \ldots, x_{N-1}$: boundaries between individual regions

$n_2, n_3, \ldots, n_N$: refractive index in individual regions. (For $N = 10$ the potential n_{10} is set to one)

l_{DASH}: dash length (in W3 units) of zero lines (default value: $1/20$ of X width of window in W3)

H: height of vertical lines indicating separating surfaces between different dielectrics (default value: 1)

Plot recommendation:
Plot a series of lines $y = k = \text{const}$, i.e., use

    NL 0 n_y

Additional plot items:

Item 5: vertical lines indicating separating surfaces between different dielectrics

Item 7: zero line for the function $|E_s|^2$ (or $\text{Re}\,E_s$ or $\text{Im}\,E_s$)

Automatically provided texts: T1, T2, TF

Example plot: Fig. 4.6

Remarks:
Note that the constituent solutions E_{sj+}, E_{sj-} apply only to region j although they are drawn in the complete x range determined by the C3 window, unless you restrict the range to region j. Note also that only the sum E_{sj} is a solution of Maxwell's equations.

4.9 Reflection and Refraction of a Harmonic Light Wave

Aim of this section: Computation and demonstration of a time-dependent harmonic wave (4.56) coming in from the left and reflected and refracted into right-moving and left-moving constituent waves.

Plot type: 0

C3 coordinates: x: position x y: time t z: $|E_c|^2$, $\text{Re}\,E_c$ or $\text{Im}\,E_c$

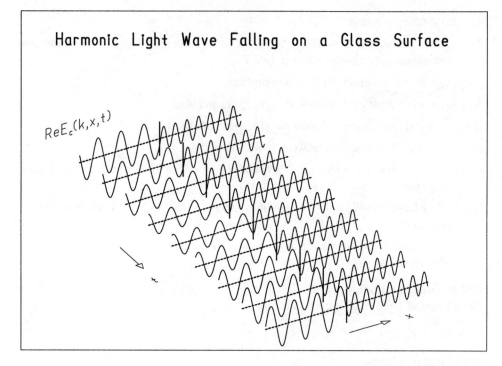

Fig. 4.7. Plot produced with descriptor 10 on file IQ011.DES

Input parameters:

```
CH 0 11 0 1
V0  f_RES                    V1  k 0 x_0
V2  N 0                      V3  f_CONSTIT
V4  x_2 x_3 x_4 x_5          V5  x_6 x_7 x_8 x_9
V6  n_2 n_3 n_4 n_5          V7  n_6 n_7 n_8 n_9
V8  l_DASH 0 H
```

with

$f_{\mathrm{RES}} = 0$: function shown is $|E_c(k, x, t)|^2$ or $|E_{cj\pm}(k, x, t)|^2$

$f_{\mathrm{RES}} = 1$: function shown is $\mathrm{Re}\, E_c(k, x, t)$ or $\mathrm{Re}\, E_{cj\pm}(k, x, t)$

$f_{\mathrm{RES}} = 2$: function shown is $\mathrm{Im}\, E_c(k, x, t)$ or $\mathrm{Im}\, E_{cj\pm}(k, x, t)$

k: wave number

All other variables: As in Sect. 4.8

Plot recommendation:
Plot a series of lines $y = t = $ const, i.e., use

```
NL 0 n_y
```

Additional plot items:

Item 5: vertical lines indicating separating surfaces between different dielectrics

Item 7: zero line for the function $|E_c|^2$ (or $\mathrm{Re}\, E_c$ or $\mathrm{Im}\, E_c$)

Automatically provided texts: T1, T2, TF

Example plot: Fig. 4.7

Remarks:
Note that the constituent solutions $E_{cj\pm}(E, x, t)$ have physical significance only in region j although they are drawn in the complete x range determined by the C3 window, unless you restrict the range to region j. Note also that only the sum E_{cj} is a solution of Maxwell's equations.

4.10 Scattering of a Wave Packet of Light

Aim of this section: Study of the time development (4.78) of a Gaussian wave packet scattered by system of different dielectrics. Study of the constituent waves.

The integration over k has to be performed numerically and is thus approximated by a sum. The larger the number of terms in that sum the better the approximation but also the longer the time needed for computation. You specify the number of terms

$$N_{\mathrm{sum}} = 2N_{\mathrm{int}} + 1$$

through the input parameter N_{int} (see below). Note that for $N_{\mathrm{int}} = 0$ you will just get the harmonic wave of the preceding section. As in Sect. 2.5, the solution obtained numerically as a sum with a finite number of terms will be periodic in x. That is, only in a limited x region will you get a good approximation to the true solution. The patterns for $E_c(x)$ (or $E_{cj+}(x)$ and $E_{cj-}(x)$) will repeat themselves periodically along the x direction, the period Δx becoming larger as N_{int} increases. You will have to make sure that in the x coordinate of your C3 window is small compared to Δx.

Plot type: 0

C3 coordinates: x: position x y: time t z: $|E_c(x, t)|^2$, $\mathrm{Re}\, E_c(x, t)$, or $\mathrm{Im}\, E_c(x, t)$

Input parameters:

```
CH 0 11 0 1
V0 fRES                    V1 k0 σq x0
V2 N Nint                  V3 fCONSTIT
V4 x2 x3 x4 x5             V5 x6 x7 x8 x9
V6 n2 n3 n4 n5             V7 n6 n7 n8 n9
V8 lDASH 0 H
```

with

$f_{\mathrm{RES}} = 0$: function shown is $|E_c(x, t)|^2$ or $|E_{cj\pm}(x, t)|^2$

$f_{\mathrm{RES}} = 1$: function shown is $\mathrm{Re}\, E_c(x, t)$ or $\mathrm{Re}\, E_{cj\pm}(x, t)$

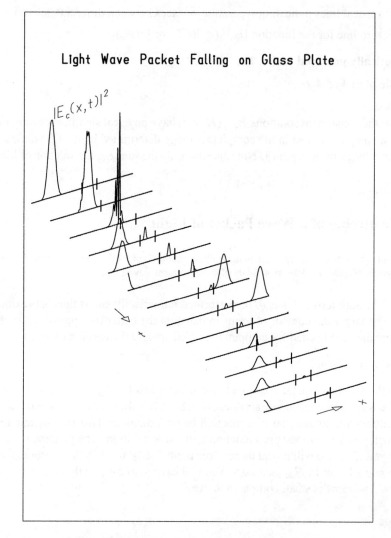

Fig. 4.8. Plot produced with descriptor 11 on file IQ011.DES

$f_{\text{RES}} = 2$: function shown is $\operatorname{Im} E_c(x, t)$ or $\operatorname{Im} E_{cj\pm}(x, t)$

k_0: expectation value of wave number

σ_q: $\sigma_k = \sigma_q k_0$ for $\sigma_q > 0$ (i.e., input parameter is given in units of p_0)
     $\sigma_k = |\sigma_q|$ for $\sigma_q < 0$ (i.e., input parameter is given in absolute units)

x_0: initial position expectation value of wave packet

N_{int}: $N_{\text{sum}} = 2N_{\text{int}} + 1$ is the number of terms in a sum approximating the momentum integration, see above. If $N_{\text{int}} < 0$ then **IQ** sets $N_{\text{int}} = 0$. If $N_{\text{int}} > 40$ then **IQ** sets $N_{\text{int}} = 40$

All other variables: As in Sect. 4.8

Plot recommendation:
Plot a series of lines $y = t = \text{const}$, i.e., use

    NL 0 n_y

Additional plot items:

Item 5: vertical lines indicating separating surfaces between different dielectrics

Item 7: zero line for the function $|E_s|^2$ (or Re E_s or Im E_s)

Automatically provided texts: TX, T1, T2, TF

Example plot: Fig. 4.8

Remarks:
Note that the constituent wave packets $E_{cj\pm}(x, t)$ have physical significance only in region j although they are drawn in the complete x range determined by the C3 window, unless you restrict the range to region j. Note that only the sum $E_{cj}(x, t)$ is a solution of Maxwell's equations.

4.11 Transmission, Reflection and Argand Diagram for a Light Wave

Aim of this section: Presentation of the complex transmission coefficient $A_N(k)$ and the complex reflection coefficient $B_1(k)$, see (4.69).

If $C(k)$ is one of these quantities, we want to illustrate its energy dependence by 4 different graphs:

- The Argand diagram Re $\{C(k)\}$ vs. Im $\{C(k)\}$
- The real part Re $\{C(k)\}$ as a function of k
- The imaginary part Im $\{C(k)\}$ as a function of k
- The absolute square $|C(k)|^2$ as a function of k

Plot type: 2

C3 coordinates: x: k or Re $\{C\}$ y: k or Im $\{C\}$ or $|C|^2$

Input parameters:

```
CH 2 11 fC 1
NL 0 fVAR
V0 fRES
V2 N
V4 x2 x3 x4 x5          V5 x6 x7 x8 x9
V6 n2 n3 n4 n5          V7 n6 n7 n8 n9
```

with

$f_C = 2$: $C(k) = A_N(k)$

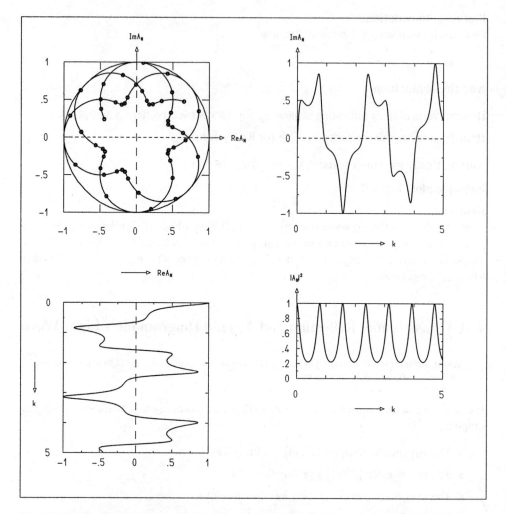

Fig. 4.9. Combined plot produced with descriptor 12 on file IQ011.DES. This descriptor quotes descriptors 13, 14, 15 and 16 to generate the individual plots situated in the top left, top right, bottom left and bottom right corners, respectively

$f_C = 3$: $C(k) = B_1(k)$

$f_{\text{VAR}} = 0$: $x = k$, $y = f(C(k))$ as given by f_{RES}

$f_{\text{VAR}} = 1$: $y = k$, $x = f(C(k))$ as given by f_{RES}

$f_{\text{VAR}} = 2$: Argand diagram ($x = \operatorname{Re}\{C(k)\}$, $y = \operatorname{Im}\{C(k)\}$), $\text{RP}(1) \le k \le \text{RP}(2)$, see Appendix A.3.4.1

$f_{\text{RES}} = 0$: $f(C(k)) = |C(k)|^2$

$f_{\text{RES}} = 1$: $f(C(k)) = \operatorname{Re}\{C(k)\}$

$f_{\text{RES}} = 2$: $f(C(k)) = \operatorname{Im}\{C(k)\}$

All other variables: As in Sect. 4.8

Additional plot items:

Item 5: polymarkers which you can place at intervals $\Delta k = $ const on the Argand diagram, see Appendix A.3.4

Item 6: unitarity circle (corresponding to $|B_1(k)|^2 = 1$ or $|B_1(k)|^2 = 0$)

Automatically provided texts: T1, T2

Example plot: Fig. 4.9

Remarks:
It is customary to draw an Argand diagram ($\mathrm{Im}\{C(k)\}$ vs. $\mathrm{Re}\{C(k)\}$) and graphs $\mathrm{Im}\{C(k)\}$ and $\mathrm{Re}\{C(k)\}$ in such a way that the graphs appear to be projections to the right and below the Argand diagram, respectively. You can do that by using a *combined plot* descriptor which in turn quotes several individual descriptors (see Appendix A.5.3) as in the example plot, Fig. 4.9.

4.12 Exercises

Please note:

(i) for the following exercises use descriptor file IQ011.DES

(ii) in many exercises we use refractive indices with numerical values much larger than available in ordinary dielectrics like glass.

4.8.1 Plot **(a)** the real part, **(b)** the imaginary part, **(c)** the absolute square of the stationary electric field strength in vacuum vertically incident on glass of refractive index $n_2 = 2$ extending from $x = 0$ to infinity. As the range for the wave numbers choose $0.1 \le k \le 5$. Start from descriptor 38. **(d)** Why is the wavelength in region 2, $x \ge 0$, shorter than in region 1, $x < 0$? **(e)** Why is the absolute square constant in region 2? **(f)** What is the origin of the wiggly pattern in region 1?

4.8.2 (a,b,c) Repeat Exercise 4.8.1 (a,b,c) for $n_2 = 10$. Start from descriptor 38. **(d)** Why is the transmission of light into region 2 close to zero?

4.8.3 (a,b,c) Repeat Exercise 4.8.1 (a,b,c) for $n_2 = 0.1$. Start from descriptor 38. This case corresponds to the reflection and refraction of a wave propagating inside an optically denser medium ($n_1 = 1$) incident on a ten-times-thinner medium ($n_2 = 0.1$). **(d)** Why is the transmission into the thinner medium large?

4.8.4 Determine the phase shift between the incident and reflected wave for **(a)** reflection at an optically denser medium $n_2 = 10$, **(b)** reflection at an optically thinner medium $n_2 = 0.1$. To this end compare the phases of the incoming and reflected constituent waves in region 1 with each other. Choose a wave number range $0.01 \le k \le 5$ in six wave number intervals of equal length. Start from descriptor 38.

4.8.5 Plot **(a)** the real part, **(b)** the imaginary part, **(c)** the absolute square of the stationary electric field strength for a sheet of glass of thickness $d = 1$ and refractive index 2. In the domain behind the glass choose $n = 1$. Divide the wave number range $0.001 \le k \le 3\pi/2$ into six intervals of length $\pi/4$. Start from descriptor 38. **(d)** What is the phenomenon behind the absence of wiggles of the absolute square in region 1 for the wave number $k = \pi/2, \pi, 3\pi/2$? **(e)** Why do the resonances in this case occur at $\lambda = 2d/m$, $m = 1, 2, 3, \ldots$?

4.8.6 (a,b,c) Repeat Exercise 4.8.5 (a,b,c), however with a denser medium with $n = 4$ behind the glass. Start from descriptor 38. **(d)** Why do the transmission resonances occur at values $\lambda = 4d/(2m+1)$? **(e)** Why is the square of the electric field in region 3 different from region 1?

4.9.1 Plot the motion of **(a)** the real part, **(b)** the imaginary part, **(c)** the absolute square of a harmonic light wave for different refractive indices in three regions:
$$n_1 = 1 \quad , \quad x < 0$$
$$n_2 = 2 \quad , \quad 0 \le x < 1$$
$$n_3 = 1 \quad , \quad 1 \le x$$
for the wave number $k = 5\pi/4$. Choose the time range $0.001 \le t \le 1$ and the x range $-4 \le x \le 4$. Start from descriptor 39. **(d)** Why are the amplitudes of the electric field strength time independent in region 3? **(e)** Why is this not so in region 1?

4.9.2 Plot the real parts of the constituent waves for Exercise 4.9.1. **(a)** E_{1+}, **(b)** E_{1-}, **(c)** E_{2+}, **(d)** E_{2-}, **(e)** E_{3+}, **(f)** E_{3-}. Start from descriptor 39. **(g)** Why is there no constituent wave E_{3-}?

4.9.3 Repeat Exercise 4.9.1 (a,b,c) for $k = 3\pi/2$.

4.9.4 (a–f) Repeat Exercise 4.9.2 (a–f) for $k = 3\pi/2$. **(g)** Why is there no constituent wave E_{1-}?

4.10.1 A light wave packet of $k_0 = 7.854$ and relative width $\sigma_k/k_0 = 0.01$ is incident on glass of refractive index $n = 4$ and thickness $d = 9.9$ mounted between $x = 0.1$ and $x = 10.1$. Its initial position is $x_0 = -15$. Plot the time dependence of the absolute square of the electric field strength of the wave packet for 10 intervals between **(a)** $0 \le t \le 60$ and **(b)** $60 \le t \le 120$. The interference pattern upon reflection is resolved only for sufficiently high accuracy AC(1). Start from descriptor 40. **(c)** By which factor is the speed of the wave packet in the glass slower than in vacuum? **(d)** Why is the amplitude of $|E_c|^2$ inside the glass plate so much smaller than in vacuum? **(e)** Why does the width of the wave packet shrink upon entering the glass?

4.10.2 (a,b) Repeat Exercise 4.10.1 (a,b), however with two additional layers of refractive index $n = 2$ for $0 \le x < 0.1$ and the other for $10 \le x < 10.1$. Start from descriptor 40. **(c)** Why does practically no reflection occur at any of the surfaces of the regions of different refractive index?

4.10.3 (a) Repeat Exercise 4.10.1 (a) for coated glass layer 1 (coating): $0 \le x < 0.1$, $n_1 = 1.2247$; layer 2 (glass): $0.1 \le x < 10$, $n_2 = 1.5$; layer 3 (coating): $10 \le x < 10.1$, $n_3 = 1.2247$. As the average wave number k_0 of the wave packet, choose 12.826. Start from descriptor 40. **(b)** Calculate the thickness of a coating of the above refractive index for visible light of a vacuum wavelength $\lambda = 550\,\text{nm}$.

4.10.4 (a) Repeat Exercise 4.10.3 (a) without the coating. Start from descriptor 40. **(b)** Why is the reflected part of the wave packet so much smaller than in Exercise 4.10.1?

4.10.5 (a) Repeat Exercise 4.10.3 (a) for a wave number 1.5 times larger. **(b)** Explain why coatings of actual optical lenses often reflect bluish light.

4.11.1 Study the transmission and reflection coefficients of an arrangement of dielectrics of three regions for a range in wave number $0.001 \le k \le 30$: region 1: $n_1 = 1$, region 2:

$0 \leq x < 0.1$, $n_2 = 2$; region 3: $0.1 \leq x$, $n_3 = 4$. Start from descriptor 41. Plot **(a)** the absolute square $|A_N|^2$ of the transmission coefficient, **(b)** the absolute square $|B_1|^2$ of the reflection coefficients, **(c)** the Argand diagram of A_N, **(d)** the Argand diagram of B_1. For (c) and (d) start from descriptor 42. **(e)** Read the wave numbers of resonant transmission off the graph and compare them with the values given by the $(2m + 1)\lambda/4 = d$ condition.

4.11.2 Repeat Exercise 4.11.1 for the choice of refractive indices $n_2 = 1.2247$, $n_3 = 1.5$.

5 A Two-Particle System: Coupled Harmonic Oscillators

Contents: Wave function of two distinguishable particles. Hamiltonian of two coupled oscillators. Separation of center-of-mass and relative coordinates. Stationary two-particle wave functions
and eigenvalues. Initial Gaussian wave packet for distinguishable particles. Time-evolution of the
Gaussian wave packet. Marginal distributions. Wave functions for distinguishable particles. Symmetrization and antisymmetrization. Pauli principle. Bosons and Fermions. Normal oscillations.

5.1 Physical Concepts

5.1.1 The Two-Particle System

The *wave function of a two-particle system* in one spatial dimension $\psi(x_1, x_2, t)$ is a function of two coordinates x_1, x_2 and the time t. The Hamiltonian

$$H = T_1 + T_2 + V(x_1, x_2) \tag{5.1}$$

consists of the two kinetic energies

$$T_i = -\frac{\hbar^2}{2m_i}\frac{\partial^2}{\partial x_i^2} \quad , \quad i = 1, 2 \quad , \tag{5.2}$$

and the potential energy $V(x_1, x_2)$ of the two particles. The time-dependent Schrödinger
equation has the usual form

$$i\hbar\frac{\partial}{\partial t}\psi(x_1, x_2, t) = H\psi(x_1, x_2, t) \quad . \tag{5.3}$$

With the separation of time and spatial coordinates

$$\psi(x_1, x_2, t) = e^{-iEt/\hbar}\varphi_E(x_1, x_2) \tag{5.4}$$

we obtain the stationary Schrödinger equation

$$H\varphi_E(x_1, x_2) = E\varphi_E(x_1, x_2) \tag{5.5}$$

for the stationary wave function $\varphi_E(x_1, x_2)$.

In the following we shall deal with two particles of equal mass $m_1 = m_2 = m$ through-
out. These particles are bound to the origin by harmonic-oscillator potentials

$$V_1(x_1) = \frac{k}{2}x_1^2 \quad , \quad V_2(x_2) = \frac{k}{2}x_2^2 \tag{5.6}$$

with the same *spring constants* $k > 0$. In addition they are coupled to each other through the harmonic two-particle potential

$$V_c(x_1 - x_2) = \frac{\kappa}{2}(x_1 - x_2)^2 \tag{5.7}$$

with the coupling constant κ. Thus the Hamiltonian reads

$$H = T_1 + T_2 + V_1(x_1) + V_2(x_2) + V_c(x_1 - x_2) \quad . \tag{5.8}$$

With the *total mass* M and the *reduced mass* μ,

$$M = 2m \quad , \quad \mu = m/2 \quad , \tag{5.9}$$

and *center-of-mass coordinate* R and *relative coordinate* r,

$$R = (x_1 + x_2)/2 \quad , \quad r = x_2 - x_1 \quad , \tag{5.10}$$

the Hamiltonian can be separated into two terms,

$$H = H_R + H_r \quad , \tag{5.11}$$

each depending on one coordinate only:

$$H_R = -\frac{\hbar^2}{2M}\frac{d^2}{dR^2} + kR^2 \quad , \quad H_r = -\frac{\hbar^2}{2\mu}\frac{d^2}{dr^2} + \frac{1}{2}\left(\frac{k}{2} + \kappa\right)r^2 \quad . \tag{5.12}$$

Both the *center-of-mass motion* and the *relative motion* are harmonic oscillations. They can be separated by factorizing the stationary wave function

$$\varphi_E(x_1, x_2) = U_N(R)u_n(r) \quad , \tag{5.13}$$

where N and n are the quantum numbers of the center-of-mass and relative motion, respectively. The factors fulfill the stationary Schrödinger equations

$$H_R U_N(R) = \left(N + \tfrac{1}{2}\right)\hbar\omega_R\, U_N(R) \quad , \quad H_r u_n(r) = \left(n + \tfrac{1}{2}\right)\hbar\omega_r\, u_n(r) \tag{5.14}$$

with the angular frequencies

$$\omega_R^2 = k/m \quad , \quad \omega_r^2 = (k + 2\kappa)/m \quad . \tag{5.15}$$

The eigenvalue in (5.5)

$$E = E_N + E_n \tag{5.16}$$

is the sum of the eigenvalues

$$E_N = \left(N + \tfrac{1}{2}\right)\hbar\omega_R \quad \text{and} \quad E_n = \left(n + \tfrac{1}{2}\right)\hbar\omega_r \tag{5.17}$$

of the center-of-mass and relative motion. The eigenfunctions $U_N(R)$ and $u_n(r)$ are the eigenfunctions (3.14) of harmonic oscillators of single particles with the angular frequencies ω_R and ω_r, respectively.

5.1.2 Initial Condition for Distinguishable Particles

For the moment we assume that the two particles are *distinguishable*, e.g., that one is a proton and the other one a neutron. As the generalization of the initial condition of the single-particle oscillator we take a Gaussian two-particle wave packet

$$\psi(x_1, x_2, 0) = \frac{1}{\sqrt{2\pi}\sigma_1\sigma_2(1 - c^2)^{1/4}} \exp\left\{-\frac{1}{4(1 - c^2)}\right.$$

$$\left. \times \left[\frac{(x_1 - \langle x_1 \rangle)^2}{\sigma_1^2} - 2c\frac{(x_1 - \langle x_1 \rangle)}{\sigma_1}\frac{(x_2 - \langle x_2 \rangle)}{\sigma_2} + \frac{(x_2 - \langle x_2 \rangle)^2}{\sigma_2^2}\right]\right\} \quad (5.18)$$

x_1, x_2: coordinates of particles 1 and 2
σ_1, σ_2: widths in x_1, x_2 of Gaussian wave packet
c : correlation between x_1 and x_2, $-1 < c < 1$
$\langle x_1 \rangle, \langle x_2 \rangle$: position expectation values

5.1.3 Time-Dependent Wave Functions and Probability Distributions for Distinguishable Particles

The time-evolution of the above initial wave function can be calculated by expanding $\psi(x_1, x_2, 0)$ into a sum over the complete set of eigenfunctions $U_N(R)u_n(r)$,

$$\psi(x_1, x_2, 0) = \sum_{N=0}^{\infty}\sum_{n=0}^{\infty} w_{Nn}U_N(R)u_n(r) \quad , \quad (5.19)$$

which determines the coefficients w_{Nn}. The time-dependent solution of the Schrödinger equation is given by

$$\psi(x_1, x_2, t) = \sum_{N=0}^{\infty}\sum_{n=0}^{\infty} w_{Nn}e^{-i(E_N+E_n)t/\hbar}U_N(R)u_n(r) \quad . \quad (5.20)$$

For brevity we discuss only the absolute square of the time-dependent wave function:

$$\varrho_D(x_1, x_2, t) = |\psi(x_1, x_2, t)|^2 = \frac{1}{2\pi\sigma_1(t)\sigma_2(t)(1 - c^2(t))^{1/2}} \exp\left\{-\frac{1}{2(1 - c^2(t))} \times\right.$$

$$\left.\left[\frac{x_1 - \langle x_1(t) \rangle)^2}{\sigma_1^2(t)} - 2c(t)\frac{(x_1 - \langle x_1(t) \rangle)}{\sigma_1(t)}\frac{(x_2 - \langle x_2(t) \rangle)}{\sigma_2(t)} + \frac{(x_2 - \langle x_2(t) \rangle)^2}{\sigma_2^2(t)}\right]\right\} . (5.21)$$

It differs from the absolute square of $\psi(x_1, x_2, 0)$ only through the time dependence of the expectation values $\langle x_1(t) \rangle, \langle x_2(t) \rangle$, the widths $\sigma_1(t), \sigma_2(t)$, and the correlation $c(t)$. For $t = 0$ they assume the values of the initial wave packet (5.18). The quantity $\varrho(x_1, x_2, t)$ is the *joint probability density* for finding at time t the distinguishable particles 1 and 2 at the locations x_1 and x_2, respectively.

5.1.4 Marginal Distributions for Distinguishable Particles

The probability distribution of particle 1, independent of the position of particle 2, is given by

$$\varrho_{D1}(x_1, t) = \int_{-\infty}^{+\infty} \varrho_D(x_1, x_2, t)dx_2 \quad . \tag{5.22}$$

Consequently, the probability density $\varrho_{D2}(x_2, t)$ of particle 2, independent of the position of particle 1, is given by integrating $\varrho_D(x_1, x_2, t)$ over x_1:

$$\varrho_{D2}(x_2, t) = \int_{-\infty}^{+\infty} \varrho_D(x_1, x_2, t)dx_1 \quad . \tag{5.23}$$

5.1.5 Wave Functions for Indistinguishable Particles. Symmetrization for Bosons. Antisymmetrization for Fermions

For *indistinguishable* particles (e.g., two protons) the Hamiltonian is symmetric under permutations of the coordinates x_1 and x_2 of the particles 1 and 2:

$$H(x_1, x_2) = H(x_2, x_1) \quad . \tag{5.24}$$

Since the particles cannot be distinguished by measurement, all measurable quantities are symmetric in the particles 1 and 2. To ensure the symmetry of the expectation values the two-particle wave functions of indistinguishable particles are *symmetric* for *bosons*:

$$\psi_B(x_1, x_2, t) = \psi_B(x_2, x_1, t) \quad ; \tag{5.25}$$

or *antisymmetric* for *fermions*:

$$\psi_F(x_1, x_2, t) = -\psi_F(x_2, x_1, t) \quad . \tag{5.26}$$

The requirement of antisymmetrization is the *Pauli principle*. Its most important physical implication is that two indistinguishable fermions cannot occupy the same state or be at the same position:

$$\psi_F(x, x, t) = 0 \quad . \tag{5.27}$$

Because of the symmetry of the Hamiltonian (5.24), time-dependent wave functions for bosons or fermions can be obtained by symmetrization or antisymmetrization of the time-dependent solution (5.20):

$$\psi_{B,F}(x_1, x_2, t) = N_{B,F}\frac{1}{\sqrt{2}}[\psi(x_1, x_2, t) \pm \psi(x_2, x_1, t)] \quad . \tag{5.28}$$

The factor $N_{B,F}$ ensures the normalization of the boson or fermion wave function. The probability density for bosons and fermions is given by

$$\varrho_B(x_1, x_2, t) = |\psi_B(x_1, x_2, t)|^2 = |N_B|^2[\varrho_S(x_1, x_2, t) + \varrho_I(x_1, x_2, t)] \tag{5.29}$$

and

$$\varrho_F(x_1, x_2, t) = |\psi_F(x_1, x_2, t)|^2 = |N_F|^2[\varrho_S(x_1, x_2, t) - \varrho_I(x_1, x_2, t)] \quad , \tag{5.30}$$

where

$$\varrho_S(x_1, x_2, t) = \tfrac{1}{2}[\varrho_D(x_1, x_2, t) + \varrho_D(x_2, x_1, t)] \tag{5.31}$$

is the *symmetrized probability density* of of distinguishable particles. The term ϱ_I is the *interference term*

$$\varrho_I(x_1, x_2, t) = \tfrac{1}{2}[\psi^*(x_1, x_2, t)\psi(x_2, x_1, t) + \psi^*(x_2, x_1, t)\psi(x_1, x_2, t)] \quad . \tag{5.32}$$

Whereas ϱ_B, ϱ_F, $\varrho_S \geq 0$, the interference term ϱ_I can assume positive and negative values. The joint probability densities $\varrho_B(x_1, x_2, t)$ and $\varrho_F(x_1, x_2, t)$ are both symmetric under permutation of x_1 and x_2.

5.1.6 Marginal Distributions of the Probability Densities of Bosons and Fermions

Since the joint probability densities ϱ_B and ϱ_F are symmetric in x_1 and x_2, there is only one marginal distribution for bosons and one for fermions:

$$\varrho_B(x, t) = \int_{-\infty}^{+\infty} \varrho_B(x, x_2, t) dx_2 \tag{5.33}$$

and

$$\varrho_F(x, t) = \int_{-\infty}^{\infty} \varrho_F(x, x_2, t) dx_2 \quad . \tag{5.34}$$

Their physical significance is that they give the probability for finding one of the two particles at the position x independent of the position of the other one. Of course it is possible to compute marginal distributions also for the densities ϱ_S and ϱ_I although these have no direct physical significance:

$$\varrho_{S,I}(x, t) = \int_{-\infty}^{\infty} \varrho_{S,I}(x, x_2, t) dx_2 \quad . \tag{5.35}$$

5.1.7 Normal Oscillations

In our system of two identical oscillators in one dimension, with a harmonic coupling, the two normal oscillations of classical mechanics are:

i) the oscillations of the center of mass with a time-independent relative coordinate, and
ii) the oscillation in the relative motion with the center of mass at rest.

The initial conditions (with the two particles initially at rest) which correspond to the two normal oscillations are:

i) identical initial positions $x_{10} = x_{20}$ for the two particles, and
ii) opposite initial positions $x_{10} = -x_{20}$ for the two particles.

In quantum mechanics also, these initial positions lead to the corresponding normal oscillations for the expectation values of the positions $\langle x_1(t) \rangle$, $\langle x_2(t) \rangle$ of the two particles.

Further Reading

Alonso, Finn: Vol. 3, Chaps. 4
Brandt, Dahmen: Chaps. 4,5
Feynman, Leighton, Sands: Vol. 3, Chaps. 4
Gasiorowicz: Chap. 8
Merzbacher: Chap. 20
Messiah: Vol. 2, Chap. 14
Schiff: Chap. 10

5.2 Stationary States

Aim of this section: Computation and presentation of the stationary wave function (5.13).

Plot type: 0

C3 coordinates: x: position x_1 y: position x_2
z: $\varphi_E(x_1, x_2)$ or $|\varphi_E(x_1, x_2)|^2 = \varrho_E(x_1, x_2)$

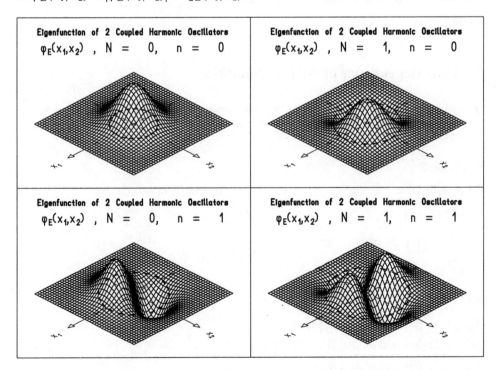

Fig. 5.1. Plot produced with descriptor 1 on file IQ040.DES

Input parameters:

CH 0 40 0

     V0 f_{RES} V1   0   0 k κ

     V2 m

with

$f_{\mathrm{RES}} = 0$: function shown is $|\varphi_E(x_1, x_2)|^2$

$f_{\mathrm{RES}} = 1$: function shown is $\varphi_E(x_1, x_2)$

k: spring constants of individual oscillators (default value: 1)

κ: spring constant of coupling

m: mass of oscillators (default value: 1)

The quantum numbers N and n are equal to the horizontal position ($h = N$) and the vertical position ($v = n$) of the graph within a multiple plot, see Appendix A.5.2.

Additional plot item:

Item 5: ellipse bounding the region in the $x_1 x_2$ plane which is accessible in classical mechanics given by $E = V(x_1, x_2)$

Automatically provided texts: T1, T2, TX

Example plot: Fig. 5.1

5.3 Time Dependence of Global Quantities

Aim of this section: Illustration of the time dependence $\langle x_1(t) \rangle$ and $\langle x_2(t) \rangle$ of the position expectation values, the time dependence $\sigma_1(t)$ and $\sigma_2(t)$ of the widths in x_1 and x_2 and of the time dependence $c(t)$ of the correlation coefficient.

Plot type: 2

C3 coordinates: x: time t y: $\langle x_1(t) \rangle$ or $\langle x_2(t) \rangle$ or $\sigma_1(t)$ or $\sigma_2(t)$ or $c(t)$

Input parameters:

     CH   2   40   0 f_{CASE}

                               V1 $\langle x_{10} \rangle$ $\langle x_{20} \rangle$ k κ

     V2 m σ_{10} σ_{20} c_0

with

$f_{\mathrm{CASE}} = 1$: function plotted is $\langle x_1(t) \rangle$

$f_{\mathrm{CASE}} = 2$: function plotted is $\langle x_2(t) \rangle$

$f_{\mathrm{CASE}} = 3$: function plotted is $\sigma_1(t)$

$f_{\mathrm{CASE}} = 4$: function plotted is $\sigma_2(t)$

$f_{\mathrm{CASE}} = 5$: function plotted is $c(t)$

$f_{\mathrm{CASE}} = 0$: if a multiple plot is asked for, the first graph is $\langle x_1(t) \rangle$, the second $\langle x_2(t) \rangle$, and so on, until the fifth graph, which is $c(t)$

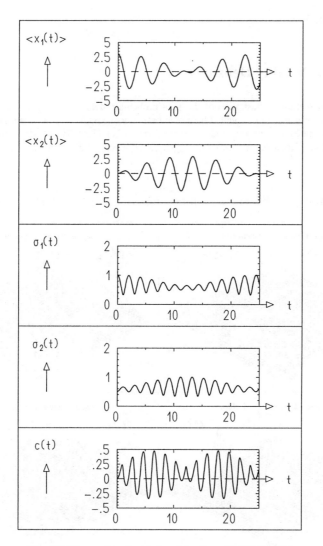

Fig. 5.2. Plot produced with descriptor 2 on file IQ040.DES

$\langle x_{10} \rangle$: initial position expectation value of particle 1

$\langle x_{20} \rangle$: initial position expectation value of particle 2

σ_{10}: initial width in position x_1

σ_{20}: initial width in position x_2

c_0: initial correlation coefficient

k: spring constant for individual oscillators (default value: 1)

κ: spring constant of coupling

m: mass of particle (default value: 1)

Automatically provided texts: 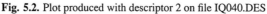 T1, T2

Example plot: Fig. 5.2

5.4 Joint Probability Densities

Aim of this section: Illustration of the joint probability densities $\varrho_D(x_1, x_2, t)$, $\varrho_B(x_1, x_2, t)$, $\varrho_F(x_1, x_2, t)$ for a system of coupled harmonic oscillators composed of two distinguishable particles, two identical bosons or two identical fermions, (5.21), (5.29), (5.30). (It is also possible to illustrate the the functions $\varrho_S(x_1, x_2, t)$ and $\varrho_I(x_1, x_2, t)$ given by (5.31) and (5.32).)

Plot type: 0

C3 coordinates: x: position x_1 y: position x_2 z: $\varrho_i(x_1, x_2, t)$, $i = D, B, F, S, I$

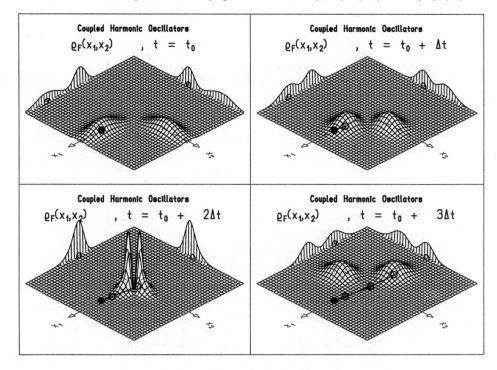

Fig. 5.3. Plot produced with descriptor 3 on file IQ040.DES

Input parameters:

CH 0 40 1 f_{CASE}

V1 $\langle x_{10} \rangle$ $\langle x_{20} \rangle$ k κ

V2 m σ_{10} σ_{20} c_0 V3 t_0 Δt

V4 f_{COV}

V6 0 R

with

$f_{CASE} = 0$: function plotted is $\varrho_D(x_1, x_2, t)$

$f_{\text{CASE}} = 1$: function plotted is $\varrho_B(x_1, x_2, t)$

$f_{\text{CASE}} = 2$: function plotted is $\varrho_F(x_1, x_2, t)$

$f_{\text{CASE}} = 3$: function plotted is $\varrho_S(x_1, x_2, t)$

$f_{\text{CASE}} = 4$: function plotted is $\varrho_I(x_1, x_2, t)$

R: radius (in W3 units) of circle indicating the positions $x_1(t)$ and $x_2(t)$ of the corresponding classical particles. (Default value $1/40$ of X window in W3.)

t_0: initial time

Δt: time interval. In a multiple plot graph $\varrho(x_1, x_2, t_0)$, $\varrho(x_1, x_2, t_0 + \Delta t)$, ... are drawn

$f_{\text{COV}} = 0$: covariance ellipse is drawn for $f_{\text{CASE}} = 0$

$f_{\text{COV}} = 1$: covariance ellipse is not drawn

All other variables: As in Sect. 5.3

Additional plot items:

Item 5: circle symbolizing position of classical particles

Item 6: covariance ellipse

Item 7: trajectory of classical system in $x_1 x_2$ plane

Automatically provided texts: `T1`, `T2`, `TX`

Example plot: Fig. 5.3

Remarks:
Over the borders $x_1 = x_{1\min}$ and $x_2 = x_{2\min}$ of the region (determined by the C3 window) of the $x_1 x_2$ plane the marginal distributions $\varrho_2(x_2, t)$ and $\varrho_1(x_1, t)$ are plotted. Moreover, the trajectory of the system of two classical coupled oscillators is shown in the $x_1 x_2$ plane for the time interval (t_0, t). The position for t_0 is indicated by a full circle, the positions for $t_0 + \Delta t, t_0 + 2\Delta t, \ldots$ by open circles.

5.5 Marginal Distributions

Aim of this section: Illustration of the marginal distributions $\varrho_{D1}(x_1, t)$, $\varrho_{D2}(x_2, t)$, $\varrho_B(x, t)$, $\varrho_F(x, t)$, $\varrho_S(x, t)$, $\varrho_I(x, t)$ discussed in Sect. 5.1, (5.22), (5.23), (5.33), (5.34), (5.35).

Plot type: 0

C3 coordinates: x: position x_1 or x_2 or x y: time t

z: ϱ_i , $i = \text{D1}, \text{D2}, \text{B}, \text{F}, \text{S}, \text{I}$

Input parameters:

```
CH 0 40 2 fCASE
                        V1 ⟨x10⟩ ⟨x20⟩ k κ
V2 m σ10 σ20 c0
V4 fMARG
V6 lDASH R
```

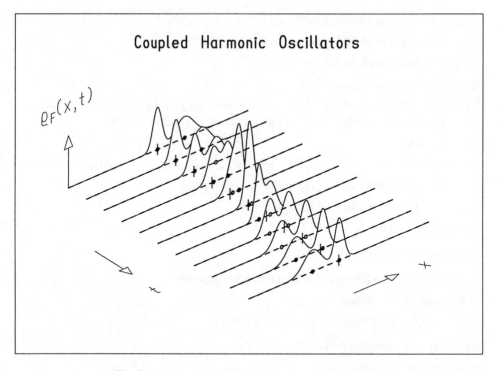

Fig. 5.4. Plot produced with descriptor 4 on file IQ040.DES

with

$f_{CASE} = 0$: function shown is $\varrho_{Di}(x_i, t)$

$f_{CASE} = 1$: function shown is $\varrho_B(x, t)$

$f_{CASE} = 2$: function shown is $\varrho_F(x, t)$

$f_{CASE} = 3$: function shown is $\varrho_S(x, t)$

$f_{CASE} = 4$: function shown is $\varrho_I(x, t)$

$f_{MARG} = 1$: for $f_{CASE} = 0$ the function $\varrho_{D1}(x_1, t)$ is shown

$f_{MARG} = 2$: for $f_{CASE} = 0$ the function $\varrho_{D2}(x_2, t)$ is shown

l_{DASH}: dash length (in W3 units) of zero lines (default value: $1/10$ of X width of window in W3)

R: radius (in W3 units) of circles indicating the positions x_1 and x_2 of the corresponding classical particles (default value $1/40$ of X window in W3)

All other variables: As in Sect. 5.3

Plot recommendation:
Plot a series of lines $y = t = $ const, i.e., use

    NL 0 n_y

Additional plot items:

Item 5: circles symbolizing positions of classical particles

Item 6: short line in z direction intersecting each zero line at $x = 0$ indicating the classical rest position of both particles

Item 7: zero lines

Automatically provided texts: T1, T2, TF

Example plot: Fig. 5.4

5.6 Exercises

Please note:

(i) you may watch a semiautomatic demonstration of the material of this chapter by typing SA D40

(ii) for the following exercises use descriptor file IQ040.DES

(iii) if not stated otherwise in the exercises the numerical values of the mass of the particle and of Planck's constant are put to 1.

5.2.1 Plot the stationary wave functions of two uncoupled oscillators with $k = 2$, $m = 1$ for the quantum numbers $N = 0, 1$, $n = 0, 1$ in a multiple plot. Start from descriptor 5.

5.2.2 Plot the stationary wave functions of two uncoupled oscillators with $k = 2$, $m = 1$, in four multiple plots for the quantum numbers **(a)** $N = 0, 1$, $n = 0, 1$, **(b)** $N = 0, 1$, $n = 2, 3$, **(c)** $N = 2, 3$, $n = 0, 1$, **(d)** $N = 2, 3$, $n = 2, 3$. Start from descriptor 5. **(e)** What determines the number and location of the node lines of the wave functions?

5.2.3 (a–d) Repeat Exercise 5.2.2 (a–d) with $m = 2$. **(e)** Which effect on the wave functions does the doubling of the mass have?

5.2.4 (a–d) Repeat Exercise 5.2.2 (a–d) with nonvanishing coupling $\kappa = 2$. **(e)** How does the coupling affect the wave function?

5.2.5 (a–d) Repeat Exercise 5.2.2 (a–d) with nonvanishing attractive coupling $\kappa = 5$. **(e)** How does the coupling affect the wave function? **(f)** Which correlation in the particle coordinates x_1, x_2 do you observe?

5.2.6 (a–d) Repeat Exercise 5.2.2 (a–d) with nonvanishing repulsive coupling $\kappa = -0.6$. **(e)** How does the repulsive coupling affect the wave function? **(f)** Which correlation in the particle coordinates x_1, x_2 do you observe?

5.2.7 (a) Plot in a multiple plot the probability density of two coupled oscillators with $k = 2$, $m = 1$ for the quantum numbers $N = 1, 2$, $n = 1, 2$ for the spring constant $\kappa = 5$ of the coupling. Start from descriptor 5. **(b)** Why are the outer maxima in the plot for $N = 2$, $n = 2$ higher than the inner ones? **(c)** Why are the widths of the inner maxima smaller than the ones of the outer maxima? **(d)** Why does the region where the probability density is essentially different from zero not in all plots evenly fill the region of possible classical orbits given by the dashed ellipse?

5.2.8 (a) Repeat Exercise 5.2.7 (a) with repulsive coupling $\kappa = -0.6$. **(b)** Why do the locations of the maxima of the probability density form a rectangular grid in the $x_1 x_2$ plane?

5.2.9 Plot in multiple plots the probability density of two coupled oscillators with $k = 1$, $m = 1$, $\kappa = 4$ for the quantum numbers **(a)** $N = 2, 3$, $n = 0, 1$, **(b)** $N = 2, 3$, $n = 2, 3$, **(c)** $N = 4, 5$, $n = 0, 1$, **(d)** $N = 4, 5$, $n = 2, 3$. Start from descriptor 5. **(e)** Calculate the total energies of the coupled oscillators for the above quantum numbers. **(f)** Compare the plots for $N = 5$, $n = 0$ with $N = 2$, $n = 2$. Explain why the graphs for high N and small n extend mainly along one of the principal axes of the dashed ellipse?

5.3.1 Consider a system of two uncoupled harmonic oscillators with mass $m = 1$ and spring constants $k = 2$, $\kappa = 0$. **(a)** Plot the time dependence of **i)** the position expectation values $\langle x_1(t) \rangle$, $\langle x_2(t) \rangle$, **ii)** the widths $\sigma_1(t)$, $\sigma_2(t)$ in x_1 and x_2, **iii)** the correlation coefficient $c(t)$ for the initial values $\langle x_{10} \rangle = 3$, $\langle x_{20} \rangle = 0$, $\sigma_{10} = 1.01$, $\sigma_{20} = 0.5$, $c_0 = 0$. Start from descriptor 6. **(b)** Why is the expectation value $\langle x_2(t) \rangle$ equal to zero? **(c)** Why do the widths $\sigma_1(t)$, $\sigma_2(t)$ vary with time? **(d)** Why is the correlation $c(t)$ equal to zero?

5.3.2 (a) Repeat Exercise 5.3.1 for nonvanishing correlation $c_0 = 0.2$. Start from descriptor 6. **(b)** Explain the time dependence of the correlation.

5.3.3 Repeat Exercise 5.3.1 for nonvanishing $x_{20} = 3$ and $\sigma_1 = \sigma_2 = \sigma_0/\sqrt{2}$, where $\sigma_0 = \sqrt{\hbar/m\omega}$ is the ground-state width of the uncoupled oscillators. Start from descriptor 6. **(a)** Calculate $\sigma_0/\sqrt{2}$. **(b)** Plot the global quantities $\langle x_1(t) \rangle$, $\langle x_2(t) \rangle$, $\sigma_1(t)$, $\sigma_2(t)$, $c(t)$. **(c)** Why are the widths $\sigma_1(t)$, $\sigma_2(t)$ time independent ? **(d)** Do you expect time-independent widths also for nonvanishing correlation?

5.3.4 (a) Repeat Exercise 5.3.3 for nonvanishing correlation $c_0 = 0.4$. Start from descriptor 6. **(b)** Why do the widths $\sigma_1(t)$, $\sigma_2(t)$ no longer remain time independent? **(c)** Why does the correlation $c(t)$ change periodically?

5.3.5 (a) Repeat Exercise 5.3.3 for the anticorrelation $c_0 = -0.4$. Start from descriptor 6. **(b)** Explain the difference between the correlation $c(t)$ of this exercise and of Exercise 5.3.4.

5.3.6 (a) Repeat Exercise 5.3.1 for nonvanishing $x_{20} = 1.5$ and $\sigma_1 = \sigma_2 = \sigma_0/\sqrt{2}$, where $\sigma_0 = \sqrt{\hbar/m\omega}$ is the ground-state width of the uncoupled oscillators. Start from descriptor 6. **(b)** Why is the result qualitatively similar to that in Exercise 5.3.3?

5.3.7 Consider the system of two coupled oscillators. Choose the same initial data as in Exercise 5.3.1, however take a nonvanishing spring constant $\kappa = 0.5$ for the coupling of the two oscillators. Start from descriptor 6. **(a)** Plot the global quantities. **(b)** Explain the behavior of the time-dependent expectation values $\langle x_1(t) \rangle$, $\langle x_2(t) \rangle$. **(c)** Why does the correlation $c(t)$ become different from zero as time increases?

5.3.8 Repeat Exercise 5.3.7, however with equal initial widths $\sigma_{10} = \sigma_{20} = 0.5$. Start from descriptor 6.

5.3.9 (a) Repeat Exercise 5.3.8, however with initial positions $x_{10} = 3$, $x_{20} = -3$. Start from descriptor 6. **(b)** What oscillation do the expectation values $\langle x_1(t) \rangle$, $\langle x_2(t) \rangle$ perform?

5.3.10 (a,b) Repeat Exercise 5.3.9 (a,b), however with strong initial anticorrelation $c_0 = -0.95$. Start from descriptor 6.

5.3.11 Repeat Exercise 5.3.10 with strong positive correlation $c_0 = 0.95$. Start from descriptor 6.

5.3.12 (a) Repeat Exercise 5.3.9 with initial position expectation values $\langle x_{10} \rangle = 3$, $\langle x_{20} \rangle = 3$ and vanishing initial correlation $c_0 = 0$. Start from descriptor 6. **(b)** What kind of oscillation do the two position expectation values perform?

5.4.1 Plot the joint probability density $\varrho_D(x_1, x_2)$ for two uncoupled harmonic oscillators of distinguishable particles of mass $m = 1$ and spring constant $k = 2$ for the initial values $\langle x_{10} \rangle = 3$, $\langle x_{20} \rangle = 0$, $\sigma_1 = 1.0$, $\sigma_2 = 0.5$, $c_0 = 0$ in two 2×2 multiple plots **(a)** for the times $t_n = n\Delta t$, $n = 0, 1, 2, 3$, $\Delta t = 0.501$ and **(b)** for the times $t_n = n\Delta t$, $n = 4, 5, 6, 7$, $\Delta t = 0.501$. Start from descriptor 7. **(c)** Why are the axes of the two-variable Gauss distributions parallel to the coordinate axes?

5.4.2 (a,b) Repeat Exercise 5.4.1 (a,b) for $\langle x_{20} \rangle = 3$, $c_0 = 0.8$, and the step width $\Delta t = 0.55536$. Start from descriptor 7. **(c)** Why are the axes of the two-dimensional Gauss distribution no longer parallel to the coordinate axes? **(d)** How does the positivity of the initial correlation c_0 show in the plots?

5.4.3 (a,b) Plot the joint probability density $\varrho_D(x_1, x_2)$ for two uncoupled harmonic oscillators of distinguishable particles of mass $m = 1$ and spring constant $k = 2$ for the initial values $\langle x_{10} \rangle = -2$, $\langle x_{20} \rangle = 3$, $\sigma_{10} = 1$, $\sigma_{20} = 0.5$, $c_0 = -0.8$ in two 2×2 multiple plots (a) for the times $t_n = n\Delta t$, $n = 0, 1, 2, 3$, $\Delta t = 0.501$ and (b) for the times $t_n = n\Delta t$, $n = 4, 5, 6, 7$, $\Delta t = 0.501$. Start from descriptor 7. **(c)** How does the negative initial correlation show in the plots?

5.4.4 (a,b) Repeat Exercise 5.4.1 (a,b) for a coupling spring constant $\kappa = 0.8$. Start from descriptor 7. **(c)** Which effect showing in the plots is due to the coupling of the two oscillators?

5.4.5 (a,b) Repeat Exercise 5.4.1 (a,b) for a coupling spring constant $\kappa = 1.5$. Start from descriptor 7. **(c)** Why does the amplitude in the variable x_2 grow faster than in the plots of Exercise 5.4.4?

5.4.6 (a,b) Repeat Exercise 5.4.1 (a,b) for a coupling spring constant $\kappa = 3$. Start from descriptor 7. **(c)** What causes the change of the initially uncorrelated Gauss distribution to a correlated one?

5.4.7 (a,b) Repeat Exercise 5.4.1 (a,b) for a coupling spring constant $\kappa = 5$. Start from descriptor 7. **(c)** Why does the expectation value (circle) in the $x_1 x_2$ plane oscillate more often than in the former exercises?

5.4.8 (a,b) Repeat Exercise 5.4.1 (a,b) for a coupling spring constant $\kappa = 10$. Start from descriptor 7. **(c)** Why is the Gaussian distribution most narrow when its position expectation values $\langle x_1 \rangle$, $\langle x_2 \rangle$ are close to zero?

5.4.9 Plot the joint probability density $\varrho_D(x_1, x_2)$ for two coupled harmonic oscillators of distinguishable particles of mass $m = 1$, spring constant $k = 2$, and coupling spring constant $\kappa = 20$ for the initial conditions $\langle x_{10} \rangle = 3$, $\langle x_{20} \rangle = -3$, $\sigma_1 = 1.0$, $\sigma_2 = 0.5$, $c_0 = 0$, in two 2×2 multiple plots **(a)** for $t_n = n\Delta t$, $n = 0, 1, 2, 3$, $\Delta t = 0.2$ and **(b)** for $t_n = n\Delta t$, $n = 4, 5, 6, 7$, $\Delta t = 0.2$. Start from descriptor 7. **(c)** What particular kind of oscillation do you observe?

5.4.10 (a,b) Repeat Exercise 5.4.9 (a,b) for a coupling spring constant $\kappa = 20$ and a step width of $\Delta t = 0.2$. Take as initial values for the position expectation values $\langle x_{10} \rangle = 3$, $\langle x_{20} \rangle = 3$. Start from descriptor 7. **(c)** What particular kind of oscillation do you observe? **(d)** Why is the motion so much slower than in Exercise 5.4.9?

5.4.11 (a,b) Repeat Exercise 5.4.1 (a,b) for a repulsive coupling spring constant $\kappa = -0.8$ and step width $\Delta t = 0.5$. Start from descriptor 7. **(c)** How does the repulsive coupling between the two oscillators make itself felt in the plots?

5.4.12 (a,b) Repeat Exercise 5.4.1 (a),(b) for a repulsive spring constant $\kappa = -0.95$ and step width $\Delta t = 2$. Start from descriptor 7. **(c)** Why does the initially uncorrelated Gauss distribution develop a correlation of the kind observed?

5.4.13 (a,b) Repeat Exercise 5.4.1 (a,b) for bosons. **(c)** Why do you initially observe two humps? **(d)** What creates the very high peak in the plots where the two bosons are close together?

5.4.14 (a,b) Repeat Exercise 5.4.2 (a,b) for bosons. **(c)** Why do you observe only one hump in the plots?

5.4.15 (a,b) Repeat Exercise 5.4.3 (a,b) for bosons. **(c)** How does the correlation show in the initial double humps?

5.4.16 Repeat Exercise 5.4.4 for bosons.

5.4.17 (a,b) Repeat Exercise 5.4.8 (a,b) for bosons. **(c)** How does the strong attractive coupling show in the plots?

5.4.18 (a,b) Repeat Exercise 5.4.9 (a,b) for bosons. **(c)** How is the difference from the graphs for indistinguishable particles explained?

5.4.19 (a,b) Repeat Exercise 5.4.10 (a,b) for bosons. **(c)** Why do you observe quick oscillations of the width in $(x_2 - x_1)$ of the Gaussian hump?

5.4.20 Repeat Exercise 5.4.11 for bosons.

5.4.21 (a,b) Repeat Exercise 5.4.1 (a,b) for fermions. **(c)** Why is the joint probability always exactly zero along the line $x_1 = x_2$?

5.4.22 Repeat Exercise 5.4.4 for fermions.

5.4.23 (a,b) Repeat Exercise 5.4.8 (a,b) for fermions.

5.4.24 (a,b) Repeat Exercise 5.4.9 (a,b) for fermions.

5.4.25 (a,b) Repeat Exercise 5.4.10 (a,b) for fermions, however for $\sigma_{10} = \sigma_{20} = 1$. **(c)** Why does the joint probability distribution vanish?

5.4.26 (a,b) Repeat Exercise 5.4.10 (a,b) for nonvanishing initial correlation. **(c)** Why does the joint probability density still vanish?

5.4.27 (a,b) Repeat Exercise 5.4.10 (a,b) for vanishing correlation and different initial widths $\sigma_1 = 1$, $\sigma_2 = 0.2$. **(c)** Why does the existence of a fermion wave function with identical expectation values for the two fermions not contradict the Pauli principle?

5.5.1 Study the marginal distributions of two distinguishable particles of equal mass $m = 1$ in two uncoupled oscillators with the initial conditions $\langle x_{10} \rangle = 3$, $\langle x_{20} \rangle = 0$, $k = 2$, $\kappa = 0.8$, $\sigma_{10} = 1$, $\sigma_{20} = 0.5$, $c_0 = 0$. **(a)** Plot the marginal distribution $\varrho_{D1}(x_1, t)$ of particle 1 for the time interval $0 \le t \le 4$ in ten steps. Start from descriptor 8. **(b)** Plot the marginal distribution $\varrho_{D2}(x_2, t)$ of particle 2 for the same interval. Start from descriptor 9. (c) For simpler comparison plot both distributions (a), (b) as a combined plot (a) above (b), see Appendix A.5.3. Start from mother descriptor 10.

5.5.2 (a–c) Repeat Exercise 5.5.1 (a–c) for a longer time interval $0 \le t \le 10$.

5.5.3 Study the marginal distributions of indistinguishable particles with the same parameters as in Exercise 5.5.1. **(a)** Plot the marginal distribution for bosons. **(b)** Plot the marginal distributions for fermions. **(c)** For comparison, plot both distributions above each other in a double plot. **(d)** What are the differences in the two plots? **(e)** Why are these differences so marginal?

6 Free Particle Motion in Three Dimensions

Contents: Description of the three-dimensional motion of a free particle of sharp momentum by a harmonic plane wave. Schrödinger equation of free motion in three dimensions. Gaussian wave packet. Angular momentum. Spherical harmonics as eigenfunctions of angular momentum. Radial Schrödinger equation of free motion. Spherical Bessel functions. Partial wave decomposition of plane wave and Gaussian wave packet.

6.1 Physical concepts

6.1.1 The Schrödinger Equation of a Free Particle in Three Dimensions. The Momentum Operator

Classical free motion in three dimensions can be viewed as three simultaneous one-dimensional motions in the coordinates x, y, z. In quantum mechanics the situation is the same. Three-dimensional free motion is viewed as three one-dimensional harmonic waves (2.1) propagating simultaneously in the coordinates x, y, z of the position vector r:

$$
\begin{aligned}
\psi_p(r, t) = \frac{1}{(2\pi\hbar)^{3/2}} &\exp\left[-\frac{i}{\hbar}\left(\frac{p_x^2}{2M}t - p_x x\right)\right] \\
\times \exp\left[-\frac{i}{\hbar}\left(\frac{p_y^2}{2M}t - p_y y\right)\right] &\exp\left[-\frac{i}{\hbar}\left(\frac{p_z^2}{2M}t - p_z z\right)\right] .
\end{aligned}
\tag{6.1}
$$

Since the kinetic energy E of the particle is given by

$$
E = (p_x^2 + p_y^2 + p_z^2)/(2M) = p^2/2M \quad ,
\tag{6.2}
$$

M: mass of particle
$p = (p_x, p_y, p_z)$: momentum of particle

the expression (6.1) can be rewritten in the form

$$
\psi_p(r, t) = \frac{1}{(2\pi\hbar)^{3/2}} e^{-i(Et - p \cdot r)/\hbar} = \frac{1}{(2\pi\hbar)^{3/2}} e^{-i(\omega t - k \cdot r)} \quad ;
\tag{6.3}
$$

$\omega = E/\hbar$: angular frequency of the wave function
$k = p/\hbar$: wave-number vector

The phase velocity of this wave is

$$
v = p/2M \quad .
\tag{6.4}
$$

The wave function (6.3) is the solution of the *free Schrödinger equation in three dimensions*

$$i\hbar \frac{\partial}{\partial t}\psi(\boldsymbol{r}, t) = T\psi(\boldsymbol{r}, t) \tag{6.5}$$

having the same formal appearance as (2.2). However, the operator T of the kinetic energy is now that of a particle in three dimensions

$$T = -\frac{\hbar^2}{2M}\left(\frac{\partial^2}{\partial x^2} + \frac{\partial^2}{\partial y^2} + \frac{\partial^2}{\partial z^2}\right) \quad . \tag{6.6}$$

With the help of the *gradient operator*

$$\nabla = \left(\frac{\partial}{\partial x}, \frac{\partial}{\partial y}, \frac{\partial}{\partial z}\right) \tag{6.7}$$

it takes the form

$$T = -\frac{\hbar^2}{2M}\nabla^2 \quad , \tag{6.8}$$

i.e., a multiple of the *Laplacian* ∇^2.

The wave function (6.3) lends itself to the factorization

$$\psi_p(\boldsymbol{r}, t) = \mathrm{e}^{-\mathrm{i}Et/\hbar}\varphi_p(\boldsymbol{r}) \tag{6.9}$$

into a time-dependent exponential and a *stationary wave function*

$$\varphi_p(\boldsymbol{r}) = \frac{1}{(2\pi\hbar)^{3/2}}\mathrm{e}^{\mathrm{i}\boldsymbol{p}\cdot\boldsymbol{r}/\hbar} = \frac{1}{(2\pi\hbar)^{3/2}}\mathrm{e}^{\mathrm{i}\boldsymbol{k}\cdot\boldsymbol{r}} \quad . \tag{6.10}$$

If we choose the unit vector $\boldsymbol{e}_z$ in the z direction parallel to the *wave-number vector* $\boldsymbol{k} = \boldsymbol{p}/\hbar$, we have $\boldsymbol{k} = k\boldsymbol{e}_z$, and the stationary wave assumes the simple form

$$\varphi_p(\boldsymbol{r}) = \frac{1}{(2\pi\hbar)^{3/2}}\mathrm{e}^{\mathrm{i}kz} \quad . \tag{6.11}$$

This is a complex function of the coordinate z only. It can be decomposed into real and imaginary parts:

$$\varphi_p(\boldsymbol{r}) = \mathrm{Re}\,\varphi_p(\boldsymbol{r}) + \mathrm{i}\,\mathrm{Im}\,\varphi_p(\boldsymbol{r}) \quad , \tag{6.12}$$

with

$$\mathrm{Re}\,\varphi_p(\boldsymbol{r}) = \frac{1}{(2\pi\hbar)^{3/2}}\cos kz \quad , \quad \mathrm{Im}\,\varphi_p(\boldsymbol{r}) = \frac{1}{(2\pi\hbar)^{3/2}}\sin kz \quad . \tag{6.13}$$

On inserting (6.9) into the time-dependent equation (6.5) we obtain

$$T\varphi_p(\boldsymbol{r}) = E\varphi_p(\boldsymbol{r}) \quad \text{or} \quad -\frac{\hbar^2}{2M}\nabla^2\varphi_p(\boldsymbol{r}) = E\varphi_p(\boldsymbol{r}) \tag{6.14}$$

as the *stationary* (time-independent) *Schrödinger equation* for the stationary wave function $\varphi_p(\boldsymbol{r})$. Equation (6.14) is also viewed as an eigenvalue equation, where E is the continuous eigenvalue of the kinetic energy and $\varphi_p(\boldsymbol{r})$ a continuum eigenfunction of the operator T of the kinetic energy.

In accordance with the classical relation for the kinetic energy $T = \boldsymbol{p}^2/2M$ of a single particle with momentum $\boldsymbol{p}$ we conclude from (6.8) that

$$\hat{p} = \frac{\hbar}{i}\nabla \quad , \quad \text{i.e.,} \quad \hat{p}_x = \frac{\hbar}{i}\frac{\partial}{\partial x} \quad , \quad \hat{p}_y = \frac{\hbar}{i}\frac{\partial}{\partial y} \quad , \quad \hat{p}_z = \frac{\hbar}{i}\frac{\partial}{\partial z} \quad , \tag{6.15}$$

is the *momentum operator*. The stationary wave function (6.10) is a continuum eigenfunction of the momentum operator

$$\hat{p}\varphi_p(r) = \frac{\hbar}{i}\nabla\varphi_p(r) = p\varphi_p(r) \tag{6.16}$$

or in components

$$\hat{p}_x\varphi_p(r) = \frac{\hbar}{i}\frac{\partial}{\partial x}\varphi_p(r) = p_x\varphi_p(r) \quad , \quad \text{etc.} \tag{6.17}$$

6.1.2 The Wave Packet. Group Velocity. Normalization. The Probability Ellipsoid

The wave function (6.3) does not correspond to an actual physical situation, since the norm of a plane wave diverges. A physical particle corresponds to a wave packet formed with a *spectral function* as in (2.4).

$$\psi(r, t) = \int f(p)e^{-iEt/\hbar}\varphi_p(r - r_0)d^3p \tag{6.18}$$

$r = (x, y, z)$: position vector
$r_0 = (x_0, y_0, z_0)$: initial position expectation value of wave packet at $t = 0$
$p = (p_x, p_y, p_z)$: momentum vector
$f(p)$: spectral function of wave packet

We choose again a *Gaussian spectral function* in three dimensions as a product

$$f(p) = f_x(p_x)f_y(p_y)f_z(p_z) \tag{6.19}$$

of three Gaussians, one for every coordinate $a = x, y, z$.

$$f_a(p_a) = \frac{1}{(2\pi)^{1/4}\sigma_{p_a}^{1/2}}\exp\frac{(p_a - p_{a_0})^2}{4\sigma_{p_a}^2} \tag{6.20}$$

p_a: a coordinate of momentum, $a = x, y, z$
p_{a_0}: expectation value of momentum
σ_{p_a}: width of Gaussian spectral function f_a

The factors in front of the exponential of (6.20) *normalize* the spectral function $f(p)$ to one:

$$\int f^2(p)d^3p = 1 \quad , \tag{6.21}$$

and thus the wave packet (6.18):

$$\int |\psi(r, t)|^2 d^3r = 1 \quad . \tag{6.22}$$

Because of the factorization of the time-dependent exponential and of the stationary wave function φ_p into factors depending on one momentum component only, the integral in (6.18) yields the wave function of the three-dimensional Gaussian wave packet

$$\psi(\mathbf{r}, t) = M_x(x, t)e^{i\phi_x(x,t)} M_y(y, t)e^{i\phi_y(y,t)} M_z(z, t)e^{i\phi_z(z,t)} \quad, \tag{6.23}$$

where the explicit expressions for the *modulus* M_a and the *phase* ϕ_a can be derived easily from (2.7) and (2.8).

The absolute square of $\psi(\mathbf{r}, t)$ yields the *probability density*

$$\varrho(\mathbf{r}, t) = \frac{1}{(2\pi)^{3/2}\sigma_x\sigma_y\sigma_z} \exp\left[-\frac{(x - \langle x\rangle)^2}{2\sigma_x^2} - \frac{(y - \langle y\rangle)^2}{2\sigma_y^2} - \frac{(z - \langle z\rangle)^2}{2\sigma_z^2}\right] \tag{6.24}$$

for a particle at the position $\mathbf{r}$ at time t. The *position expectation value* is given by

$$\langle \mathbf{r}(t)\rangle = (\langle x(t)\rangle, \langle y(t)\rangle, \langle z(t)\rangle) \quad,$$

with

$$\langle \mathbf{r}(t)\rangle = \mathbf{r}_0 + \mathbf{v}t \quad, \quad a = x, y, z \quad. \tag{6.25}$$

This represents the motion of a particle with constant velocity

$$\mathbf{v} = \mathbf{p}_0/M \tag{6.26}$$

along a straight line, starting at $t = 0$ with the initial position $\mathbf{r}_0$. The velocity $\mathbf{v}$ is called the *group velocity* since it determines the propagation of a wave packet or wave group. It is different from the phase velocity (6.4).

The *width* of the Gaussian is time dependent:

$$\sigma_a^2(t) = \sigma_{a_0}^2 + \left(\frac{\hbar t}{2M} \frac{1}{\sigma_{a_0}}\right)^2 \quad, \quad a = x, y, z \tag{6.27}$$

$\sigma_{a_0} = \hbar/2\sigma_{pa}$: initial width of wave packet in the coordinate $a = x, y, z$

The plots produced with **IQ** refer to the two-dimensional distribution

$$\varrho(x, y, t) = \int_{-\infty}^{+\infty} \varrho(x, y, z, t)dz \tag{6.28}$$

which represents the probability density for a particle having at time $t = 0$ the coordinates x and y irrespective of z. The explicit result for $\varrho(x, y, t)$ is

$$\varrho(x, y, t) = \frac{1}{2\pi\sigma_x\sigma_y} \exp\left\{-\frac{(x - \langle x\rangle)^2}{2\sigma_x^2} - \frac{(y - \langle y\rangle)^2}{2\sigma_y^2}\right\} \quad. \tag{6.29}$$

6.1.3 Angular Momentum. Spherical Harmonics

Angular momentum is a vector $\mathbf{L} = (L_x, L_y, L_z)$ of the form

$$\mathbf{L} = \mathbf{r} \times \mathbf{p} \quad; \tag{6.30}$$

in components

$$L_x = yp_z - zp_y \quad, \quad L_y = zp_x - xp_z \quad, \quad L_z = xp_y - yp_x \quad. \tag{6.31}$$

In quantum mechanics the momentum $\boldsymbol{p}$ is a multiple of the del or nabla operator, see (6.15), so that the *operator of angular momentum* $\hat{\boldsymbol{L}}$ is given by

$$\hat{\boldsymbol{L}} = \hat{\boldsymbol{r}} \times \hat{\boldsymbol{p}} = \frac{\hbar}{i} \boldsymbol{r} \times \nabla \quad . \tag{6.32}$$

Its three components

$$\hat{L}_x = \frac{\hbar}{i}\left(y\frac{\partial}{\partial z} - z\frac{\partial}{\partial y}\right) \ , \ \hat{L}_y = \frac{\hbar}{i}\left(z\frac{\partial}{\partial x} - x\frac{\partial}{\partial z}\right) \ , \ \hat{L}_z = \frac{\hbar}{i}\left(x\frac{\partial}{\partial y} - y\frac{\partial}{\partial x}\right) \tag{6.33}$$

do not commute with each other. Instead, they satisfy the *commutation relations*

$$[\hat{L}_x, \hat{L}_y] = i\hbar\hat{L}_z \quad , \quad [\hat{L}_y, \hat{L}_z] = i\hbar\hat{L}_x \quad , \quad [\hat{L}_z, \hat{L}_x] = i\hbar\hat{L}_y \quad . \tag{6.34}$$

Each of these components does commute, however, with the square

$$\hat{\boldsymbol{L}}^2 = \hat{L}_x^2 + \hat{L}_y^2 + \hat{L}_z^2 \tag{6.35}$$

of the angular momentum vector $\hat{\boldsymbol{L}}$:

$$[\hat{\boldsymbol{L}}^2, \hat{L}_a] = 0 \quad , \quad a = x, y, z \quad . \tag{6.36}$$

Polar coordinates (radius r, polar angle ϑ, azimuth φ) are related to Cartesian coordinates (x, y, z) by

$$x = r\sin\vartheta\cos\varphi \quad , \quad y = r\sin\vartheta\sin\varphi \quad , \quad z = r\cos\vartheta \quad . \tag{6.37}$$

Using these polar coordinates the components and the square of $\hat{\boldsymbol{L}}$ have the representations

$$\hat{L}_x = i\hbar\left(\sin\varphi\frac{\partial}{\partial\vartheta} + \cotan\vartheta\cos\varphi\frac{\partial}{\partial\varphi}\right)$$

$$\hat{L}_y = i\hbar\left(\cos\varphi\frac{\partial}{\partial\vartheta} - \cotan\vartheta\sin\varphi\frac{\partial}{\partial\varphi}\right)$$

$$\hat{L}_z = -i\hbar\frac{\partial}{\partial\varphi} \tag{6.38}$$

$$\hat{\boldsymbol{L}}^2 = -\hbar^2\left[\frac{1}{\sin\vartheta}\frac{\partial}{\partial\vartheta}\left(\sin\vartheta\frac{\partial}{\partial\vartheta}\right) + \frac{1}{\sin^2\vartheta}\frac{\partial^2}{\partial\varphi^2}\right] \quad . \tag{6.39}$$

Thus, the eigenfunctions of $\hat{\boldsymbol{L}}^2$ and $\hat{L}_z$ are the *spherical harmonics* $Y_{\ell m}(\vartheta, \varphi)$ depending on ϑ and φ only, with the two indices relating to the eigenvalues of the square and the z component of angular momentum

$$\hat{\boldsymbol{L}}^2 Y_{\ell m} = \ell(\ell+1)\hbar^2 Y_{\ell m} \quad , \quad \ell = 0, 1, 2, \ldots \quad , \tag{6.40}$$

$$\hat{L}_z Y_{\ell m} = m\hbar Y_{\ell m} \quad , \quad -\ell \le m \le \ell \quad . \tag{6.41}$$

The *angular-momentum quantum number* ℓ is interpreted as the modulus of angular momentum and m – usually called *magnetic quantum number* – as its z component. Together with (6.39), (6.40) is up to a factor $\hbar^2$ identical with (9.15) of Chap. 9, which deals with mathematical functions. The details of the spherical harmonics $Y_{\ell m}$ are given there.

6.1.4 The Stationary Schrödinger Equation in Polar Coordinates. Separation of Variables. Spherical Bessel Functions. Continuum Normalization. Completeness

The stationary Schrödinger equation (6.14) of a free particle can be expressed in polar coordinates if the kinetic energy is

$$T\varphi(\mathbf{r}) = \left[-\frac{\hbar^2}{2M} \frac{1}{r} \frac{\partial^2}{\partial r^2} r + \frac{1}{2Mr^2} \hat{\mathbf{L}}^2 \right] \varphi(\mathbf{r}) = E\varphi(\mathbf{r}) \quad , \tag{6.42}$$

where the square $\hat{\mathbf{L}}^2$ of angular momentum is given by (6.39). Separation of the radial variable r and the angles ϑ and φ is achieved by factorization:

$$\varphi(\mathbf{r}) = R_\ell(r) Y_{\ell m}(\vartheta, \varphi) \quad . \tag{6.43}$$

Then, the radial wave function $R_\ell(r)$ satisfies the *free radial Schrödinger equation*

$$-\frac{\hbar^2}{2M} \left[\frac{1}{r} \frac{d^2}{dr^2} r - \frac{\ell(\ell+1)}{r^2} \right] R_\ell = E R_\ell \quad , \quad r > 0 \quad . \tag{6.44}$$

It is equivalent to, $k^2 = 2ME/\hbar^2$,

$$r^2 \frac{d^2}{dr^2} R_\ell(k, r) + 2r \frac{d}{dr} R_\ell(k, r) + [k^2 r^2 - \ell(\ell+1)] R_\ell(k, r) = 0 \quad . \tag{6.45}$$

Choosing $x = kr$ as dimensionless variable and setting $z_\ell(x) \sim R_\ell(k, r)$, we arrive at the differential equation (9.30) of Chap. 9 for the spherical Bessel functions $z_\ell(x)$.

The kinetic energy of radial motion $T_r = -(\hbar^2/2M)r^{-1}(d^2/dr^2)r$ is a Hermitian operator only for wave functions which are not singular at $r = 0$. This requirement restricts the $R_\ell(k, r)$ to be proportional to the *spherical Bessel functions of the first kind j_ℓ*:

$$R_\ell(k, r) = \sqrt{\frac{2}{\pi}} k j_\ell(kr) \quad . \tag{6.46}$$

The factor in front of j_ℓ in (6.46) ensures a continuum normalization of the kind

$$\int_0^\infty R_\ell(k, r) R_\ell(k', r) r^2 dr = \delta(k' - k) \quad . \tag{6.47}$$

The eigenfunctions $\varphi_{\ell m}(k, r)$ of the kinetic-energy operator T belonging to the energy eigenvalue $E = \hbar^2 k^2/2M$ and to the angular-momentum quantum numbers ℓ, m are called *free partial waves*,

$$\varphi_{\ell m}(k, \mathbf{r}) = R_\ell(k, r) Y_{\ell m}(\vartheta, \varphi) \quad . \tag{6.48}$$

These eigenfunctions exhibit a continuum normalization in k:

$$\int \varphi_{\ell' m'}^*(k', \mathbf{r}) \varphi_{\ell m}(k, \mathbf{r}) dV = \delta(k' - k) \delta_{\ell' \ell} \delta_{m' m} \quad . \tag{6.49}$$

Their completeness relation (9.18) reads

$$\sum_{\ell, m} \int_0^\infty \varphi_{\ell m}^*(k, \mathbf{r}') \varphi_{\ell m}(k, \mathbf{r}) k^2 dk = \delta^3(\mathbf{r}' - \mathbf{r}) \quad . \tag{6.50}$$

It allows a decomposition of wave functions into free partial waves.

6.1.5 Partial-Wave Decomposition of the Plane Wave

The stationary plane wave of momentum $p = \hbar k$ has the form

$$e^{ip \cdot r/\hbar} = e^{ik \cdot r} = e^{ikr \cos \vartheta} \quad . \tag{6.51}$$

In a system of polar coordinates with the z axis in the direction of k, the above formula shows that there is no φ dependence. Thus a decomposition into free partial waves $\varphi_{\ell 0}(k, r)$ (6.48), containing only the spherical harmonics (9.16), with $m = 0$

$$Y_{\ell 0}(\vartheta, \varphi) = \sqrt{\frac{2\ell + 1}{4\pi}} P_\ell(\cos \vartheta) \tag{6.52}$$

is possible. Here P_ℓ is a simple Legendre polynomial. One obtains

$$e^{ik \cdot r} = e^{ikr \cos \vartheta} = \sum_{\ell=0}^{\infty} \varphi_\ell = \sum_{\ell=0}^{\infty} i^\ell (2\ell + 1) j_\ell(kr) P_\ell(\cos \vartheta) \tag{6.53}$$

r: radius vector
k: wave vector of plane wave
$\cos \vartheta = k \cdot r/kr$: cosine of polar angle
ℓ: angular-momentum quantum number
$P_\ell(\cos \vartheta)$: Legendre polynomial of order ℓ
$j_\ell(kr)$: spherical Bessel function of first kind

6.1.6 Partial-Wave Decomposition of the Gaussian Wave Packet

The Gaussian wave packet (6.18) at time $t = 0$ is decomposed into free partial waves starting from the completeness relation (6.50):

$$\psi(r, 0) = \frac{2}{\pi} \sum_{\ell=0}^{\infty} \sum_{m=-\ell}^{\ell} \int b_{\ell m}(k) j_\ell(kr) Y_{\ell m}(\vartheta, \varphi) k^2 dk \quad . \tag{6.54}$$

The probability $W_{\ell m}$ of finding a contribution of angular momentum ℓ, m irrespective of the wave number k is given by

$$W_{\ell m} = \frac{2}{\pi} \int b^*_{\ell m}(k) b_{\ell m}(k) k^2 dk \quad . \tag{6.55}$$

For a Gaussian wave packet with a probability sphere, i.e., $\sigma_x = \sigma_y = \sigma_z = \sigma(t)$, the probability $W_{\ell m}$ is given by

$$W_{\ell m} = \exp\left\{-\sigma_0^2 \left(k_0^2 + \frac{r_0^2}{2\sigma_0^4}\right)\right\} \sqrt{\frac{\pi}{2\sigma_0^2 k^2}} I_{\ell+\frac{1}{2}}(\sigma_0^2 k^2)(2\ell + 1)$$
$$\times \frac{(\ell - |m|)!}{(\ell + |m|)!} |P_\ell^m(\cos \varphi)|^2 \exp\left\{-\frac{m}{2} \ln |\zeta|^4\right\} \tag{6.56}$$

$r_0 = y\hat{k}_0 + b\hat{b}$: initial position expectation value
$\hat{k}_0$: unit vector in the direction of wave-vector expectation value
$\hat{b}$: unit vector, perpendicular to $\hat{k}_0$, $\hat{k}_0 \cdot \hat{b} = 0$

$$k^4 = \left(k_0^2 - r_0^2/4\sigma_0^4\right)^2 + (k_0 \cdot r_0)^2/\sigma_0^4$$

P_ℓ^m: associated Legendre function

$I_{\ell+\frac{1}{2}}$: modified Bessel function

Consider a free Gaussian wave packet, the probability ellipsoid of which is a sphere, i.e., $\sigma_x(t) = \sigma_y(t) = \sigma_z(t) = \sigma(t)$. At a given initial time $t = 0$ the width is σ_0. The momentum expectation value $p_0 = \hbar k_0$ is time independent. We decompose the initial position expectation value r_0 into vectors parallel and perpendicular to k_0, i.e.,

$$r_0 = y\hat{k}_0 + b\hat{b} \quad , \quad \hat{k}_0 \cdot \hat{b} = 0 \quad .$$

Then b is the expectation value of the *impact parameter* with respect to the origin. The probability that a particle described by the wave packet has angular-momentum quantum numbers ℓ,m is

$$W_{\ell m} = \exp\left\{-\sigma_0^2\left(k_0^2 + \frac{r_0^2}{2\sigma_0^4}\right)\right\}\sqrt{\frac{\pi}{2\sigma_0^2 k^2}}I_{\ell+\frac{1}{2}}\left(\sigma_0^2 k^2\right)(2\ell + 1)$$

$$\times \frac{(\ell - |m|)!}{(\ell + |m|)!}\left|P_\ell^m(\cos\varphi')\right|^2\exp\left\{-\frac{m}{2}\ln|\zeta|^4\right\} \quad , \tag{6.57}$$

with

$$k^4 = \left(k_0^2 - \frac{r_0^2}{4\sigma_0^4}\right)^2 + \frac{(k_0 \cdot r_0)^2}{\sigma_0^4}$$

P_ℓ^m: associated Legendre functions with complex argument

$I_{\ell+\frac{1}{2}}$: modified Bessel function

The complex vector

$$\kappa = k_0 - i\frac{1}{2\sigma_0^2}r_0 \tag{6.58}$$

is decomposed into

$$\kappa = \kappa\{e_1\cos\varphi' + e_2\sin\varphi'\} \quad . \tag{6.59}$$

The quantity κ is the complex square root

$$\kappa = \left[\left(k_0 - \frac{i}{2\sigma_0^2}y\right)^2 - \frac{1}{4\sigma_0^4}b^2\right]^{1/2} \quad . \tag{6.60}$$

The complex angle φ' is defined by

$$\cos\varphi' = \frac{1}{\kappa}(e_1 \cdot \kappa) \quad , \quad \sin\varphi' = \frac{1}{\kappa}(e_2 \cdot \kappa) \quad . \tag{6.61}$$

So far m was the quantum number of the z component of angular momentum, i.e., the *quantization axis* was the z axis. For an arbitrary quantization axis we rotate the coordinate system

$$e_1 = \hat{k}_0 \quad , \quad e_2 = \hat{b} \quad , \quad e_3 = e_1 \times e_2$$

into the arbitrary system η_1, η_2, η_3 with the transformation

$$e_j = \sum_{i=1}^{3} \eta_i R_{ij} \quad .$$

The matrix R_{ij} represents a rotation with the angle α about the axis

$$\hat{\alpha} = \eta_1 \sin \beta \cos \gamma + \eta_2 \sin \beta \sin \gamma + \eta_3 \cos \beta \quad , \tag{6.62}$$

i.e., β and γ are the polar and azimuthal angles which define the rotation axis in the original system e_1, e_2, e_3. With this we define the complex angle φ'' through its cosine

$$\cos \varphi'' = R_{13} \cos \varphi' + R_{23} \sin \varphi' \quad ,$$

and the quantity

$$\zeta = \cos \varphi'' + i \sin \varphi'' \quad .$$

Here

$$R_{13} = (1 - \cos \alpha) \cos \beta \sin \beta \sin \gamma + \sin \alpha \sin \beta \sin \gamma \quad ,$$

$$R_{23} = (1 - \cos \alpha) \cos \beta \sin \beta \cos \gamma - \sin \alpha \sin \beta \cos \gamma \quad .$$

The quantization axis with respect to which m is defined is η_3. The marginal distribution

$$W_\ell = \sum_{m=-\ell}^{\ell} W_{\ell m} = e^{-\sigma_0^2 \lambda^2} \sqrt{\frac{\pi}{2\sigma_0^2 k^2}} I_{\ell+1/2}(\sigma_0^2 k^2)(2\ell + 1) P_\ell \left(\frac{1}{\sqrt{1 - \xi^2}} \right) \tag{6.63}$$

describes the weight of the contribution of the angular momentum ℓ in the wave packet. Here

$$\lambda^2 = k_0^2 + \frac{r_0^2}{4\sigma_0^2} \tag{6.64}$$

and

$$\sqrt{1 - \xi^2}^{-1} = \frac{\lambda^2}{k^2} \quad . \tag{6.65}$$

Further Reading

Alonso, Finn: Vol. 3, Chap. 3
Berkeley Physics Course: Chaps. 7,8
Brandt, Dahmen: Chaps. 9,10
Feynman, Leighton, Sands: Vol. 3, Chap. 18
Flügge: Vol. 1, Chaps. 2B,2C
Gasiorowicz: Chaps. 9,10,11
Merzbacher: Chap. 9
Messiah: Vol. 1, Chap. 9, Vol.2, Chap. 13
Schiff: Chaps. 2,4

6.2 The 3D Harmonic Plane Wave

Aim of this section: Illustration of the stationary harmonic wave (6.11). (The illustration can be presented in the form of a Cartesian or a polar 3D plot.)

6.2.1 A Cartesian 3D Plot

Plot type: 0

C3 coordinates: x: position coordinate z y: position coordinate x
z: Re $\{e^{ikz}\}$ or Im $\{e^{ikz}\}$ or $|e^{ikz}|^2$

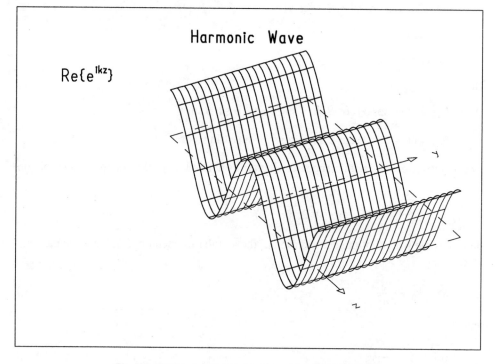

Fig. 6.1. Plot produced with descriptor 1 on file IQ032.DES

Input parameters:

```
CH 0 32 1 0
VO fRES                              V1 k
```

with

$f_{\text{RES}} = 0$: function plotted is $|e^{ikz}|^2$

$f_{\text{RES}} = 1$: function plotted is Re $\{e^{ikz}\}$

$f_{\text{RES}} = 2$: function plotted is Im $\{e^{ikz}\}$

k: wave number

Automatically provided texts: `T1, T2, TF`

Example plot: Fig. 6.1

6.2.2 A Polar 3D Plot

Plot type: 1

C3 coordinates: x: position coordinate z $\quad$ y: position coordinate x
r: radial coordinate r $\quad$ φ: polar angle ϑ $\quad$ z: $\mathrm{Re}\{e^{ikz}\}$ or $\mathrm{Im}\{e^{ikz}\}$ or $|e^{ikz}|^2$

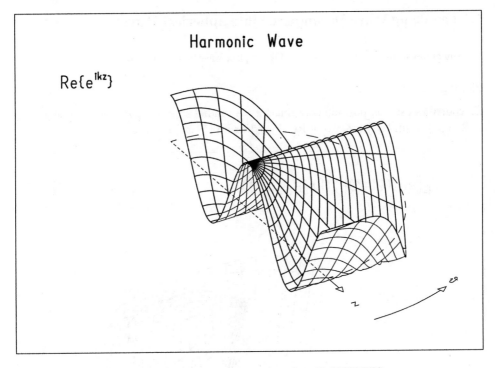

Fig. 6.2. Plot produced with descriptor 2 on file IQ032.DES

Input parameters:

```
CH 1 32 1 0
VO fRES                    V1 q 0 0 fkE
```

with

$f_{RES} = 0$: function plotted is $|e^{ikz}|^2$

$f_{RES} = 1$: function plotted is $\mathrm{Re}\{e^{ikz}\}$

$f_{RES} = 2$: function plotted is $\mathrm{Im}\{e^{ikz}\}$

q: wave number is $k = q$ for $f_{kE} = 0$

q: wave number is $k = \sqrt{2Mq}/\hbar$ for $f_{kE} = 1$

$f_{kE} = 0$: input value q is interpreted as wave number

$f_{kE} = 1$: input value q is interpreted as kinetic energy

Automatically provided texts: T1, TP, TF

Example plot: Fig. 6.2

Remarks:
Note that the numerical values of M and $\hbar$ are fixed at $M = 1$ and $\hbar = 1$ in this section.

6.3 The Plane Wave Decomposed into Spherical Waves

Aim of this section: Decomposition (6.53) of a plane wave into spherical waves.

Plot type: 1

C3 coordinates: x: position coordinate z r: radial coordinate r φ: polar angle ϑ
z: $\mathrm{Re}\{\varphi_\ell\}$ or $\mathrm{Im}\{\varphi_\ell\}$ or $|\varphi_\ell|^2$ or $\mathrm{Re}\{\sum \varphi_\ell\}$ or $\mathrm{Im}\{\sum \varphi_\ell\}$ or $|\sum \varphi_\ell|^2$

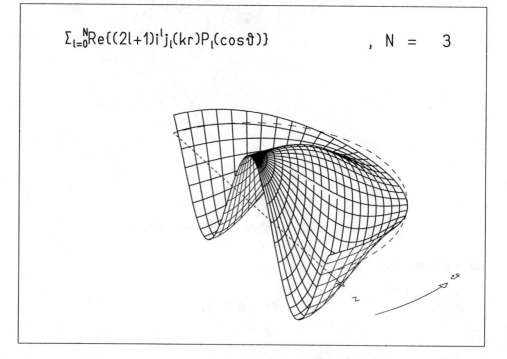

$$\sum_{l=0}^{N} \mathrm{Re}\{(2l+1)i^l j_l(kr) P_l(\cos\vartheta)\} \qquad , \quad N \; = \quad 3$$

Fig. 6.3. Plot produced with descriptor 3 on file IQ032.DES

Input parameters:

```
CH 1 32 0 0
VO fRES                                    V1 q 0 0 fkE
                                           V3 fFUNC L
```

with

$f_{RES} = 0$: function plotted is absolute square of function computed (see below)

$f_{RES} = 1$: function plotted is real part of function computed (see below)

$f_{RES} = 2$: function plotted is imaginary part of function computed (see below)

q: wave number is $k = q$ for $f_{kE} = 0$

q: wave number is $k = \sqrt{2Mq}/\hbar$ (with $M = 1$, $\hbar = 1$) for $f_{kE} = 1$

$f_{kE} = 0$: input value q is interpreted as wave number

$f_{kE} = 1$: input value q is interpreted as kinetic energy

L: $N = L$ for $f_{FUNC} = 1$

L: $\ell = L$ for $f_{FUNC} = 2$

$f_{FUNC} = 1$: function computed is $\sum_{\ell=0}^{N}(2\ell + 1)\mathrm{i}^{\ell} j_{\ell}(kr)P_{\ell}(\cos \vartheta)$

$f_{FUNC} = 2$: function computed is $(2\ell + 1)\mathrm{i}^{\ell} j_{\ell}(kr)P_{\ell}(\cos \vartheta)$

If you ask for a multiple plot the input value of L is taken only for the first plot and successively increased by one for each subsequent plot.

Automatically provided texts: T1, TP, TX

Example plot: Fig. 6.3

6.4 The 3D Gaussian Wave Packet

Aim of this section: Illustration of the probability density $\varrho(x, y, t)$, (6.29), and the corresponding wave function $\Psi(x, y, t)$.

Plot type: 0

C3 coordinates: x: position coordinate x y: position coordinate y
z: $\varrho(x, y, t) = |\Psi(x, y, t)|^2$ or Re $\{\Psi(x, y, t)\}$ or Im $\{\Psi(x, y, t)\}$

Input parameters:

```
CH 0 32 0
VO fRES                                    V1 x0 y0
V2 vx0 vy0                                 V3 σx0 σy0
V4 Δt
V6 R
```

with

$f_{RES} = 0$: function plotted is $|\Psi(x, y, t)|^2$

$f_{\text{RES}} = 1$: function plotted is Re $\Psi(x, y, t)$

$f_{\text{RES}} = 1$: function plotted is Im $\Psi(x, y, t)$

x_0, y_0: initial position expectation values ($t = 0$)

v_{x0}, v_{y0}: initial velocities

σ_{x0}, σ_{y0}: initial widths

Δt: in a multiple plot $\varrho(x_1, x_2, t)$ is plotted for $t = 0$, $t = \Delta t$, $t = 2\Delta t$, ... (default value: $\Delta t = 1$)

R: radius of point symbolizing position of corresponding classical particle. (Default value: $1/50$ of width in X of W3 window)

Additional plot items:

Item 5: circles indicating position of classical particle at times 0, Δt, ..., t

Item 6: trajectory of classical particle

Item 7: probability ellipse

Automatically provided texts: T1, T2, TF, TX

Example plot: Fig. 6.4

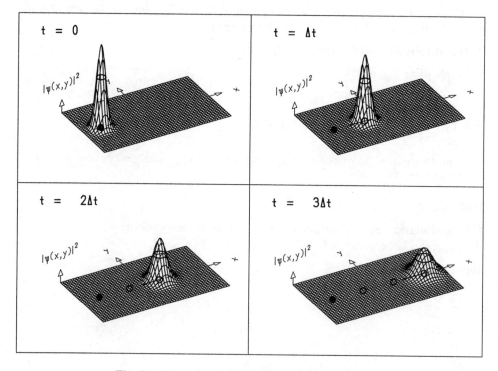

Fig. 6.4. Plot produced with descriptor 4 on file IQ032.DES

6.5 The Probability Ellipsoid

Aim of this section: Drawing for times $t_0, t_0 + \Delta t, \ldots t_0 + (N-1)\Delta t$ the ellipsoid having the principal axes of lengths $\sigma_x, \sigma_y, \sigma_z$ parallel to the coordinate axes, see (6.27), which characterizes a 3D Gaussian wave packet with (uncorrelated) widths $\sigma_x, \sigma_y, \sigma_z$.

Plot type: 10

C3 coordinates: x: position coordinate x y: position coordinate y
z: position coordinate z

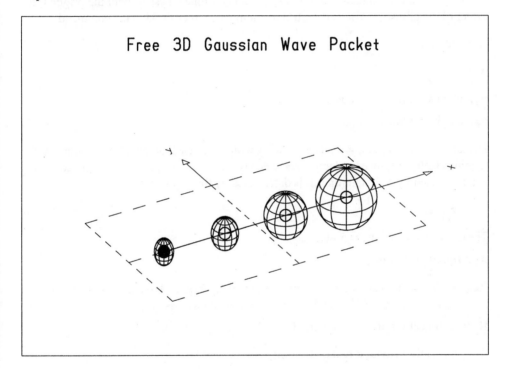

Fig. 6.5. Plot produced with descriptor 5 on file IQ032.DES

Input parameters:

```
CH 10 30 1
                          V1 x0 y0 z0
V2 vx0 vy0 vz0            V3 σx0 σy0 σz0
V6 R                     V7 0 t0 Δt N
```

with

x_0, y_0, z_0: initial position expectation values ($t = t_0$)

v_{x0}, v_{y0}, v_{z0}: velocities

$\sigma_{x0}, \sigma_{y0}, \sigma_{z0}$: initial widths

R: radius (in W3 units) of point indicating position of corresponding classical particle (center of ellipsoid)

t_0: initial time

Δt: time elapsed between two instants for which ellipsoid is drawn

N: total number of ellipsoids drawn

Plot recommendation:
Each ellipsoid is drawn with two sets of lines $\vartheta = $ const and $\varphi = $ const where ϑ and φ are the polar and azimuthal angle of a spherical coordinate system with the origin at the center of the ellipsoid. You can control the number of lines through the command

> NL n_φ n_ϑ

with

n_φ: number of lines $\varphi = $ const

n_ϑ: number of lines $\vartheta = $ const

Defaults are $n_\varphi = 12$, $n_\vartheta = 5$. The accuracy with which each of these lines is drawn is determined by the distance (in degrees) between two consecutive points which are computed and connected by a straight line. It is determined by the command

> AC $\Delta\alpha$

The default value is $\Delta\alpha = 5$ (degrees).

Additional plot items:

Item 5: circles indicating the position of the corresponding classical particle at times t_0, $t_0 + \Delta t, \ldots, t_0 + (N-1)\Delta t$. For $t = t_0$ a filled circle is drawn

Item 6: trajectory of classical particle

Automatically provided texts: T1, T2, TF

Example plot: Fig. 6.5

6.6 Angular-Momentum Decomposition of a Wave Packet

Aim of this section: Computation and presentation of the probabilities $W_{\ell m}$ and W_ℓ that a particle represented by a wave packet of spherical symmetry has the angular-momentum quantum numbers ℓ and m, respectively; see (6.55) and (6.63).

Plot type: 3

C3 coordinates: x: angular-momentum quantum number ℓ
y: magnetic quantum number m z: probabilities $W_{\ell m}$ and W_ℓ

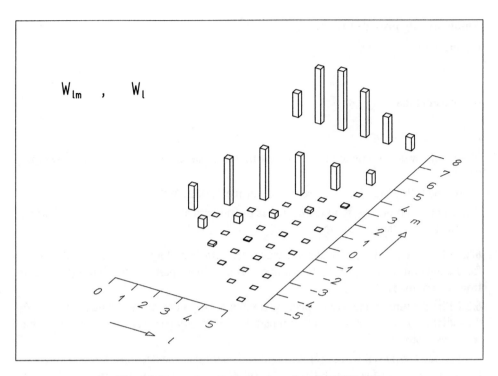

Fig. 6.6. Plot produced with descriptor 6 on file IQ032.DES

Input parameters:

    CH 3 32

                                    V1 p_0 σ_0 y b

        V2 α β γ

with

p_0: momentum expectation value

σ_0: initial width

y: component of initial position expectation value parallel to p_0

b: expectation value of impact parameter

α, β, γ: angles (in degrees) defining quantization axis

Plot recommendation:
Using the commands **XX** and **YY** in Appendix A.2.2.3 choose the parameter for the C3 windows to be

$$x_a = 0 \quad , \quad x_b = \ell_{\max} \quad , \quad y_a = -\ell_{\max} \quad , \quad y_b = \ell_{\max} + 3$$

where $\ell_{\max}$ is some positive integer. To get blocks plotted at all meaningful points (ℓ, m) use the **NL** command in the form

$$\text{NL} \;\; \ell_{\max} + 1 \;\; , \;\; 2\ell_{\max} + 4$$

The marginal distribution is plotted as a series of blocks along the line $m = \ell_{\max} + 3$.

Automatically provided texts: T1, T2, TF

Example plot: Fig. 6.6

6.7 Exercises

Please note:

(i) you may watch a semiautomatic demonstration of the material of this chapter by typing SA D32

(ii) for the following exercises use descriptor file IQ032.DES

(iii) if not stated otherwise in the exercises the numerical values of the particle mass and of Planck's constant are put to 1.

6.2.1 Plot the three dimensional plane wave in a Cartesian plot for the wave number $k = 2$ for the interval $0 \le x \le 2\pi$: **(a)** real part, **(b)** imaginary part, **(c)** absolute square. Start from descriptor 1.

6.2.2 Plot the three-dimensional plane wave in a polar plot for the wave number $k = 2$ for the radial interval $0 \le r \le 2\pi$, **(a)** real part, **(b)** imaginary part, **(c)** absolute square. Start from descriptor 2.

6.3.1 Study the partial waves for different ℓ appearing in the decomposition of the plane harmonic wave for **(a)** $\ell = 0$, **(b)** $\ell = 1$, **(c)** $\ell = 2, \ldots$, **(l)** $\ell = 13$, **(m)** $\ell = 20$. Start from descriptor 3. For even (odd) values of ℓ the real (imaginary) part of the partial waves are nonvanishing. **(n)** Calculate the minimal classical angular momentum in units $\hbar$ of a particle of wave number $k = 2$ that does not enter the radial domain $0 \le r \le 2\pi$. **(o)** Plot the partial wave for $\ell = 13$. **(p)** Why does the partial wave become more and more suppressed in the central region (r close to zero) for increasing ℓ?

6.3.2 Study the sums of partial waves for different N approximating the plane harmonic wave for **(a)** $N = 0$, **(b)** $N = 1, \ldots$, **(k)** $N = 10$, **(l)** $N = 13$, **(m)** $N = 20$. Start from descriptor 3. **(n)** How fast does the region in which the plotted partial sum resembles a plane harmonic wave grow in radius with increasing N?

6.4.1 A three-dimensional Gaussian wave packet in the xy plane has the initial position $x_0 = -2$, $y_0 = -2$, initial velocity $v_{x0} = 4$, $v_{y0} = 4$, initial width $\sigma_{x0} = 0.5$, $\sigma_{y0} = 0.5$. Plot for the times $t = 0$ and $t = 1$ **(a)** the absolute square, **(b)** the real part, **(c)** the imaginary part of the wave function. Start from descriptor 4. **(d)** Calculate the spatial widening of the wave packet with time ($M = 1$, $\hbar = 1$).

6.4.2 (a–d) Repeat Exercise 6.4.1 for the initial position $x_0 = -2$, $y_0 = -2$, initial velocity $v_{x0} = 4$, $v_{y0} = 2$, initial width $\sigma_{x0} = 0.25$, $\sigma_{y0} = 0.8$. **(e)** Calculate the vector of classical angular momentum of a particle described by the wave packet ($M = 1$, $\hbar = 1$).

6.4.3 (a,b,c) Repeat Exercise 6.4.1 (a,b,c) for the initial position $x_0 = -4$, $y_0 = -3$, initial velocity $v_{x0} = 4$, $v_{y0} = 3$, initial width $\sigma_{x0} = 0.1$, $\sigma_{y0} = 0.3$, for $\Delta t = 1.5$. **(d)** Explain the different speeds by which the main axes of the covariance ellipse grow. **(e)** Which direction do the ripples in the real and imaginary part of the wave function possess?

6.4.4 A wave packet at rest is to be plotted with the initial parameters $x_0 = 0$, $y_0 = 0$, $v_{x0} = 0$, $v_{y0} = 0$, $\sigma_{x0} = 1$, $\sigma_{y0} = 1$, for $\Delta t = 2.5$, plot **(a)** the absolute square, **(b)** the real part, (c) the imaginary part. Start from descriptor 4. (d) Why does this wave packet exhibit a rotationally invariant structure?

6.5.1 (a) Plot the probability ellipsoids of a particle with the initial conditions $x_0 = -3$, $y_0 = -2$, $z_0 = -1$, $v_{x0} = 5$, $v_{y0} = 1.5$, $v_{z0} = 1$, $\sigma_{x0} = 1.0$, $\sigma_{y0} = 1.05$, $\sigma_{z0} = 2$, for 3 instants in time $t = 0, 1, 2$. Start from descriptor 5. **(b)** Calculate the angular-momentum vector of the classical particle with the above initial data ($M = 1$, $\hbar = 1$).

6.5.2 Plot the probability ellipsoids of a particle with the initial conditions $x_0 = -3$, $y_0 = -2$, $z_0 = -1$, $v_{x0} = 5$, $v_{y0} = 1.5$, $v_{z0} = 1$, $\sigma_{x0} = 0.25$, $\sigma_{y0} = 0.35$, $\sigma_{z0} = 0.5$ for three instants in time $t = 0, 0.8, 1.6$. Start from descriptor 5.

6.5.3 (a) Plot the probability ellipsoids of a particle with the initial conditions $x_0 = -5$, $y_0 = -2$, $z_0 = -1$, $v_{x0} = 5$, $v_{y0} = 1.5$, $v_{z0} = 1$, $\sigma_{x0} = 1.8$, $\sigma_{y0} = 1.5$, $\sigma_{z0} = 1$, for the three instants in time $t = 0, 1, 2$. **(b)** Why does the ellipsoid enlarge much more slowly in time relative to its initial size than in Exercise 6.5.2? Start from descriptor 5.

6.5.4 (a) Plot the probability ellpsoids for a spherical wave packet at rest with the widths $\sigma_{x0} = \sigma_{y0} = \sigma_{z0} = 0.2$ for four instants in time $t = 0, 0.2, 0.4, 0.6$. **(b)** Why does the radius of the sphere grow almost linearly over time only for later times? Start from descriptor 5.

6.6.1 (a) Plot the probabilities $W_{\ell m}$, W_ℓ of the partial wave decomposition for the quantization axis $\hat{n}$ pointing in the z direction of a wave packet with initial conditions $p_0 = 1.5$, $\sigma_0 = 1$, $y = 2$, $b = 0.6667$. Start from descriptor 6. **(b)** Calculate the angular-momentum vector of a classical particle possessing the above initial data ($\hbar = 1$). **(c)** Calculate the angular momentum for a particle with the impact parameter $b' = b + \sigma_0$ and for $b'' = b - \sigma_0$. **(d)** Interpret the two results of b and c in terms of the partial probabilities $W_{\ell m}$ as plotted in (a).

6.6.2 (a) Plot the probabilities $W_{\ell m}$, W_ℓ of the partial wave decomposition for the quantization axis $\hat{n}$ pointing in the z direction for a wave packet with initial conditions $p_0 = 3$, $\sigma_0 = 0.5$, $y = 2$, $b = 0.3333$. Start from descriptor 6. **(b)** Why does the plot look the same as in Exercise 6.6.1 (a)?

6.6.3 (a) Plot the probabilities $W_{\ell m}$, W_ℓ of the partial-wave decomposition for the quantization axis $\hat{n}$ pointing in the z direction for a wave packet with initial conditions $p_0 = 3$, $\sigma_0 = 1$, $y = 2$, $b = 0.3333$. Start from descriptor 6. **(b)** Why are the $W_{\ell m} = 0$ for $m = \ell - 1, \ell - 3, \ldots, -\ell + 1$ for the wave packets investigated so far? **(c)** Why has the distribution of the $W_{\ell m}$'s and the W_ℓ's widened compared to Exercise 6.6.2?

7 Bound States in Three Dimensions

Contents: Introduction of the Schrödinger equation with potential. Partial wave decomposition. Spherical harmonics as eigenfunctions of angular momenta. Radial Schrödinger equation. Centrifugal barrier. Normalization and orthogonality of bound state wave functions. Infinitely deep square-well potential. Spherical step potential. Harmonic oscillator. Coulomb potential. Harmonic particle motion.

7.1 Physical Concepts

7.1.1 The Schrödinger Equation for a Particle under the Action of a Force. The Centrifugal Barrier. The Effective Potential

The Schrödinger equation (6.5) of a free particle of mass M introduced in Sect. 6.1 contains the kinetic energy T as the only term in the Hamiltonian operator. Under the action of a conservative force $\boldsymbol{F}(\boldsymbol{r}) = -\nabla V(\boldsymbol{r})$ the Hamiltonian H contains both the kinetic energy T and the potential energy $V(\boldsymbol{r})$

$$H = T + V(\boldsymbol{r}) = -\frac{\hbar^2}{2M}\nabla^2 + V(\boldsymbol{r}) \quad . \tag{7.1}$$

The Schrödinger equation for the three-dimensional motion under the action of a force then reads

$$i\hbar\frac{\partial}{\partial t}\psi(\boldsymbol{r}, t) = \left(-\frac{\hbar^2}{2M}\nabla^2 + V(\boldsymbol{r})\right)\psi(\boldsymbol{r}, t) \quad . \tag{7.2}$$

With a separation of time and space coordinates,

$$\psi(\boldsymbol{r}, t) = e^{-iEt/\hbar}\varphi_E(\boldsymbol{r}) \quad , \tag{7.3}$$

the stationary Schrödinger equation is an eigenvalue equation:

$$H\varphi_E(\boldsymbol{r}) = E\varphi_E(\boldsymbol{r}) \quad \text{or} \quad \left(-\frac{\hbar^2}{2M}\nabla^2 + V(\boldsymbol{r})\right)\varphi_E(\boldsymbol{r}) = E\varphi_E(\boldsymbol{r}) \quad . \tag{7.4}$$

Again E is the energy eigenvalue, $\varphi_E(\boldsymbol{r})$ the corresponding eigenfunction.

For a *spherically symmetric potential*

$$V(\boldsymbol{r}) = V(r)$$

a further separation of radial and angular coordinates by means of an eigenfunction corresponding to the energy eigenvalue E and the angular-momentum quantum numbers ℓ, m,

$$\varphi_{E\ell m}(r) = R_{E\ell}(r)\, Y_{\ell m}(\vartheta, \varphi) \quad , \tag{7.5}$$

is carried out along the same lines as in Sect. 6.1. We arrive at the *radial Schrödinger equation* for the *radial wave function* $R_{E\ell}(r)$

$$\left(-\frac{\hbar^2}{2M} \frac{1}{r} \frac{d^2}{dr^2} r + V_\ell^{\text{eff}}(r) \right) R_{E\ell}(r) = E R_{E\ell}(r) \quad , \quad r > 0 \quad , \tag{7.6}$$

with the *effective potential*

$$V_\ell^{\text{eff}}(r) = \frac{\hbar^2}{2M} \frac{\ell(\ell+1)}{r^2} + V(r) \quad . \tag{7.7}$$

The first term of the left-hand side of (7.6) represents the *kinetic energy of the radial motion*

$$T^{\text{rad}} = \hat{p}_r^2 / 2M \quad , \tag{7.8}$$

with the *operator of radial momentum*

$$\hat{p}_r = \frac{\hbar}{i} \left(\frac{1}{r} + \frac{\partial}{\partial r} \right) \quad . \tag{7.9}$$

The first term of (7.7) is the *centrifugal barrier*. It corresponds to the *rotational energy* relative to the origin of the coordinate frame

$$T^{\text{rot}} = \hat{\boldsymbol{L}}^2 / 2\Theta \tag{7.10}$$

of a particle with squared angular momentum $\hat{\boldsymbol{L}}^2 Y_{\ell m} = \hbar^2 \ell(\ell+1) Y_{\ell m}$ and a *moment of inertia* $\Theta = Mr^2$ with respect to the origin. The second term is the spherically symmetric potential $V(r)$ of the force $\boldsymbol{F}(r) = -\nabla V(r) = -e_r dV(r)/dr$ acting on the particle. Since the centrifugal barrier is a repulsive potential (for $\ell \geq 1$) it tends to push the particle away from the origin $r = 0$.

The solutions of (7.6) are physical for $r > 0$. The radial kinetic energy is a Hermitian operator, i.e., a physical observable, only for wave functions free of singularities at $r = 0$. This requirement represents the *boundary condition* for solutions of the Schrödinger equation at $r = 0$.

7.1.2 Bound States. Scattering States. Discrete and Continuous Spectra

We denote by V_∞ the value of the spherically symmetric potential far out,

$$V_\infty = \lim_{r \to \infty} V(r) \quad . \tag{7.11}$$

We consider potentials only for which the intervals in r, for which $V(r) \leq E < V_\infty$ for any given energy value have a finite total length. Then there are two types of solution:

i) *Bound states* for $E < V_\infty$: there exist only solutions at *discrete energy eigenvalues* $E_{n\ell}$; the integer n is the principal quantum number used to enumerate the eigenvalue, ℓ is the quantum number of angular momentum

ii) *Scattering states* for $E \geq V_\infty$: there is a *continuous spectrum of eigenvalues* $E_\ell(k)$; it fills the domain $V_\infty \leq E_\ell(k)$

In this chapter we deal with bound states only.

The radial wave function for a bound state with energy eigenvalue $E_{n\ell}$ will be denoted by $R_{n\ell}(r)$. It satisfies the radial Schrödinger equation

$$\left(-\frac{\hbar^2}{2M} \frac{1}{r} \frac{d^2}{dr^2} r + V_\ell^{\text{eff}}(r) \right) R_{n\ell}(r) = E_{n\ell} R_{n\ell}(r) \quad . \tag{7.12}$$

In the domains in r where $V_\ell^{\text{eff}}(r) > E_{n\ell}$ the integral over the absolute square of the wave function must be finite, otherwise the contributions to the potential energy coming from these domains would diverge. Thus the integral over the absolute square of the bound-state wave function over the range $0 \le r < \infty$ must be finite so that the integral over the absolute square of the radial bound-state function can be normalized to one:

$$\int_0^\infty R_{n\ell}^*(r) R_{n\ell}(r) r^2 dr = 1 \quad . \tag{7.13}$$

Since the radial kinetic energy and the effective potential energy in (7.12) are Hermitian operators, the eigenfunctions $R_{n\ell}$ are orthogonal for different principal quantum numbers, so that together with (7.13) we have the *orthonormality relation* for the radial bound-state wave functions

$$\int_0^\infty R_{n'\ell}^*(r) R_{n\ell}(r) r^2 dr = \delta_{n'n} \quad . \tag{7.14}$$

The total bound-state wave functions

$$\varphi_{n\ell m}(\boldsymbol{r}) = R_{n\ell}(r) Y_{\ell m}(\vartheta, \varphi) \tag{7.15}$$

are orthonormal in all three quantum numbers n, ℓ, m,

$$\int \varphi_{n'\ell'm'}^*(\boldsymbol{r}) \varphi_{n\ell m}(\boldsymbol{r}) dV = \delta_{n'n} \, \delta_{\ell'\ell} \, \delta_{m'm} \quad , \tag{7.16}$$

because of the orthonormality of the radial wave functions (7.14) and of the spherical harmonics (9.18).

The probability density

$$\varrho_{n\ell m}(\boldsymbol{r}) = |\varphi_{n\ell m}(\boldsymbol{r})|^2 = |R_{n\ell}(r)|^2 |Y_{\ell m}(\vartheta, \varphi)|^2 \tag{7.17}$$

of a bound state described by the wave function $\varphi_{n\ell m}(\boldsymbol{r})$ is a function of r and ϑ only. Because of (9.16) and (9.17) the φ dependence vanishes upon taking the absolute square of the spherical harmonics

$$|Y_{\ell m}(\vartheta, \varphi)|^2 = \frac{2\ell + 1}{4\pi} \frac{(\ell - m)!}{(\ell + m)!} \left(P_\ell^{|m|}(\cos \vartheta) \right)^2 \quad . \tag{7.18}$$

Moreover, the probability density is a function of the modulus $|m|$ of the magnetic quantum number. Altogether we have

$$\varrho_{n\ell m}(r, \vartheta) = |R_{n\ell}(r)|^2 \frac{2\ell + 1}{4\pi} \frac{(\ell - m)!}{(\ell + m)!} \left(P_\ell^{|m|}(\cos \vartheta) \right)^2 \tag{7.19}$$

r: radial variable

ϑ: polar angle

$R_{n\ell}(r)$: radial wave function

$P_\ell^{|m|}(\cos\vartheta)$: associated Legendre function

n: principal quantum number

ℓ: quantum number of angular momentum

m: magnetic quantum number

The zeros of the radial wave function $R_{n\ell}(r_i) = 0$ appear in a plot of $\varrho_{n\ell m}(r, \vartheta)$ over the coordinate plane of r and ϑ as *circular node lines* of radius r_i. The zeros of the spherical harmonics $Y_{\ell m}(\vartheta_j, \varphi) = 0$ appear as *node rays* originating at the origin under the polar angles ϑ_j.

7.1.3 The Infinitely Deep Square-Well Potential

The *infinitely deep square-well potential*

$$V(r) = \begin{cases} 0 & , \quad r \le a \\ \infty & , \quad r > a \end{cases}$$

confines the particle to a sphere of radius a. The solutions $R_{n\ell}(r)$ of the radial Schrödinger equation have to vanish for values $r > a$, otherwise they would give rise to infinite contributions of the potential energy for $r > a$. Thus we are looking for solutions of the radial Schrödinger equation

$$\left(-\frac{\hbar^2}{2M} \frac{1}{r} \frac{d^2}{dr^2} r + V_\ell^{\text{eff}}(r) \right) R_{n\ell}(r) = E_{n\ell}(r) \quad , \quad 0 < r \le a \tag{7.20}$$

in the range $0 \le r \le a$ only. The solution has to be free of singularities at $r = 0$ and has to vanish at $r = a$. This allows only for spherical Bessel functions of the first kind $j_\ell(kr)$. The wave number k has to be determined in accordance with the boundary condition at $r = a$. Thus $k = k_{n\ell}$ has to be chosen so that $j_\ell(k_{n\ell}a) = 0$ is a zero of the Bessel function j_ℓ. With the normalization to one the solution is given by

$$R_{n\ell} = \begin{cases} \left[\frac{a^3}{2}(j_{\ell+1}(k_{n\ell}a))^2 \right]^{-1/2} j_\ell(k_{n\ell}r) & , \quad 0 \le r \le a \\ 0 & , \quad r > a \end{cases} \tag{7.21}$$

The energy eigenvalue is determined by the wave number

$$E_{n\ell} = \hbar^2 k_{n\ell}^2/2M \quad . \tag{7.22}$$

The *principal quantum number* $n = 0, 1, 2, \ldots$ is equal to the number of nodes (zeros) of the spherical Bessel function $j_\ell(k_{n\ell}r)$ in the domain $0 < r < a$. The wave numbers $k_{n\ell}$ increase monotonously with n, as does the energy eigenvalue $E_{n\ell}$. A simple heuristic argument behind this is that the radial kinetic-energy contribution increases with increasing curvature of the wave function. The curvature itself grows monotonously with the number of nodes.

7.1.4 The Spherical Step Potential

For many applications a potential with stepwise constant values with n regions is a sufficient approximation:

$$V(r) = \begin{cases} V_1 & , \ 0 \le r < r_1 & \text{region 1} \\ V_2 & , \ r_1 \le r < r_2 & \text{region 2} \\ \vdots & & \vdots \\ V_{N-1} & , \ r_{N-2} \le r < r_{N-1} & \text{region } N-1 \\ V_N & , \ r_{N-1} \le r & \text{region } N \end{cases} \quad . \tag{7.23}$$

The eigenfunction

$$\varphi_{n\ell m}(\boldsymbol{r}) = R_{n\ell}(r)Y_{\ell m}(\vartheta, \varphi)$$

belonging to the energy eigenvalue $E_{n\ell}$ is determined by the radial wave function $R_{n\ell}(r)$, which is a solution of the radial Schrödinger equation (7.12). Since the potential (7.23) is a stepwise constant function with N regions, $R_{n\ell}(r)$ consists of N pieces $R_{n\ell q}$, $q = 1, \ldots, N$,

$$R_{n\ell}(r) = \begin{cases} R_{n\ell 1}(r) & , \ 0 \le r < r_1 & \text{region 1} \\ R_{n\ell 2}(r) & , \ r_1 \le r < r_2 & \text{region 2} \\ \vdots & & \vdots \\ R_{n\ell N-1}(r) & , \ r_{N-2} \le r < r_{N-1} & \text{region } N-1 \\ R_{n\ell N}(r) & , \ r_{N-1} \le r & \text{region } N \end{cases} \quad . \tag{7.24}$$

The piece $R_{n\ell q}(r)$ fulfills the free radial Schrödinger equation

$$\left(-\frac{\hbar^2}{2M} \frac{1}{r} \frac{d^2}{dr^2} r + \frac{\ell(\ell+1)\hbar^2}{2Mr^2} \right) R_{n\ell q}(r) = (E_{n\ell} - V_q)R_{n\ell q}(r) \quad . \tag{7.25}$$

For the solution $R_{n\ell q}$ two cases have to be distinguished:

i) In regions with $E_{n\ell} > V_q$ we obtain a real wave number

$$k_{n\ell q} = \left| \sqrt{2M(E_{n\ell} - V_q)}/\hbar \right| \quad . \tag{7.26}$$

The solution is a linear superposition of the two linearly independent spherical Bessel functions j_ℓ and n_ℓ, (9.31), (9.32),

$$R_{n\ell q}(r) = A_{n\ell q}j_\ell(k_{n\ell q}r) + B_{n\ell q}n_\ell(k_{n\ell q}r) \quad . \tag{7.27}$$

ii) In regions with $E_{n\ell} < V_q$ the wave number is purely imaginary:

$$k_{n\ell q} = i\kappa_{n\ell q} \quad , \quad \kappa_{n\ell q} = \left| \sqrt{2M(V_q - E_{n\ell})}/\hbar \right| \quad . \tag{7.28}$$

The solution is a linear combination of the two linearly independent Hankel functions $h_\ell^{(\pm)}$, (9.33), (9.34),

$$R_{n\ell q}(r) = A_{n\ell q}h_\ell^{(+)}(i\kappa_{n\ell q}r) + B_{n\ell q}h_\ell^{(-)}(i\kappa_{n\ell q}r) \quad . \tag{7.29}$$

For the imaginary argument the Hankel functions are products of real polynomials $C_\ell^\pm(i\kappa_{n\ell q}r)$ and of decreasing and increasing exponential functions

$$h_\ell^{(\pm)}(i\kappa_{n\ell q}r) = C_\ell^\pm \frac{\exp(\mp\kappa_{n\ell q}r)}{\kappa_{n\ell q}r} \quad , \tag{7.30}$$

see (9.35), (9.36).

The solution (7.24) has to satisfy two boundary conditions:

i) At the origin $r = 0$: there is an absence of singularities, as already discussed in this section. Since the spherical Bessel function n_ℓ is singular at $r = 0$, see (9.40), this requires

$$B_{n\ell 1} = 0 \quad , \quad \text{i.e.,} \quad R_{n\ell 1}(r) = A_{n\ell 1}j_\ell(k_{n\ell 1}r) \quad . \tag{7.31}$$

ii) In region N, $r_{N-1} \leq r$: for bound states the energy eigenvalue $E_{n\ell}$ is smaller than V_∞, i.e., $E_{n\ell} < V_N$, so that the wave function (7.29) in this region is given by (7.30). The radial wave function has to vanish sufficiently fast for $r \to \infty$ to allow for the normalization (7.13) of the radial wave function. This is taken care of by putting

$$B_{n\ell N} = 0 \quad , \quad \text{i.e.,} \quad R_{n\ell N}(r) = A_{n\ell N}h^{(+)}(i\kappa_{n\ell q}r) \quad . \tag{7.32}$$

The remaining discussion runs very much parallel to Sect. 3.1. The continuity of the function $R_{n\ell}(r)$ and its derivative at the positions $r_1, \ldots, r_{N-1}$ poses $2(N-1)$ conditions analogous to (3.22)

$$R_{n\ell q}(r_q) = R_{n\ell q+1}(r_q) \tag{7.33}$$

and

$$\frac{dR_{n\ell q}}{dr}(r_q) = \frac{dR_{n\ell q+1}}{dr}(r_q) \tag{7.34}$$

for the $2N - 2$ unknown coefficients $A_{n\ell 1}, A_{n\ell 2}, B_{n\ell 2}, \ldots, A_{n\ell N-1}, B_{n\ell N-1}, A_{n\ell N}$. For every value ℓ of angular momentum this is a system of $2N - 2$ linear equations for an equal number of unknowns. It has a nontrivial solution only if its determinant

$$D_\ell = D_\ell(E) \tag{7.35}$$

vanishes. This leads to a transcendental equation for the energy eigenvalues $E_{n\ell}$

$$D_\ell(E_{n\ell}) = 0 \quad . \tag{7.36}$$

In general, its solutions can only be found numerically; they are calculated by the computer. Once the eigenvalue $E_{n\ell}$ is determined as a single zero of (7.36) the system of linear equations (7.33), (7.34) can be solved yielding the coefficients $A_{n\ell q}, B_{n\ell q}$ as a function of one of them. This last undetermined coefficient is then fixed by the normalization condition (7.13).

The set of eigenfunctions of step potentials with $V_N < \infty$ is finite, thus they do not form a complete set. In Chap. 8 we present the continuum eigenfunctions supplementing the discrete ones to a complete set of functions.

7.1.5 The Harmonic Oscillator

The *three-dimensional harmonic oscillator* is described by the spherically harmonic potential

$$V(r) = \frac{1}{2}M\omega^2 r^2 \tag{7.37}$$

r: radius
ω: angular frequency
M: mass of particle

The radial eigenfunctions of the harmonic oscillator are, see (9.48),

$$R_{n_r\ell} = N_{n_r\ell}\left(\frac{r^2}{\sigma_0^2}\right)^{\ell/2} \exp\left(-\frac{r^2}{2\sigma_0^2}\right) L_{n_r}^{\ell+1/2}\left(\frac{r^2}{\sigma_0^2}\right) \tag{7.38}$$

$\sigma_0^2 = \hbar/(M\omega)$: ground-state width of oscillator
$n_r = (n - \ell)/2$: radial quantum number
n: principal quantum number
ℓ: angular-momentum quantum number
$L_{n_r}^{\ell+1/2}$: associated Laguerre polynomial, see (9.45)

$$N_{n_r\ell} = \sqrt{\frac{n_r!2^{n_r+\ell+2}}{\sqrt{\pi}(2(n_r+\ell)+1)!!\sigma_0^3}} \quad : \quad \text{normalization constant} \tag{7.39}$$

The *principal quantum number*

$$n = 2n_r + \ell \quad , \quad n = 0, 1, 2, \ldots \tag{7.40}$$

determines the energy eigenvalues of the bound states

$$E_{n\ell} = \left(n + \tfrac{3}{2}\right)\hbar\omega \quad . \tag{7.41}$$

The eigenfunctions (7.38) form a complete set of functions of the radial variable r. The full three-dimensional wave function is again obtained as a product of the radial wave function $R_{n_r\ell}$ and the spherical harmonic $Y_{\ell m}$:

$$\varphi_{n\ell m}(\mathbf{r}) = R_{n_r\ell}(r)Y_{\ell m}(\vartheta, \varphi) \quad , \quad n = 2n_r + \ell \quad . \tag{7.42}$$

They form a complete set of functions of the radius vector $\mathbf{r}$.

Of course, the Hamiltonian of the three-dimensional harmonic oscillator can be treated as a sum of three Hamiltonians of one-dimensional oscillators (3.13) in the Cartesian coordinates x, y and z. This leads to eigenfunctions

$$\varphi'_{n_1 n_2 n_3}(\mathbf{r}) = \varphi_{n_1}(x)\varphi_{n_2}(y)\varphi_{n_3}(z) \quad , \tag{7.43}$$

which are products of one-dimensional oscillator wave functions φ_n (3.14). Clearly, they belong to energy eigenvalues

$$E_n = \left(n + \tfrac{3}{2}\right)\hbar\omega \quad , \quad n = n_1 + n_2 + n_3 \quad , \tag{7.44}$$

determined by the principal quantum number n which is now simply the sum of the three principal quantum numbers n_1, n_2, n_3 of the three one-dimensional oscillators. Also the set (7.43) of eigenfunctions is complete. In fact, the eigenfunctions $\varphi_{n\ell m}(\mathbf{r})$, (7.42), to a given eigenvalue $E_{n\ell}$ can be superimposed as a linear combination of eigenfunctions $\varphi'_{n_1 n_2 n_3}(\mathbf{r})$ belonging to the same energy eigenvalue. Thus their indices n_1, n_2, n_3 have to satisfy the relation $n_1 + n_2 + n_3 = n$.

7.1.6 The Coulomb Potential. The Hydrogen Atom

In the hydrogen atom an electron of mass M_e carrying elementary charge $(-e)$ moves under the attractive Coulomb force of a proton of a mass M_p about 2000 times as heavy as the electron. The *Coulomb potential*

$$U(r) = \frac{e}{4\pi\varepsilon_0} \frac{1}{r} \tag{7.45}$$

yields the potential energy of the electron upon multiplication with the charge of the electron

$$V(r) = -eU(r) = -\frac{e^2}{4\pi\varepsilon_0} \frac{1}{r} \quad ; \tag{7.46}$$

r: radial variable
e: elementary charge
ε_0: electric field constant

The constant $e^2/(4\pi\varepsilon_0)$, having the dimension of action times velocity, can be expressed in units $\hbar c$ having the same dimension:

$$\frac{e^2}{4\pi\varepsilon_0} = \alpha\hbar c \quad ; \tag{7.47}$$

$\hbar = h/2\pi$: Planck's constant
c: velocity of light

The proportionality factor

$$\alpha = 1/137 \tag{7.48}$$

is Sommerfeld's *fine-structure constant*. The Coulomb potential energy now reads

$$V(r) = -\hbar c \frac{\alpha}{r} \quad . \tag{7.49}$$

The bound-state solutions $R_{n\ell}(r)$ of the radial Schrödinger equation (7.12), fulfilling the boundary condition of absence of singularities at $r = 0$ and sufficient decrease for $r \to \infty$, are given by

$$R_{n\ell}(r) = \frac{2}{n^2 a^{3/2}} \sqrt{\frac{(n-\ell-1)!}{(n+\ell)!}} \left(\frac{2r}{na}\right)^\ell \exp\left(-\frac{r}{na}\right) L_{n-\ell-1}^{2\ell+1}\left(\frac{2r}{na}\right) \tag{7.50}$$

r: radial variable
n: principal quantum number $n = 1, 2, 3, \ldots$
ℓ: angular-momentum quantum number $\ell = 0, 1, \ldots, n - 1$,
$a = \hbar/(\alpha M c) = 0.5292 \times 10^{-10}$m: Bohr radius
$M = M_p M_e/(M_p + M_e) \approx M_e$: reduced mass of electron
M_p: proton mass
M_e: electron mass
$\alpha = e^2/(4\pi\epsilon_0\hbar c) = 1/137$: Sommerfeld's fine-structure constant
e: elementary charge

$\hbar$: Planck's constant

ε_0: electric-field constant

L_p^k: associated Laguerre polynomial, see (9.45), (9.50)

The energy eigenvalues depend solely on the *principal quantum number n*:

$$E_n = -\frac{1}{2}Mc^2\frac{\alpha^2}{n^2} \quad . \tag{7.51}$$

The factor in front of n^{-2} has the numerical value $Mc^2\alpha/2 = 13.65$ eV.

7.1.7 Harmonic Particle Motion

In Sect. 3.1 we introduced the three-dimensional Gaussian wave packet of momentum expectation value p_0. Its probability distribution can be characterized by the probability ellipsoid, as discussed in Chap. 6. We calculate the motion of a Gaussian wave packet with uncorrelated initial widths $\sigma_{x0}, \sigma_{y0}, \sigma_{z0}$ of initial momentum expectation value p_0 under the action of a harmonic force (7.37). The center of the probability ellipsoid, which is initially at rest, moves like

$$
\begin{aligned}
x(t) &= x_0 \cos\omega t + \frac{p_{x0}}{M\omega}\sin\omega t, \\
y(t) &= y_0 \cos\omega t + \frac{p_{y0}}{M\omega}\sin\omega t, \\
z(t) &= z_0 \cos\omega t + \frac{p_{z0}}{M\omega}\sin\omega t,
\end{aligned}
\tag{7.52}
$$

which represents the motion on an ellipse about the origin in the plane containing the point $r = 0$ and the initial position $r_0 = (x_0, y_0, z_0)$ and being tangential to the initial momentum $p_0 = (p_{x0}, p_{y0}, p_{z0})$. As we have learned in Sect. 3.1, (3.31), the width of a one-dimensional harmonic oscillator oscillates with 2ω, twice the oscillator frequency.

Further Reading

Alonso, Finn: Vol. 3, Chap. 3
Berkeley Physics Course: Vol. 4, 8
Brandt, Dahmen: Chaps. 10,12
Feynman, Leighton, Sands: Vol. 3, Chap. 19
Flügge: Vol. 1, Chap. 1D
Gasiorowicz: Chaps. 12,17
Merzbacher: Chap. 10
Messiah: Vol. 1, Chaps. 9,11
Schiff: Chaps. 3,4

7.2 Radial Wave Functions in Simple Potentials

Aim of this section: Illustration of the radial wave function $R_{n\ell}(r)$, the energy eigenvalue spectrum $E_{n\ell}$, the potential $V(r)$ and the effective potential $V^{\text{eff}}(r)$ for four types of potential; the infinitely deep square well ((7.21), (7.22)), the square well of finite depth ((7.27) for $N = 2$), the harmonic oscillator ((7.38), (7.41)), and the Coulomb potential ((7.50), (7.51)).

Plot type: 2

C3 coordinates: x: radial coordinate r
y: $R_{n\ell}(r)$ or $R_{n\ell}^2(r)$ or $rR_{n\ell}(r)$ or $r^2R_{n\ell}^2(r)$ and $E_{n\ell}$ and $V(r)$ and $V^{\text{eff}}(r)$

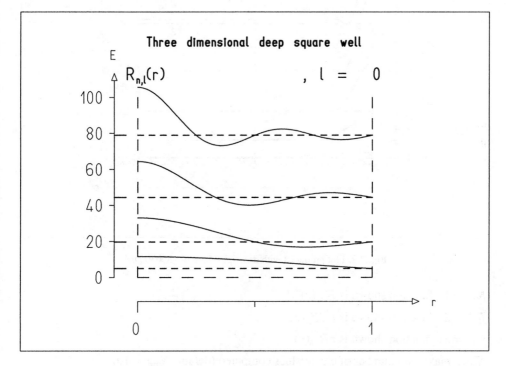

Fig. 7.1. Plot produced with descriptor 1 on file IQ030.DES

Input parameters:

CH 2 30 f_{CASE}
V0 f_{RES}
V2 ℓ
V6 l_{DASH} s
V8 Δx_- Δx_+ Δy_- Δy_+

V1 N_{max} E_{max}
V3 p_1 p_2
V7 f_{POT} f_{EF} f_{TS} f_{EV}

with

$f_{\text{CASE}} = 1$: infinitely deep square well

$f_{\text{CASE}} = 2$: square well of finite depth

$f_{\text{CASE}} = 3$: harmonic oscillator

$f_{\text{CASE}} = 4$: Coulomb potential

$f_{\text{RES}} = 0$: function shown is $R_{n\ell}^2(r)$

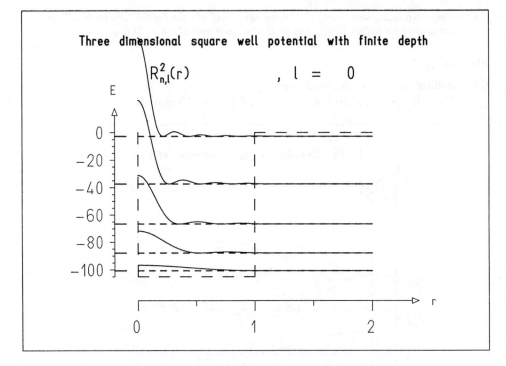

Fig. 7.2. Plot produced with descriptor 2 on file IQ030.DES

$f_{RES} = 1$: function shown is $R_{n\ell}(r)$

$f_{RES} = 2$: function shown is $r^2 R_{n\ell}^2(r)$

$f_{RES} = 3$: function shown is $r R_{n\ell}(r)$

N_{max}: maximum number of eigenvalues computed (default $N_{max} = 10$)

E_{max}: only eigenvalues $E_{n\ell} < E_{max}$ are looked for. (Default value: maximum of energy scale, i.e., YY(4))

ℓ: angular-momentum quantum number. If you ask for a multiple plot the input value of ℓ is taken for the first plot and increased successively by one for every plot

p_1, p_2: parameters for specific potential:

    for $f_{CASE} = 1$ (infinitely deep square well):
        a=p_1: radius of square well (default $a = 1$)

    for $f_{CASE} = 2$ (square well of finite depth):
        $a = p_1$: radius of square well (default $a = 1$)
        $V_0 = p_2$: depth of square well (default $V_0 = -1$)
        (If by mistake a positive sign is given to the input value of V_0 the sign is set negative by the program.)

    for $f_{CASE} = 3$ (harmonic oscillator):
        $\omega = p_1$: angular frequency of the oscillator (default $\omega = 1$)

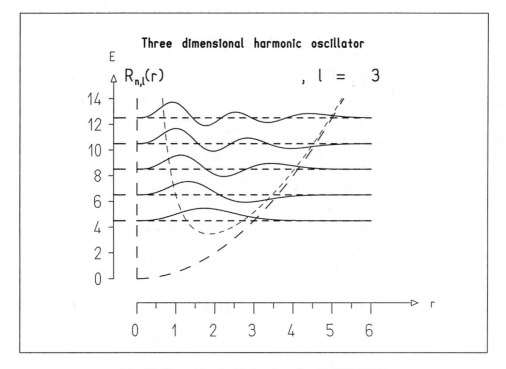

Fig. 7.3. Plot produced with descriptor 3 on file IQ030.DES

for $f_{CASE} = 4$ (Coulomb potential):

$\alpha = e^2/4\pi\varepsilon_0 = p_1^2$: (default $\alpha = 1$)

with $\alpha = \dfrac{e^2}{4\pi\varepsilon_0}\dfrac{1}{\hbar c}$ being Sommerfeld's fine-structure constant (e = elementary charge, c = velocity of light in vacuum)

l_{DASH}: dash length (in W3 units) for potential (default value: $1/10$ of X width of W3 window). Dash length for effective potential and for eigenvalues is $l_{DASH}/2$

s: scale factor for plotting function (e.g., $R_{n\ell}(r)$), default value: 1. (Since the function $R_{n\ell}(r)$ is shown with the eigenvalue serving as zero line, actually $f(r) = E_{n\ell} + sR_{n\ell}(r)$ is plotted.)

$f_{POT} = 0$: normally

$f_{POT} = 1$: if plot of potential $V(r)$ and of effective potential $V^{\text{eff}}(r)$ is *not* wanted

$f_{EF} = 0$: normally

$f_{EF} = 1$: if plot of radial eigenfunction $R_{n\ell}(r)$ or related function is *not* wanted

$f_{TS} = 0$: normally

$f_{TS} = 1$: if term scheme on left side of plot is *not* wanted

$f_{EV} = 0$: normally

$f_{EV} = 1$: if eigenvalues are *not* wanted

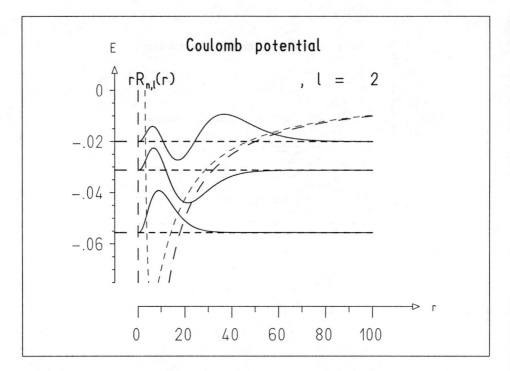

Fig. 7.4. Plot produced with descriptor 4 on file IQ030.DES

Δx_-, Δx_+, Δy_-, Δy_+: if these parameters are left zero the lines showing the potential and the effective potential are drawn inside the "box" defined by XX and YY, see Appendix A.2.2.3. You can extend the box to the left by the fraction f_x of its original length in x by setting Δx_- equal to f_x, e.g., $\Delta x_- = 0.1$. Similarly Δx_+, Δy_-, Δy_+ extend the box to the right, bottom and top, respectively

Additional plot items:

Item 6: potential

Item 7: eigenvalues serving as zero lines and in term scheme

Item 8: effective potential (for $\ell > 0$)

Automatically provided texts: T1, T2, TX

Example plots: Figs. 7.1–7.4

Remarks:

1. The y scale (controlled by the commands Y1 and Y2 is used as scale for the term scheme. The lines in the term scheme extend in X from the energy scale half the distance to the lower boundary in X of the W3 window.

2. Note that the numerical values of $\hbar$ and M are fixed in this section to the values $\hbar = 1$, $M = 1$.

7.3 Radial Wave Functions in the Step Potential

Aim of this section: Computation and presentation of the radial wave function $R_{n\ell}$, (7.27), the energy eigenvalue spectrum $E_{n\ell}$, the potential $V(r)$ and the effective potential $V^{\text{eff}}(r)$ for the spherical step potential (7.23).

Plot type: 2

C3 coordinates: x: distance r
y: $R_{n\ell}(r)$ or $R_{n\ell}^2(r)$ or $rR_{n\ell}(r)$ or $r^2 R_{n\ell}^2(r)$, and $E_{n\ell}$ and $V(r)$ and $V^{\text{eff}}(r)$

Input parameters:

```
CH 2 30 5
V0 fRES              V1 Nmax Emax Nsearch
V2 ℓ                 V3 N
V4 r1 r2 r3 r4       V5 V1 V2 V3 V4
V6 lDASH s           V7 fPOT fEF fTS fEV
V8 Δx- Δx+ Δy- Δy+   V9 fd sd xD
```

with

$f_{\text{RES}} = 0$: function shown is $R_{n\ell}^2(r)$

$f_{\text{RES}} = 1$: function shown is $R_{n\ell}(r)$

$f_{\text{RES}} = 2$: function shown is $r^2 R_{n\ell}^2(r)$

$f_{\text{RES}} = 3$: function shown is $rR_{n\ell}(r)$

N_{max}: maximum number of eigenvalues computed (default: $N_{\text{max}} = 10$)

E_{max}: only eigenvalues $E_{n\ell} < E_{\text{max}}$ are looked for. (Default value: maximum of energy scale, i.e., min(YY(4), 0).)

N_{search}: the interval $(E_{\text{min}}, E_{\text{max}})$ is divided into N_{search} subintervals and each is searched for an eigenvalue. If neighboring eigenvalues vary little from one another a large value has to be chosen for N_{search} (default: $N_{\text{search}} = 100$)

ℓ: angular-momentum quantum number. If you ask for a multiple plot the input value of ℓ is taken for the first plot and increased successively by one for every plot

N: number of regions ($2 \leq N \leq 5$)

$r_1, \ldots, r_{N-1}$: values of r separating regions

$V_1, \ldots, V_{N-1}$: potentials in regions $1, \ldots, N-1$ ($V_N = 0$)

$l_{\text{DASH}}, s, f_{\text{POT}}, f_{\text{EF}}, f_{\text{TS}}, f_{\text{EV}}, \Delta x_-, \Delta x_+, \Delta y_-, \Delta y_+$: as in Sect. 7.2

$f_D \leq 0$: function $D_\ell(E)$ not shown

$f_D > 0$: the function $D_\ell = D_\ell(E)$, (7.35), which has zeros for the eigenvalue $E = E_n$ is shown. The function value is plotted along the x axis, the independent variable along the y axis. To allow convenient scaling actually $f_D \sinh^{-1}[s_D D_\ell(E)]$ is plotted, where s_D is a scale factor. Usually $f_D = 1$, $s_D = 0$ will suffice

s_D: scale factor (default value $s_D = 1$) for plotting the function $D_\ell = D_\ell(E)$, see above

x_D: x position (in C3 coordinates) of zero line for function $D_\ell = D_\ell(E)$

Additional plot items:

Item 5: function $D_\ell = D_\ell(E)$

Item 6: potential

Item 7: eigenvalues serving as zero lines and in term scheme

Item 8: effective potential (for $\ell > 0$)

Automatically provided texts: T1, T2, TX

Example plot: Fig. 7.5

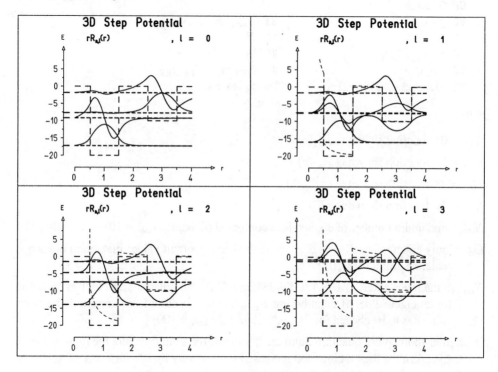

Fig. 7.5. Plot produced with descriptor 5 on file IQ030.DES

Remarks:

1. The y scale (controlled by the commands Y1 and Y2) is used as scale for the term scheme. The lines in the term scheme extend in X from the energy scale half the distance to the lower boundary in X of the W3 window.

2. Note that the numerical values of $\hbar$ and M are fixed in this section to the values $\hbar = 1$, $M = 1$.

7.4 Probability Densities

Aim of this section: Illustration of the probability density (7.17) describing a particle in an eigenstate of a spherically symmetric potential.

The functions $\varrho_{n\ell m}(r, \vartheta)$ are drawn as a surface over a half plane bounded by the z axis for the four types of potential of Sect. 7.2, i.e., the infinitely deep square well, the square well of finite depth, the harmonic oscillator and the Coulomb potential.

Plot type: 1

C3 coordinates: x: position coordinate z φ: polar angle ϑ
r: radial coordinate r z: $\varrho_{n\ell m}(r, \vartheta)$

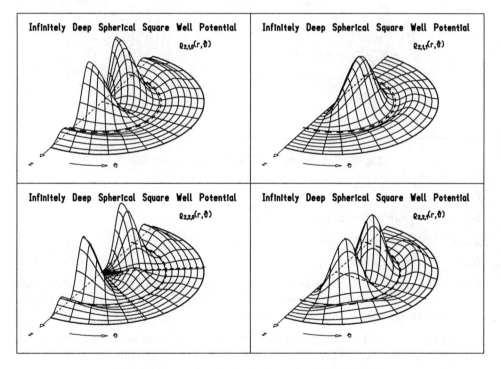

Fig. 7.6. Plot produced with descriptor 6 on file IQ030.DES

Input parameters:

CH 1 30 f_{CASE}

V1 N_{search} ε
V3 n ℓ m

V4 p_1 V5 p_2
V6 s

with

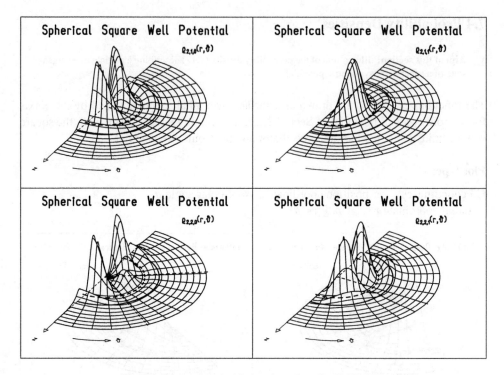

Fig. 7.7. Plot produced with descriptor 7 on file IQ030.DES

$f_{CASE} = 1$: infinitely deep square well

$f_{CASE} = 2$: square well of finite depth

$f_{CASE} = 3$: harmonic oscillator

$f_{CASE} = 4$: Coulomb potential

N_{search} (for $f_{CASE} = 2$ only): the interval (E_{min}, E_{max}) is divided into N_{search} subintervals and each is searched for an eigenvalue. If neighboring eigenvalues vary little from one another a large value has to be chosen for N_{search} (default: $N_{search} = 100$)

ε (for $f_{CASE} = 1, 2$ only): an accuracy parameter for the solution of the eigenvalue problem. The smaller the value of ε the higher the accuracy (default: $\varepsilon = 1$)

n: principal quantum number (for $f_{CASE} = 1, 2, 4$), radial quantum number n_r (for $f_{CASE} = 3$)

ℓ, m: angular-momentum and magnetic quantum number, respectively. If you ask for a multiple plot the input values are taken for the first (top left) plot only. The input value of ℓ is increased by one for every row, the value of m by one for every column

p_1, p_2: parameters for specific potential:

   for $f_{CASE} = 1$ (infinitely deep square well):
      $a = p_1$: radius of square well (default $a = 1$)

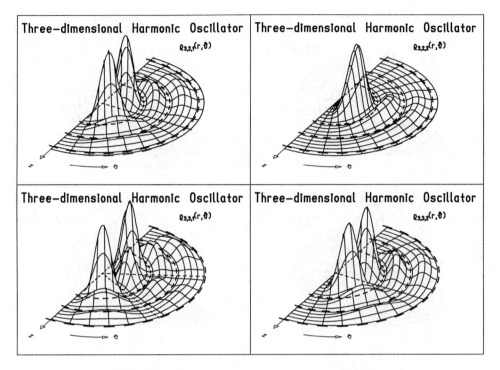

Fig. 7.8. Plot produced with descriptor 8 on file IQ030.DES

for $f_{CASE} = 2$ (square well of finite depth):
  $a = p_1$: radius of square well (default $a = 1$)
  $V_0 = p_2$: depth of square well (default $V_0 = -1$)
  (If by mistake a positive sign is given to the input value of V_0 the sign is set negative by the program.)

for $f_{CASE} = 3$ (harmonic oscillator):
  no specific input parameter is needed. The angular frequency of the oscillator is fixed ot $\omega = 1$

for $f_{CASE} = 4$ (Coulomb potential):
  no specific input parameter is needed. The fine structure constant is fixed to $\alpha = e^2/4\pi\varepsilon_0\hbar c = 1$

s: scale factor (default $s = 1$), see remark below

Additional plot items:

Item 5: radial node lines

Item 6: polar node lines

Item 7: half circle indicating edge of square-well potential

Automatically provided texts: T1, TP, TX, CA

Example plot: Fig. 7.6–7.9

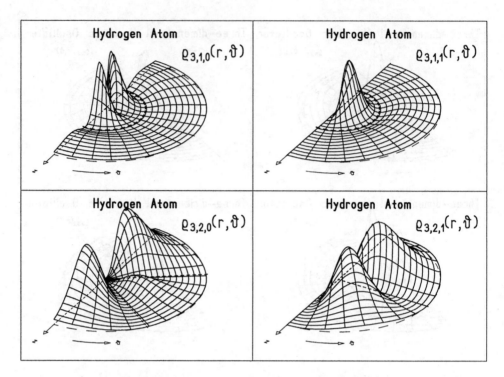

Fig. 7.9. Plot produced with descriptor 9 on file IQ030.DES

Remark:
Since the maximum value which a function $\varrho_{n\ell m}$ can take depends not only on the potential but also strongly on the quantum numbers n, ℓ, m, it takes some experience to choose the appropriate window in z and viewport in Z with the ZZ command. **IQ** does this automatically if you type

    NL(3) 1

Then the variables ZZ(3) and ZZ(4) are set to ZZ(3)=0 and ZZ(4)= $s f_{max}$, where f_{max} is the maximum value of the probability density. The variables ZZ(1) and ZZ(2) are still under your control.

7.5 Harmonic Particle Motion

Aim of this section: Illustration of the motion (7.52) of a Gaussian wave packet in a harmonic-oscillator potential by plotting the probability ellipsoid at times $t_0, t_0 + \Delta t, \ldots$. (The illustration is restricted to wave packets with uncorrelated widths $\sigma_x, \sigma_y, \sigma_z$. The probability density is then simply a product of three one-dimensional densities as treated in Sect. 3.4.)

Plot type: 10

C3 coordinates: x: position coordinate x y: position coordinate y
z: position coordinate z

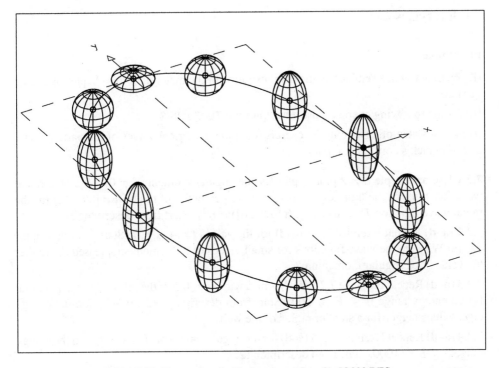

Fig. 7.10. Plot produced with descriptor 10 on file IQ030.DES

Input parameters:

    CH 10 30 0
                                    V1 x_0 y_0 z_0
    V2 p_{x0} p_{y0} p_{z0} V3 σ_{x0} σ_{y0} σ_{z0}
    V6 R V7 T t_0 Δt N

with

x_0, y_0, z_0: initial position expectation values ($t = t_0$)

p_{x0}, p_{y0}, p_{z0}: initial momentum expectation values

σ_{x0}, σ_{y0}, σ_{z0}: initial widths

R: radius (in W3 units) of point indicating position of corresponding classical particle (center of ellipsoid)

$T = 2\pi/\omega$: period of oscillator (default $T = 1$)

t_0: initial time

Δt: time elapsed between two instants for which ellipsoid is drawn

N: total number of ellipsoids drawn

Note that the numerical values of $\hbar$ and M are set to 1 in this section. For additional plot items, automatically provided texts and plot recommendation see Sect. 6.5.

Example plot: Fig. 7.10

7.6 Exercises

Please note:

(i) you may watch a semiautomatic demonstration of the material of this chapter by typing
SA D30

(ii) for the following exercises use descriptor file IQ030.DES

(iii) if not stated otherwise in the exercises the the numerical values of the particle mass
and of Planck's constant are put to 1.

7.2.1 Use a multiple 2×2 plot to plot for the angular momenta $\ell = 0, 1, 2, 3$, the radial
wave functions of the infinitely deep square-well potential of radius $a = 0.9$. Plot **(a)** the
radial wave function $R_{n\ell}$, **(b)** $r R_{n\ell}$, **(c)** $R_{n\ell}^2$, **(d)** $r^2 R_{n\ell}^2$. Start from descriptor 11.

7.2.2 (a–d) Repeat Exercise 7.1.1 (a–d) for the values $\ell = 4, 5, 6, 7$. Start from descriptor
12. **(e)** Why are the wave functions for small values of r the more suppressed the higher
the values ℓ of angular momentum?

7.2.3 (a–d) Repeat Exercise 7.2.1 (a–d) with a width $a = 0.5$ of the infinitely deep potential
for an energy range $0 \leq E \leq 200$. Start from descriptor 11. **(e)** Why are the energy
eigenvalues bigger for a smaller radius of the well?

7.2.4 (a–d) Repeat Exercise 7.2.3 (a–d) for the angular momenta $\ell = 4, 5, 6, 7$ in the energy
range $0 \leq E \leq 1000$. Start from descriptor 12.

7.2.5 (a–d) Repeat Exercise 7.2.1 (a–d) for a potential depth $V_0 = -100$. Start from
descriptor 11. **(e)** Why are the differences $\Delta_i = E_i - V_0$ of the energy eigenvalues E_i and
the potential depth V_0 in this potential smaller than in the infinitely deep square well of
Exercise 7.2.1 in the range $0 \leq E \leq 100$?

7.2.6 (a–d) Repeat Exercise 7.2.2 (a–d) for a potential of finite depth $V_0 = -100$. Start
from descriptor 11. **(e)** Why do the wave functions exhibit for small r values the same
behavior as those of the infinitely deep square well?

7.2.7 Plot for a harmonic-oscillator potential $V = M\omega^2 r^2/2$ with the oscillator frequency
$\omega = 1$ in the energy range $0 \leq E \leq 15$ **(a)** the radial wave function $R_{n\ell}$, **(b)** $r R_{n\ell}$, **(c)** $R_{n\ell}^2$,
(d) $r^2 R_{n\ell}^2$. Start from descriptor 3. **(e)** Explain in classical terms why the probability
density per radial shell $r^2 R_{n\ell}^2$ for large ℓ is largest close to the maximal elongation of the
radial oscillator.

7.2.8 (a–d) Repeat Exercise 7.2.7 (a–d) for the angular momenta $\ell = 4, 5, 6, 7$. **(e)** Why
do the lowest energy eigenvalues no longer occur for larger ℓ?

7.2.9 (a–d) Repeat Exercise 7.2.7 (a–d) for the angular momenta $\ell = 20, 21, 22, 23$ for the
lowest energy eigenvalues.

7.2.10 (a–d) Repeat Exercise 7.2.7 for the angular momenta $\ell = 40, 41, 42, 43$. Plot for
the few lowest energy eigenvalues **(a)** the radial wave functions $R_{n\ell}$, **(b)** $r R_{n\ell}$, **(c)** $R_{n\ell}^2$,
(d) $r^2 R_{n\ell}^2$. **(e)** Approximate the effective potential for higher values ℓ of the angular momentum by an oscillator potential having its minimum at the minimum of $V_{\text{eff}}(r)$. **(f)** Explain why these wave functions look very similar to oscillator wave functions. **(g)** What
is the effective oscillator frequency ω_{eff} for the shifted approximate oscillator? **(h)** How
are the ℓ independence of ω_{eff} and the equal spacing of the energy levels related?

7.2.11 Plot for a Coulomb potential with $\alpha = 1$ in the range $0 \le r \le 30$ for the lowest three energy levels **(a)** the radial wave function $R_{n\ell}$, **(b)** $rR_{n\ell}$, **(c)** $R_{n\ell}^2$, **(d)** $r^2R_{n\ell}^2$. Start from descriptor 4. Use double plots for the presentation of the angular momenta $\ell = 0, 1$. **(e)** Calculate the energies of the lowest three levels for the two angular momenta.

7.2.12 (a–e) Repeat Exercise 7.2.11 (a–e) for $\ell = 2, 3$.

7.2.13 (a–d) Repeat Exercise 7.2.11 (a–d) for $\ell = 20$. **(e)** Calculate the equivalent oscillator potential by an expansion of the effective potential about its minimum. **(f)** Calculate the energy levels in the approximating oscillator potential for $n_r \ll \ell$ and compare them with the exact Balmer formula. **(g)** Compare the position of the minimum with the value according to the formula for the Bohr radii.

7.2.14 (a–d) Repeat Exercise 7.2.11 (a–d) for $\ell = 40$.

7.3.1 The potential has the form
$$V(r) = \begin{cases} -20 \;, & 0 \le r < 1.5 \\ 0 \;, & 1.5 \le r < 2.5 \\ -10 \;, & 2.5 \le r < 3.5 \\ 0 \;, & 3.5 \le r \end{cases} .$$
Plot for $\ell = 0$ **(a)** the wave function $R_{n\ell}$, **(b)** $rR_{n\ell}$, **(c)** $R_{n\ell}^2$, **(d)** $r^2R_{n\ell}^2$. Start from descriptor 5. **(e)** Why is the wave function of the third state essentially different from zero only in the second well?

7.3.2 (a–d) Repeat Exercise 7.3.1 (a–d) for $\ell = 1$.

7.3.3 (a–d) Repeat Exercise 7.3.1 (a–d) for $\ell = 2$.

7.3.4 For the potential
$$V(r) = \begin{cases} 100 \;, & 0 \le r < 0.5 \\ -20 \;, & 0.5 \le r < 1.5 \\ 0 \;, & 1.5 \le r \end{cases}$$
plot in a 2×2 plot for the angular momenta $\ell = 0, 1, 2, 3$ **(a)** the wave function $R_{n\ell}$, **(b)** $R_{n\ell}^2$. Start from descriptor 5. **(c)** Why does the energy of the second state vary so much less with increasing angular momentum than in a single potential well without the repulsive hard core below $r = 0.5$?

7.4.1 (a) Plot the probability densities $\varrho_{n\ell m}$ in an infinitely deep square well of radius $a = 0.9$ with a multiple plot for $n = 1, \ell = 0, m = 0$ and $n = 1, \ell = 1, m = 0, 1$. Start from descriptor 13. **(b)** Explain the significance of the node lines in the polar angle ϑ.

7.4.2 Repeat 7.4.1 (a) for $n = 2$.

7.4.3 (a) Repeat 7.4.1 (a) for $n = 3$. **(b)** What are the node half circles correlated to?

7.4.4 Repeat 7.4.1 (a) for $n = 1, \ell = 2, m = 0, 1$ and $n = 1, \ell = 3, m = 0, 1$. Which is the correlation between ϑ node lines and quantum numbers?

7.4.5 (a) Repeat Exercise 7.4.1 (a) for $n = 1, \ell = 2, m = 2$ and $n = 1, \ell = 3, m = 2, 3$. **(b)** At which angles do the ϑ node lines occur?

7.4.6 Repeat Exercise 7.4.1 (a) for $n = 1, \ell = 15, m = 15$ in a single plot.

7.4.7 (a) Repeat Exercise 7.4.1 (a) for $n = 1, \ell = 15, m = 0$ in a single plot. **(b)** Explain the difference in this plot and the one of Exercise 7.4.6 (a) in terms of classical angular momenta.

7.4.8 (a) Plot the probability densities $\varrho_{n\ell m}$ in a three-dimensional harmonic oscillator with a multiple plot for $n_r = 1, \ell = 0, m = 0$ and $n_r = 1, \ell = 1, m = 0, 1$. Start from descriptor 8. **(b)** Why is the decrease of the wave functions with growing r slower than for Exercise 7.4.1?

7.4.9 Repeat Exercise 7.4.8 for $n_r = 2$.

7.4.10 Repeat Exercise 7.4.8 (a) for $n_r = 0$, $\ell = 15$, $m = 15$.

7.4.11 Repeat Exercise 7.4.8 (a) for $n_r = 0$, $\ell = 15$, $m = 0$.

7.4.12 (a) Plot the probability densities $\varrho_{n\ell m}$ in a Coulomb potential for the quantum numbers $n = 1$, $\ell = 0$, $m = 0$. Start from descriptor 9. **(b)** Why do no states exist for $n = 1$, $\ell = 1$, $m = 1$?

7.4.13 (a) Repeat Exercise 7.4.12 using a multiple plot for the probability densities for the quantum numbers $n = 2$, $\ell = 0$, $m = 0$ and $n = 2$, $\ell = 1$, $m = 0, 1$. **(b)** How do the nodes and the quantum numbers correspond to each other?

7.4.14 (a) Repeat Exercise 7.4.12 (a) for $n = 15$, $\ell = 14$, $m = 14$. Make sure you extend your plot to large values of r by changing the C3 window in x and y. **(b)** Why do you need a very large scale factor in z (unless you use the autoscale facility) to see the peak in the probability density? **(c)** Why does the region of large values of the wave function occur at large r?

7.4.15 Repeat 7.4.12 (a) for $n = 15$, $\ell = 14$, $m = 0$.

7.5.1 (a) Plot the motion of a three-dimensional Gaussian wave packet in a spherically symmetric harmonic-oscillator potential. As initial conditions use $x_0 = 3$, $y_0 = 1$, $z_0 = 2$, $p_{x0} = 1$, $p_{y0} = 2$, $p_{z0} = 3$, $T = 1$, $t_0 = 0$, $\Delta t = 0.1$, $N = 1$. Start from descriptor 10. **(b)** Calculate the classical angular momentum ($\hbar = 1$) for the initial conditions under (a).

7.5.2 Repeat Exercise 7.5.1 (a) for $T = 0.5$.

7.5.3 (a) Repeat Exercise 7.5.1 (a) for the initial conditions $x_0 = 5$, $y_0 = 0$, $z_0 = 0$, $p_{x0} = -2$, $p_{y0} = 0$, $p_{z0} = 0$ for the time intervals $\Delta t = 0.1667$ for $N = 4$ positions. **(b)** Calculate the classical angular momentum.

7.5.4 Repeat Exercise 7.5.1 (a) for the initial conditions $x_0 = 0$, $y_0 = 0$, $z_0 = 0$, $p_{x0} = 5$, $p_{y0} = 0$, $p_{z0} = 0$ and the time intervals $\Delta t = 0.0833$ for $N = 10$ positions.

7.5.5 (a) Repeat Exercise 7.5.4 for a Gaussian wave packet at rest $x_0 = 0$, $y_0 = 0$, $z_0 = 0$, $p_{x0} = 0$, $p_{y0} = 0$, $p_{z0} = 0$, $\sigma_{x0} = 0.2$, $\sigma_{y0} = 0.3$, $\sigma_{z0} = 1.4$ for the time interval $\Delta t = 0.25$ for two positions ($N = 2$). **(b)** Why does the shape of the wave packet change from prolate to oblate?

8 Scattering in Three Dimensions

Contents: Radial scattering wave functions. Boundary and continuity conditions. Solutions for step potentials. Scattering of plane harmonic waves. Scattering matrix element. Partial scattering amplitude. Scattered wave as sum over partial waves. Scattering amplitude as sum over partial scattering amplitudes. Differential cross section. Total cross section. Partial cross sections. Scattering amplitude and phase. Unitarity and the Argand diagram.

8.1 Physical Concepts

8.1.1 Radial Scattering Wave Functions

Besides the bound states as discussed in Chap. 7 there are continuum states for potentials with $V_\infty < \infty$, (7.11). These states will be studied in this chapter for a spherically symmetric step potential

$$
V(r) = \begin{cases}
V_1 & , \quad 0 \le r < r_1 \quad \text{region 1} \\
V_2 & , \quad r_1 \le r < r_2 \quad \text{region 2} \\
\vdots & \qquad\qquad \vdots \\
V_N & , \quad r_{N-1} \le r \qquad \text{region } N
\end{cases}
\tag{8.1}
$$

Scattering states are continuum eigenfunctions of the stationary Schrödinger equation (7.12) for eigenvalues $E \ge V_\infty$, i.e., in the case of the step potential (8.1) for all the values $E \ge V_N$.

The wave number in the N regions is

i) if $E > V_q$:

$$
k_q = \left| \sqrt{2M(E - V_q)}/\hbar \right| \quad ,
\tag{8.2}
$$

or

ii) if $E < V_q$:

$$
k_q = i\kappa_q \quad , \quad \kappa_q = \left| \sqrt{2M(V_q - E)}/\hbar \right| \quad .
\tag{8.3}
$$

The radial wave function of angular momentum ℓ consists again of N pieces

$$
R_\ell(k, r) = \begin{cases}
R_{\ell 1}(k_1, r) & , \quad 0 \le r < r_1 & \text{region 1} \\
R_{\ell 2}(k_2, r) & , \quad r_1 \le r < r_2 & \text{region 2} \\
\vdots & \qquad\quad \vdots & \qquad \vdots \\
R_{\ell N-1}(k_{N-1}, r) & , \quad r_{N-2} \le r < r_{N-1} & \text{region } N-1 \\
R_{\ell N}(k_N, r) & , \quad r_{N-1} \le r & \text{region } N
\end{cases}
\tag{8.4}
$$

Here we have put

$$k = k_N \quad , \quad 0 \le k < \infty$$

as the wave number in the region N of incident and reflected wave.

The pieces $R_{\ell q}(k_q, r)$, $q = 1, \dots, N$ of the wave function in the N regions have the form

$$R_{\ell q}(k_q, r) = A_{\ell q} j_\ell(k_q r) + B_{\ell q} n_\ell(k_q r) \quad . \tag{8.5}$$

For k_q real, i.e., $E > V_q$, the spherical Bessel functions j_ℓ and n_ℓ are real so that $R_{\ell q}(k_q, r)$ is a real function if $A_{\ell q}$ and $B_{\ell q}$ are real coefficients, as will turn out in the following. Alternatively, with the help of (9.38), (9.39) the pieces $R_{\ell q}(k_q, r)$, $q = 1, \dots, N$, of the wave function can be expressed in terms of spherical Hankel functions, see Sect. 9.1.6

$$R_{\ell q}(k_q, r) = D_{\ell q} h_\ell^{(-)}(k_q r) + F_{\ell q} h_\ell^{(+)}(k_q r) \tag{8.6}$$

with the coefficients

$$D_{\ell q} = -\frac{1}{2i}(A_{\ell q} - iB_{\ell q}) \quad \text{and} \quad F_{\ell q} = \frac{1}{2i}(A_{\ell q} + iB_{\ell q}) \quad . \tag{8.7}$$

For k_q real, i.e., $E > V_q$,

$$h_\ell^{(-)}(k_q r) = C_\ell^- \frac{e^{-ik_q r}}{k_q r} \tag{8.8}$$

is an *incoming spherical wave*, i.e., a spherical wave propagating from large values of the radial distance r in region q towards the origin $r = 0$. Analogously, for k_q real

$$h_\ell^{(+)}(k_q r) = C_\ell^+ \frac{e^{ik_q r}}{k_q r} \tag{8.9}$$

is an *outgoing spherical wave*, i.e., a spherical wave propagating from small values r in region q outward. Thus, in analogy to the one-dimensional case, the wave function in a region q with k_q real can be interpreted as a superposition of an incoming $h_\ell^{(-)}$ and an outgoing $h_\ell^{(+)}$ complex spherical wave.

For scattering wave functions we have $E > V_N$, so that in region N the wave number k_N is real and $h_\ell^{(-)}(k_N r)$ is the incident spherical wave, whereas $h_\ell^{(+)}(k_N r)$ is the reflected spherical wave, of angular momentum ℓ. For imaginary wave numbers $k_q = i\kappa_p$, i.e., $E < V_q$, $q \ne N$, the scattering wave $R_{\ell q}(i\kappa_q, r)$ in region q is a linear superposition of the real spherical Hankel functions

$$h_\ell^{(-)}(i\kappa_q r) = C_\ell^- \frac{e^{\kappa_q r}}{r} \quad , \quad h_\ell^{(+)}(i\kappa_q r) = C_\ell^+ \frac{e^{-\kappa_q r}}{r} \quad . \tag{8.10}$$

Thus for real coefficients $i(A_{\ell q} \pm iB_{\ell q})/2$ in

$$R_{\ell q}(i\kappa_q, r) = \frac{i}{2}\left[(A_{\ell q} - iB_{\ell q})h_\ell^{(-)}(i\kappa_q r) - (A_{\ell q} + iB_{\ell q})h_\ell^{(+)}(i\kappa_q r)\right] \quad , \tag{8.11}$$

the radial wave functions $R_{\ell q}(i\kappa_q, r)$ are real. The physical interpretation of $R_{\ell q}(i\kappa_q, r)$ in a region q with $E < V_q$ is again the tunnel effect. Even though the total radial energy of the particle is lower than the potential barrier the wave penetrates the wall of height V_q.

8.1.2 Boundary and Continuity Conditions. Solution of the System of Inhomogeneous Linear Equations for the Coefficients

The two *boundary conditions* for scattering solutions are

i) At $r = 0$: $R_{\ell 1}(k_1, r)$ is free of singularities, Sect. 7.1. This requires

$$B_{\ell 1} = 0 \quad , \quad \text{i.e.,} \quad R_{\ell 1}(k_1, r) = A_{\ell 1} j_\ell(k_1 r) \quad , \tag{8.12}$$

since the $n_\ell(k_1 r)$ possess a singularity at $r = 0$.

ii) At $r \to \infty$, i.e., in region N we have to have an incoming and an outgoing spherical wave. This is fulfilled by (8.6) in region N. For given boundary conditions for large r we may assume that the coefficients $A_{\ell N}$ of the Bessel function $j_\ell(kr)$ in region N are known quantities. Further below in this section we shall discuss the choice of the $A_{\ell N}$ for the boundary condition posed by an incoming plane wave.

As in Sect. 7.1, for the scattering solutions, the radial wave function has also to be *continuous* and *continuously differentiable* at the points $r_1, r_2, \ldots, r_{N-1}$ where the pieces $R_{\ell q}(k_q, r)$ have to be matched. The continuity conditions pose $2(N - 1)$ inhomogeneous linear algebraic equations for the $2(N - 1)$ coefficients $A_{\ell 1}, A_{\ell 2}, B_{\ell 2}, \ldots, A_{\ell N-1}, B_{\ell N-1}$, $B_{\ell N}$. The coefficient $A_{\ell N}$ given by the boundary condition constitutes the inhomogeneity of the system. Thus the $2(N - 1)$ unknown coefficients are uniquely determined by the inhomogeneous linear equations in terms of the coefficient $A_{\ell N}$. The coefficient $A_{\ell N}$ can be chosen to be real. The functions j_ℓ and n_ℓ are real in the regions with real k_q, i.e., $E > V_q$, and the $h_\ell^{(\pm)}$ are real in the regions with imaginary $k_q = \mathrm{i}\kappa_q$, i.e., $E < V_q$. Since the coefficients of the linear system of $2(N - 1)$ equations are real, the solutions $A_{\ell 1}, A_{\ell 2}, B_{\ell 2}, \ldots, A_{\ell N-1}, B_{\ell N-1}, B_{\ell N}$ are real coefficients and thus real functions of the incoming wave number k. Thus, the radial wave function $R_\ell(k, r)$ is a real function of k and r.

8.1.3 Scattering of a Plane Harmonic Wave

For the usual scattering experiment the particles possess a momentum p sufficiently sharp to describe the incoming particles by a three-dimensional plane harmonic wave with wave vector $k = p/\hbar$. We choose the z direction e_z of the polar coordinate frame parallel to the momentum, i.e., $p = \hbar k = \hbar k e_z$. The plane wave has the form ($\cos \vartheta = k \cdot r / kr$)

$$\varphi(k, r) = \mathrm{e}^{\mathrm{i}k \cdot r} = \mathrm{e}^{\mathrm{i}kz} = \mathrm{e}^{\mathrm{i}kr \cos \vartheta} \quad . \tag{8.13}$$

According to (6.53) it can be decomposed into partial waves

$$\varphi(k, r) = \mathrm{e}^{\mathrm{i}kr \cos \vartheta} = \sum_{\ell=0}^{\infty} (2\ell + 1)\mathrm{i}^\ell j_\ell(kr) P_\ell(\cos \vartheta) \quad . \tag{8.14}$$

This means that the incoming plane wave in region N is equivalently well described by a set of partial waves of angular momentum ℓ and magnetic quantum number $m = 0$. The radial wave function $j_\ell(kr)$ is a superposition (9.38),

$$j_\ell(kr) = \frac{1}{2\mathrm{i}} \left[h_\ell^{(+)}(kr) - h_\ell^{(-)}(kr) \right] \quad , \tag{8.15}$$

of incoming and outgoing spherical waves $\frac{1}{2i}h_\ell^{(-)}(kr)$ and $\frac{1}{2i}h_\ell^{(+)}(kr)$. The incoming radial wave in region N is thus

$$R_{\ell N}^{in}(k,r) = -\frac{1}{2i}h_\ell^{(-)}(kr) \quad , \tag{8.16}$$

for the moment leaving aside the weight factor $(2\ell+1)i^\ell$ in the partial wave decomposition (8.14). To have the term (8.16) as the incoming spherical wave in the solution of the Schrödinger equation we divide $R_\ell(k,r)$, (8.6), by $(A_{\ell N} - iB_{\ell N})$ and obtain for the ℓ-th radial wave function

$$R_\ell^{(+)}(k,r) = \frac{1}{A_{\ell N} - iB_{\ell N}}R_\ell(k,r) \quad . \tag{8.17}$$

Its piece in region N is then given by

$$R_{\ell N}^{(+)}(k,r) = -\frac{1}{2i}h_\ell^{(-)}(kr) + \frac{1}{2i}S_\ell(k)h_\ell^{(+)}(kr) \quad . \tag{8.18}$$

The coefficient

$$S_\ell(k) = \frac{A_{\ell N} + iB_{\ell N}}{A_{\ell N} - iB_{\ell N}} \quad , \tag{8.19}$$

being a function of the incoming wave number $k = p/\hbar$, is called the *scattering-matrix element* S_ℓ of the ℓ-th partial wave. It is the angular momentum projection of the S matrix. The function $R_{\ell N}^{(+)}(k,r)$ can also be rephrased in terms of $j_\ell(kr)$:

$$R_{\ell N}^{(+)}(k,r) = j_\ell(kr) + \frac{1}{2i}(S_\ell(k) - 1)h_\ell^{(+)}(kr) \quad . \tag{8.20}$$

The coefficient

$$f_\ell(k) = \frac{1}{2i}(S_\ell(k) - 1) \tag{8.21}$$

is called the *partial scattering amplitude*. It determines the effect of the potential $V(r)$ on the ℓ-th partial wave $j_\ell(kr)$ in the decomposition (8.14) of the incoming plane wave. The representation (8.20) is the appropriate form for the construction of the full three-dimensional solution

$$\varphi^{(+)}(\boldsymbol{k},\boldsymbol{r}) = \sum_{\ell=0}^{\infty} \varphi_\ell(\boldsymbol{k},\boldsymbol{r}) = \sum_{\ell=0}^{\infty}(2\ell+1)i^\ell R_\ell^{(+)}(k,r)P_\ell(\cos\vartheta) \tag{8.22}$$

representing the superposition

$$\varphi^{(+)}(\boldsymbol{k},\boldsymbol{r}) = e^{i\boldsymbol{k}\cdot\boldsymbol{r}} + \eta(\boldsymbol{k},\boldsymbol{r}) \tag{8.23}$$

of the incoming three-dimensional plane harmonic wave and the *scattered wave*

$$\eta(\boldsymbol{k},\boldsymbol{r}) = \sum_{\ell=0}^{\infty}\eta_\ell(\boldsymbol{k},\boldsymbol{r}) \quad . \tag{8.24}$$

The ℓ-th *scattered partial wave* $\eta_\ell(\boldsymbol{k},\boldsymbol{r})$ is given by

$$\eta_\ell(\boldsymbol{k},\boldsymbol{r}) = (2\ell+1)i^\ell[R_\ell^{(+)}(k,r) - j_\ell(kr)]P_\ell(\cos\vartheta) \quad . \tag{8.25}$$

In region N the piece of the scattered partial wave $\eta_{\ell N}(\mathbf{k}, \mathbf{r})$ has the explicit representation

$$\eta_{\ell N}(\mathbf{k}, \mathbf{r}) = (2\ell + 1)i^\ell f_\ell(k)h_\ell^{(+)}(k, r)P_\ell(\cos \vartheta). \tag{8.26}$$

For far out distances $kr \gg 1$ in region N the function $\eta_{\ell N}(\mathbf{k}, \mathbf{r})$, and thus $\eta_\ell(\mathbf{k}, \mathbf{r})$, is dominated by the asymptotically leading term of the Hankel function

$$h_\ell^{(+)}(kr) \sim \frac{e^{ikr}}{kr} \quad . \tag{8.27}$$

This leads to the asymptotic representation

$$\eta_\ell(\mathbf{k}, \mathbf{r}) = (2\ell + 1)\frac{i^\ell}{k}f_\ell(k)\frac{e^{ikr}}{r}P_\ell(\cos \vartheta) \quad . \tag{8.28}$$

In region N the total scattered wave is given by (8.24) and (8.28):

$$\eta(\mathbf{k}, \mathbf{r}) = \sum_{\ell=0}^\infty \eta_\ell(\mathbf{k}, \mathbf{r}) = \sum_{\ell=0}^\infty (2\ell+1)i^\ell f_\ell(k)h_\ell^{(+)}(kr)P_\ell(\cos \vartheta) \quad , \quad r_{N-1} \leq r \quad . \tag{8.29}$$

For asymptotic r values in region N, i.e., for $kr \gg 1$, the asymptotic representation is again obtained from (8.27) yielding

$$\eta(\mathbf{k}, \mathbf{r}) \approx f(k, \vartheta)\frac{e^{ikr}}{r} \tag{8.30}$$

with the *scattering amplitude* read off (8.28):

$$f(k, \vartheta) = \frac{1}{k}\sum_{\ell=0}^\infty (2\ell + 1)i^\ell f_\ell(k)P_\ell(\cos \vartheta) \quad . \tag{8.31}$$

This determines the modulation of the scattered spherical wave $r^{-1}\exp(ikr)$ in the polar angle ϑ.

The density of particles driven by the potential out of the original beam into the direction ϑ is given by $|\eta(\mathbf{k}, \mathbf{r})|^2$ which is asymptotically ($kr \gg 1$)

$$|\eta(\mathbf{k}, \mathbf{r})|^2 = |f(k, \vartheta)|^2/r^2 \quad . \tag{8.32}$$

The current ΔI of particles passing through a small area Δa vertical to the ray from the scattering potential to the position of Δa at angle ϑ and distance r is

$$\Delta I = |\eta(\mathbf{k}, \mathbf{r})|^2 v\Delta a = |f(k, \vartheta)|^2 v\frac{\Delta a}{r^2} = |f(k, \vartheta)|^2 v\Delta\Omega \quad . \tag{8.33}$$

The quantity

$$\Delta\Omega = \frac{\Delta a}{r^2} \tag{8.34}$$

is the solid angle under which Δa appears, seen from the origin.

The incident current density is the incident particle density times its velocity

$$j = |e^{i\mathbf{k}\cdot\mathbf{r}}|^2 v = v \quad . \tag{8.35}$$

Thus the current ΔI of particles scattered into the solid angle $\Delta\Omega$ can be written as

$$\Delta I = |f(k, \vartheta)|^2 \Delta\Omega j \quad . \tag{8.36}$$

The proportionality constant between the initial current density j and the current through the solid angle element $\Delta\Omega$ is the *differential scattering cross section* $d\sigma/d\Omega$:

$$\Delta I = \frac{d\sigma}{d\Omega}\Delta\Omega j \quad . \tag{8.37}$$

Thus we identify

$$\frac{d\sigma}{d\Omega} = |f(k, \vartheta)|^2 \tag{8.38}$$

as the differential cross section for particles of momentum $p = \hbar k$ on a scatterer described by the potential $V(r)$. The scattering amplitude $f(k, \vartheta)$ is given by (8.31).

The *total scattering cross section* is the integral of (8.38) over the total solid angle 4π:

$$\sigma_{\text{tot}} = \int \frac{d\sigma}{d\Omega}d\Omega = 2\pi \int_{-1}^{+1} |f(k, \vartheta)|^2 d\cos\vartheta \quad . \tag{8.39}$$

Using the orthogonality (9.10) of the Legendre polynomials

$$\int_{-1}^{+1} P_{\ell'}(\cos\vartheta)P_\ell(\cos\vartheta)d\cos\vartheta = \frac{2}{2\ell+1}\delta_{\ell'\ell} \tag{8.40}$$

we get, using (8.31),

$$\sigma_{\text{tot}} = \frac{4\pi}{k^2}\sum_{\ell=0}^{\infty}(2\ell+1)|f_\ell(k)|^2 = \sum_{\ell=0}^{\infty}\sigma_\ell \tag{8.41}$$

with the *partial cross section* of angular momentum ℓ

$$\sigma_\ell = \frac{4\pi}{k^2}(2\ell+1)|f_\ell(k)|^2 \quad . \tag{8.42}$$

8.1.4 Scattering Amplitude and Phase. Unitarity. The Argand Diagram

In Chap. 4 we derived current conservation as the basis for the conservation of probability. In three dimensions the same chain of arguments is valid. However, simple physical arguments lead to the same conclusion without calculations. Elastic scattering of particles on a spherically symmetric potential conserves particle number, energy and angular momentum. Thus the magnitude of the velocity remains unaltered in the scattering process. Since the particle number and angular momentum are conserved, this leads to conservation of the current density of the spherical waves of angular momentum ℓ. Thus the incoming current of a spherical wave is only reflected upon scattering but keeps its magnitude. Therefore the complex scattering matrix element S_ℓ determining the relative factor between incoming and outgoing spherical waves in region N must have the absolute value one. It can only be a complex phase. This is the *unitarity relation for the scattering matrix element S_ℓ*:

$$S_\ell^* S_\ell = 1 \quad . \tag{8.43}$$

It can only be a complex phase factor which is conventionally written as

$$S_\ell = e^{2i\delta_\ell} \tag{8.44}$$

with the *scattering phase* δ_ℓ of the ℓ-th partial wave. This is directly verified by (8.19) which also shows that S_ℓ has modulus one:

$$S_\ell = e^{2i\delta_\ell} = \frac{A_{\ell N} + iB_{\ell N}}{A_{\ell N} - iB_{\ell N}} \quad . \tag{8.45}$$

The phase itself is then given by

$$\cos\delta_\ell = \frac{A_{\ell N}}{\sqrt{A_{\ell N}^2 + B_{\ell N}^2}} \quad , \quad \sin\delta_\ell = \frac{B_{\ell N}}{\sqrt{A_{\ell N}^2 + B_{\ell N}^2}} \quad . \tag{8.46}$$

The partial scattering amplitude $f_\ell(k)$ is given by (8.21) in terms of the scattering matrix element S_ℓ. With (8.44) one easily expresses $f_\ell(k)$ in terms of the scattering phase δ_ℓ:

$$f_\ell(k) = e^{i\delta_\ell} \sin\delta_\ell \quad . \tag{8.47}$$

Starting from (8.43), (8.21) yields the *unitarity relation for the scattering amplitude*

$$\text{Im } f_\ell(k) = |f_\ell(k)|^2 \quad . \tag{8.48}$$

This relation is easily verified by using (8.47) directly. The unitarity relation is easily interpreted in the complex plane spanned by the real and imaginary parts of $f_\ell(k)$:

$$(\text{Re } f_\ell(k))^2 + \left(\text{Im } f_\ell(k) - \frac{1}{2}\right)^2 = \frac{1}{4} \quad . \tag{8.49}$$

This represents a circle of radius $1/2$ about the center $(0, 1/2)$ in that plane and is again referred to as the *Argand diagram* of elastic potential scattering. The analogy to Sect. 4.1 is obvious.

Further Reading

Alonso, Finn: Vol. 3, Chap. 7
Berkeley Physics Course: Vol. 4, Chaps. 8,9
Brandt, Dahmen: Chaps. 10,11,13
Feynman, Leighton, Sands: Vol. 3, Chaps. 19
Flügge: Vol. 1, Chap. 1D
Gasiorowicz: Chaps. 11,24
Merzbacher: Chaps. 11,19
Messiah: Vol. 1, Chaps. 10,11
Schiff: Chap. 5

8.2 Radial Wave Functions

Aim of this section: Presentation of the spherical step potential $V(r)$, (8.1), and of the radial wave function $R_\ell(k, r)$, (8.4), in that potential.

Plot type: 0

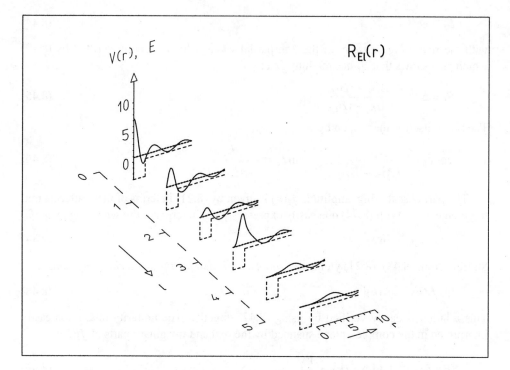

Fig. 8.1. Plot produced with descriptor 1 on file IQ031.DES

C3 coordinates: x: radial coordinate r

y: energy E or angular-momentum quantum number ℓ z: $R_\ell(k, r)$ and $V(r)$ and E

Input parameters:

CH 0 31 f_{CASE}

V1 q 0 0 f_{kE}

V2 N
V3 ℓ

V4 r_1 r_2 r_3 r_4
V5 V_1 V_2 V_3 V_4

V6 l_{DASH} s

with

$f_{\text{CASE}} = 1$: the energy E is fixed at the value determined by V1. The C3 coordinate y
determines the angular-momentum quantum number ℓ. For an example plot see
Fig. 8.1

$f_{\text{CASE}} = 2$: the angular-momentum quantum number is fixed at the value determined by
V3. The C3 coordinate y corresponds to the energy E. For an example plot see
Fig. 8.2

q: wave number $k = q$ and momentum $p = \hbar k$ for $f_{kE} = 0$. If $f_{kE} = 1$ then the input
value q is interpreted as energy, i.e., $p = \sqrt{2ME} = \sqrt{2Mq}$. Note that the numerical
values of $\hbar$ and the particle mass M are $\hbar = 1$, $M = 1$ in this section

$f_{kE} = 0$: input value q is interpreted as wave number

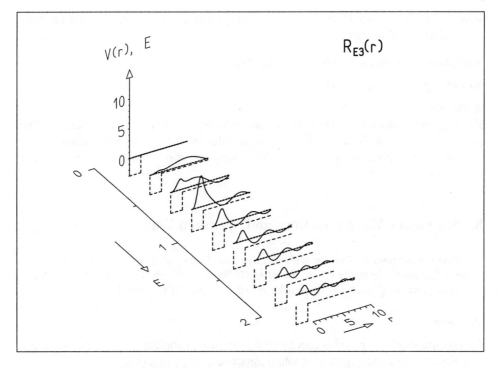

Fig. 8.2. Plot produced with descriptor 2 on file IQ031.DES

$f_{kE} = 1$: input value q is interpreted as kinetic energy

N: number of regions, $1 \leq N \leq 5$. (For input values < 1 (> 5) the minimum (maximum) permissible value is used)

ℓ: angular-momentum quantum number. If you ask for a multiple plot the input value of ℓ is taken only for the first plot and increased successively by one for every further plot

$r_1, \ldots, r_{N-1}$: boundaries between individual regions

$V_1, \ldots, V_{N-1}$: potential in individual regions ($V_N = 0$)

l_{DASH}: dash length (in W3 units) of lines indicating the potential $V(r)$ (default value: $1/10$ of X width of window in W3). Dash length for zero line is $l_{\mathrm{DASH}}/2$

s: scale factor for plotting $R_\ell(k, r)$ (default value: $s = 1$). (Since $R_\ell(k, r)$ is shown with the energy E serving as zero line, actually the function $f(r) = E + s R_\ell(k, r)$ is plotted)

Plot recommendation:
Plot a series of lines $y = $ const, i.e., use

    NL 0 n_y

Additional plot items:

Item 6: line indicating potential $V(r)$

Item 7: line indicating the energy $z = E$ = const and serving as zero line for the radial wave function $R_\ell(k, r)$

Automatically provided texts: T1, T2, TF, TX

Example plots: Figs. 8.1 and 8.2

Remarks:
For f_{CASE} = 1 make sure you use the commands YY and NL in such a manner that the lines y = const which you plot correspond to integer values of the angular momentum $y = \ell$. If they do not the nearest integer is taken. This, however, may result in an uneven spacing of the ℓ values.

8.3 Stationary Wave Functions and Scattered Waves

Aim of this section: Presentation of the stationary wave function $\varphi^{(+)}(r, \vartheta)$, (8.22), approximated as $\sum_{\ell=0}^{L} \varphi_\ell(r, \vartheta)$, the partial waves $\varphi_\ell(r, \vartheta)$, (8.22), the scattered spherical wave $\eta(r, \vartheta)$, (8.29), approximated as $\sum_{\ell=0}^{L} \eta_\ell(r, \vartheta)$, the partial scattered waves $\eta_\ell(r, \vartheta)$, (8.25), (8.28).

Plot type: 1

C3 coordinates: x: position coordinate x r: radial coordinate r φ: polar angle ϑ
z: real part or imaginary part or absolute square of φ or φ_ℓ or η or η_ℓ

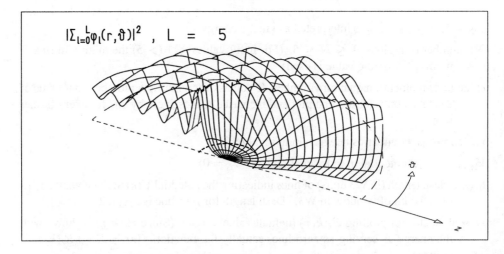

Fig. 8.3. Plot produced with descriptor 3 on file IQ031.DES

Input parameters:

```
CH 1 31 f_FUNC
V0 f_RES              V1 q 0 0 f_kE
V2 N                 V3 ℓ_1
V4 r_1 r_2 r_3 r_4    V5 V_1 V_2 V_3 V_4
```

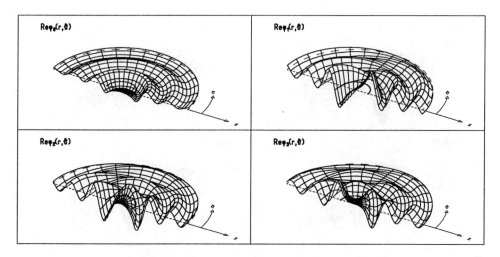

Fig. 8.4. Plot produced with descriptor 4 on file IQ031.DES

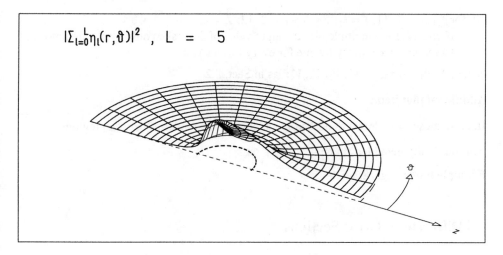

Fig. 8.5. Plot produced with descriptor 5 on file IQ031.DES

with

$f_{FUNC} = 1$: function computed is $\sum_{\ell=0}^{L} \varphi_\ell$

$f_{FUNC} = 2$: function computed is φ_ℓ

$f_{FUNC} = 11$: function computed is $\sum_{\ell=0}^{L} \eta_\ell$

$f_{FUNC} = 12$: function computed is η_ℓ

$f_{RES} = 0$: absolute square of function computed is plotted

$f_{RES} = 1$: real part of function computed is plotted

$f_{RES} = 2$: imaginary part of function computed is plotted

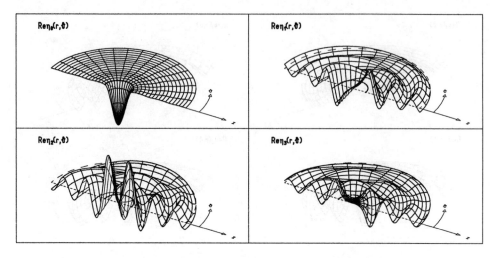

Fig. 8.6. Plot produced with descriptor 6 on file IQ031.DES

ℓ_1: for $f_{FUNC} = 1, 11$: $\ell = \ell_1$; for $f_{FUNC} = 2, 12$: $L = \ell_1$.
   If you ask for a multiple plot the input value of ℓ_1 is taken only for the first plot and increased successively by one for every further plot

q, f_{kE}, N, r_1, r_2, r_3, r_4, V_1, V_2, V_3, V_4: as in Sect. 8.2

Additional plot item:

Item 5: arcs $r = r_i$ drawn as dashed lines indicating boundaries between regions

Automatically provided texts: T1, T2, TX

Example plots: Figs. 8.3–8.6

8.4 Differential Cross Sections

   Aim of this section: Illustration of the differential cross section $d\sigma/d\Omega$, (8.38).

Plot type: 0

C3 coordinates: x: cosine of scattering angle $\cos\vartheta$
y: energy $E = \sqrt{2ME}/\hbar$ (note that the numerical values of M and $\hbar$ are set to 1 in this section) z: differential cross section $d\sigma(\cos\vartheta)/d\Omega$

Input parameters:

```
CH 0 31 3
V2 N                          V3 L
V4 r₁ r₂ r₃ r₄                V5 V₁ V₂ V₃ V₄
V6 l_DASH
```

with

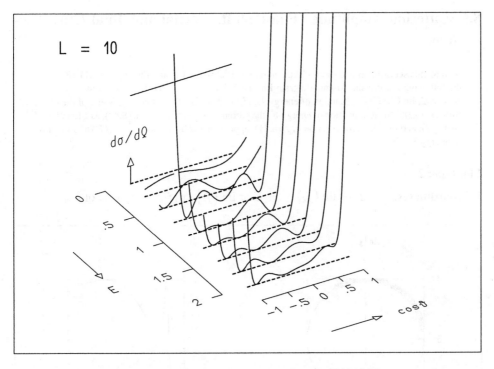

Fig. 8.7. Plot produced with descriptor 7 on file IQ031.DES

L: upper index of sum by which $f(\cos \vartheta)$ is approximated. If you ask for a multiple plot the input value of L is taken only for the first plot and increased successively by one for every further plot

l_{DASH}: dash length (in W3 units) of lines ($z = 0$, $y = $ const) serving as zero lines (default value: $1/20$ of X width of window in W3)

$N, r_1, r_2, r_3, r_4, V_1, V_2, V_3, V_4$: as in Sect. 8.2

Plot recommendation:
Plot a series of lines $y = E = $ const, i.e., use

       NL 0 n_y

Additional plot item:

Item 7: zero lines

Automatically provided texts: T1, T2, TF, TX

Example plot: Fig. 8.7

8.5 Scattering Amplitude. Phase Shift. Partial and Total Cross Sections

Aim of this section: Illustration of the complex partial scattering amplitude f_ℓ, (8.21), (8.47), in the following graphs: the Argand diagram Im $\{f_\ell(E)\}$ vs. Re $\{f_\ell(E)\}$; Re $\{f_\ell(E)\}$ as function of energy E; Im $\{f_\ell(E)\}$ as function of energy E; $|f_\ell(E)|^2$ as function of energy E and of the phase shift $\delta_\ell(E)$, (8.46), as a function of energy E; the partial cross section $\sigma_\ell(E)$, (8.42), as a function of energy E; and the total cross section σ_{tot}, (8.41), approximated by $\sigma_{\text{tot}} \approx \sum_{\ell=0}^{L} \sigma_\ell(E)$ as a function of energy E.

Plot type: 2

C3 coordinates: x: E or Re $\{f_\ell\}$ y: E or Im $\{f_\ell\}$ or $|f_\ell|^2$ or δ_ℓ or σ_ℓ or σ_{tot}

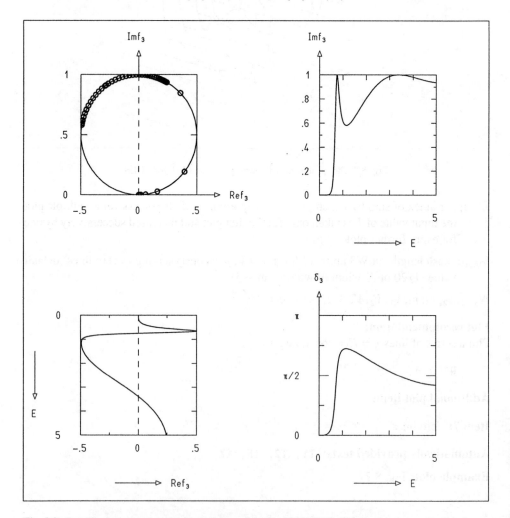

Fig. 8.8. Combined plot produced with descriptor 8 on file IQ031.DES. This descriptor quotes descriptors 9, 10, 11 and 12 which generate the four individual plots

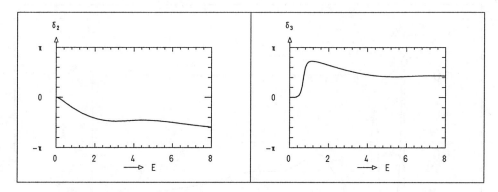

Fig. 8.9. Part of plot produced with descriptor 13 on file IQ031.DES

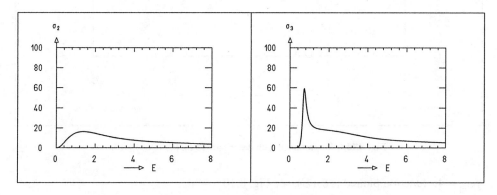

Fig. 8.10. Part of plot produced with descriptor 14 on file IQ031.DES

Input parameters:

```
CH 2 31
NL 0 fVAR
VO fRES
V2 N              V3 ℓ1
V4 r1 r2 r3 r4    V5 V1 V2 V3 V4
```

with

$f_{VAR} = 0$: $x = E$, y is function given by f_{RES}

$f_{VAR} = 1$: $y = E$, x is function given by f_{RES}

$f_{VAR} = 2$: Argand diagram ($x = \text{Re}\{f_\ell(E)\}$, $y = \text{Im}\{f_\ell(E)\}$, RP(1)$\leq E \leq$RP(2), see Appendix A.3.4.1, for an example plot see top-left part of Fig. 8.8)

$f_{RES} = 0$: function plotted is $|f_\ell(E)|^2$

$f_{RES} = 1$: function plotted is $\text{Re}\{f_\ell(E)\}$, see bottom-left part of Fig. 8.8

$f_{RES} = 2$: function plotted is $\text{Im}\{f_\ell(E)\}$, see top-right part of Fig. 8.8

$f_{RES} = 3$: function plotted is $\delta_\ell(E)$, see Fig. 8.9 and bottom-right part of Fig. 8.8

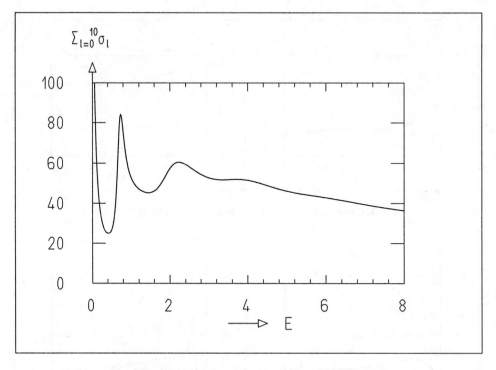

Fig. 8.11. Plot produced with descriptor 15 on file IQ031.DES

$f_{RES} = 4$: function plotted is $\sigma_\ell(E)$, see Fig. 8.10

$f_{RES} = 5$: function plotted is $\sum_{\ell=0}^{L} \sigma_\ell(E)$, see Fig. 8.11

ℓ_1: angular-momentum quantum number $\ell = \ell_1$ for $f_{RES} < 5$, upper index in sum $L = \ell_1$ for $f_{RES} = 5$. If you ask for a multiple plot the input value of ℓ_1 is taken for the first plot and increased successively by one for every plot

$N, r_1, r_2, r_3, r_4, V_1, V_2, V_3, V_4$: as in Sect. 8.2

Additional plot items:

Item 5: polymarkers which you can place at intervals $\Delta E = $ const on the Argand diagram, see Appendix A.3.4

Item 6: unitarity circle for Argand diagrams

Automatically provided texts: T1, T2

Example plots: Figs. 8.8–8.11

Remarks:
It is customary to draw an Argand diagram (Im $\{f_\ell(E)\}$ vs. Re $\{f_\ell(E)\}$) and graphs Im $\{f_\ell(E)\}$ and Re $\{f_\ell(E)\}$ in such a way that the graphs appear to be projections to the right and below the the Argand diagram, respectively. You can do that by using a *combined plot* descriptor which in turn quotes several individual descriptors (see Appendix A.5.3) as in the example plot, Fig. 8.8.

8.6 Exercises

Please note:

(i) you may watch a semiautomatic demonstration of the material of this chapter by typing
`SA D31`

(ii) for the following exercises use descriptor file `IQ031.DES`

(iii) if not stated otherwise in the exercises the numerical values of the particle mass and of Planck's constant are put to 1.

8.2.1 Plot the free radial wave functions for the momentum $p = 0.75$ and the angular momenta **(a)** $\ell = 0, 1, 2, \ldots, 10$, **(b)** $\ell = 11, 12, \ldots, 20$, **(c)** $\ell = 21, 22, \ldots, 30$. Start from descriptor 16. **(d)** Why is the wave function close to zero in a range adjacent to $r = 0$ and why does this range grow with increasing ℓ?

8.2.2 Plot the free radial wave functions for the energy range $0.01 \leq E \leq 10$ for **(a)** $\ell = 0$, **(b)** $\ell = 1$, **(c)** $\ell = 2$, **(d)** $\ell = 3$, **(e)** $\ell = 10$. Start from descriptor 17.

8.2.3 Repeat Exercise 8.2.2 for the energy range $0.01 \leq E \leq 1$ and for the angular momenta **(a)** $\ell = 6$, **(b)** $\ell = 8$, **(c)** $\ell = 10$. **(d)** Why does the range of very small values of the wave function decrease with increasing energy if ℓ is kept fixed?

8.2.4 Plot the wave functions with angular momentum $\ell = 0$ for the repulsive potential
$$V(r) = \begin{cases} 10 & , \quad 0 \leq r < 6 \\ 0 & , \quad 6 \leq r \end{cases}$$
in ten energy intervals for the range **(a)** $1 \leq E \leq 10$, **(b)** $9 \leq E \leq 12$. Start from descriptor 17.

8.2.5 **(a)** Plot the phase shift δ_ℓ for the repulsive potential of Exercise 8.2.4 for the energy range $0.001 \leq E \leq 30$ for the angular momentum $\ell = 0$. Start from descriptor 18. **(b)** Read the energy values of the first two resonances off the plot.

8.2.6 Plot the radial wave functions for the angular momentum $\ell = 0$ for an energy range about **(a)** the first, **(b)** the second resonance as determined in Exercise 8.2.5 **(b)** for the potential of Exercise 8.2.4. Start from descriptor 17.

8.2.7 Repeat Exercise 8.2.5 for angular momentum $\ell = 1$.

8.2.8 Repeat Exercise 8.2.6 for angular momentum $\ell = 1$.

8.2.9 **(a,b)** Repeat Exercise 8.2.5 **(a)** and **(b)** for the lowest resonance in angular momentum $\ell = 2$. **(c)** Why do the resonance energies for resonance wave functions with the same number of radial nodes increase with increasing angular momentum?

8.2.10 Repeat Exercise 8.2.6 for the lowest resonance in angular momentum $\ell = 2$.

8.2.11 Plot the radial wave functions for the potential
$$V(r) = \begin{cases} 0 & , \quad 0 \leq r < 1.5 \\ 10 & , \quad 1.5 \leq r < 2 \\ 0 & , \quad 2 \leq r \end{cases}$$
in the energy range $0.01 \leq E \leq 20$ for **(a)** $\ell = 0$, **(b)** $\ell = 1$, **(c)** $\ell = 2$, **(d)** $\ell = 10$. Start from descriptor 17.

8.2.12 Plot the radial wave functions for the potential of Exercise 8.2.11 for the values $0 \leq \ell \leq 10$ for the energies **(a)** $E = 0.1$, **(b)** $E = 1$, **(c)** $E = 9.9$, **(d)** $E = 20$. Start from descriptor 16.

8.2.13 Plot the phase shift δ_ℓ for the potential of Exercise 8.2.11 for the energy range $0.01 \leq E \leq 40$ for the angular momentum (a) $\ell = 0$, (b) $\ell = 1$, (c) $\ell = 2$, (d) $\ell = 3$, (e) $\ell = 6$, (f) $\ell = 8$, and for the energy range $0.01 \leq E \leq 60$ for the angular momenta (g) $\ell = 10$, (h) $\ell = 13$. Start from descriptor 18.

8.3.1 Plot the absolute square of the wave function $|\varphi(r, \vartheta)|^2$ as a function of the radial variable r and the polar angle ϑ for the repulsive potential

$$V(r) = \begin{cases} 6 & , \quad 0 \leq r < 2 \\ 0 & , \quad 2 \leq r \end{cases}$$

for the energy values (a) $E = 3$, (b) $E = 6.5$, (c) $E = 10$. Start from descriptor 19.

8.3.2 Plot the absolute square of the wave function $|\varphi(r, \vartheta)|^2$ for the potential of Exercise 8.3.1 for the energy of the lowest resonance in (a) $\ell = 0$ ($E_{\text{res}} = 7$), (b) $\ell = 1$ ($E_{\text{res}} = 8.5$), (c) $\ell = 2$ ($E_{\text{res}} = 10$), (d) $\ell = 3$ ($E_{\text{res}} = 11.9$). Start from descriptor 19.

8.3.3 Plot the partial wave functions $\varphi_\ell(r, \vartheta)$ for the potential of Exercise 8.3.1 for the energy $E = 3$ and the angular momenta (a) $\ell = 0$, (b) $\ell = 1$, (c) $\ell = 2$, (d) $\ell = 3$, (e) $\ell = 4$, (f) $\ell = 5$, (g) $\ell = 6$, (h) $\ell = 7$, (i) $\ell = 8$. Start from descriptor 19.

8.3.4 Repeat Exercise 8.3.3 for a free particle (i.e., $V(r) = 0$ everywhere) and compare the results with the ones of 8.3.3.

8.3.5 Repeat Exercise 8.3.3 for the energy $E = 9$ and compare the results with the ones of 8.3.4.

8.3.6 Repeat Exercise 8.3.1 for the absolute square of the scattering wave $|\eta(r, \vartheta)|^2$.

8.3.7 Repeat Exercise 8.3.2 for the absolute square of the scattering wave $|\eta(r, \vartheta)|^2$.

8.3.8 Repeat Exercise 8.3.3 for the partial scattering waves $\eta_\ell(r, \vartheta)$.

8.3.9 Repeat Exercise 8.3.5 for the partial scattering waves $\eta_\ell(r, \vartheta)$.

8.3.10 Plot the partial scattering waves $\eta_\ell(r, \vartheta)$ for the potential of Exercise 8.3.1 for (a) $\ell = 0$, $E = 7$, (b) $\ell = 1$, $E = 7$, (c) $\ell = 2$, $E = 7$, (d) $\ell = 3$, $E = 7$, (e) $\ell = 0$, $E = 8.5$, (f) $\ell = 1$, $E = 8.5$, (g) $\ell = 2$, $E = 8.5$, (h) $\ell = 3$, $E = 8.5$, (i) $\ell = 0$, $E = 10$, (j) $\ell = 1$, $E = 10$, (k) $\ell = 2$, $E = 10$, (l) $\ell = 3$, $E = 10$, (m) $\ell = 0$, $E = 11.9$, (n) $\ell = 1$, $E = 11.9$, (o) $\ell = 2$, $E = 11.9$, (p) $\ell = 3$, $E = 11.9$. Start from descriptor 19.

8.4.1 For the repulsive spherically symmetric potential

$$V(r) = \begin{cases} 3 & , \quad 0 \leq r < 2 \\ 0 & , \quad 2 \leq r \end{cases}$$

plot the differential cross section $d\sigma/d\Omega$ for the energy range $0.001 \leq E \leq 6$ divided into 10 intervals for the summation of angular momenta up to (a) $L = 0$, (b) $L = 1$, (c) $L = 2$, $\ldots$, (i) $L = 8$. Start from descriptor 7. (j) Why is the differential cross section obtained in (a) independent of $\cos \vartheta$?

8.4.2 Repeat Exercise 8.4.1 for a summation up to $L = 30$ for the energy ranges (a) $40 \leq E \leq 50$ and (b) $4000 \leq E \leq 5000$. Start from descriptor 7. (c) Calculate the wavelengths ($M = 1$, $\hbar = 1$) for $E = 50$ and $E = 5000$. (d) How does the cross section in the forward direction decrease for increasing energy?

8.4.3 For the attractive spherically symmetric potential

$$V(r) = \begin{cases} -3 & , \quad 0 \leq r < 2 \\ 0 & , \quad 2 \leq r \end{cases}$$

plot the differential cross section $d\sigma/d\Omega$ for the energy range $0.001 \leq E \leq 6$ divided into

10 intervals for the summation of angular momenta up to (a) $L = 0$, (b) $L = 1$, (c) $L = 2$, ..., (i) $L = 8$. Start from descriptor 7.

8.4.4 Repeat Exercise 8.4.3 for a summation up to $L = 30$ for the energy ranges (a) $40 \leq E \leq 50$ and (b) for $4000 \leq E \leq 5000$. (c) Compare the result with those in Exercise 8.4.2. Start from descriptor 7.

8.4.5 Compare the differential cross sections for the repulsive and attractive potentials of Exercises 8.4.1 and 8.4.3 for low energies $0.001 \leq E \leq 0.1$. Plot for the repulsive potential (a) $L = 0, \ldots,$ (d) $L = 3$ and for the attractive potential (e) $L = 0, \ldots,$ (h) $L = 3$. Start from descriptor 7. (i) Why are the cross sections for the attractive potential at low energies larger than those for the repulsive potential?

8.5.1 For the potential
$$V(r) = \begin{cases} 3 & , \quad 0 \leq r < 2 \\ 0 & , \quad 2 \leq r \end{cases}$$
plot for $\ell = 0$ and $0.001 \leq E \leq 6$ (a) the Argand plot $f_0(E)$, (b) Im $f_0(E)$, (c) Re $f_0(E)$, (d) $\delta_0(E)$, (e) the combined plot of (a–d). Start from the descriptors (a) 9, (b) 10, (c) 11, (d) 12, (e) 8. (f) Read the lowest resonance energy off the plot of the phase δ_0.

8.5.2 (a–e) Repeat Exercise 8.5.1 (a–e) for the energy range $0.001 \leq E \leq 20$. (f) Read the two lowest resonance energies E_1, E_2 and the corresponding phase values δ_{01}, δ_{02} off the plot for δ_0. (g) Calculate the values $f_0(E_1)$, $f_0(E_2)$ from the phases δ_{01}, δ_{02}.

8.5.3 (a–e) Repeat Exercise 8.5.2 (a–e) for $\ell = 1$. (f) Read the lowest resonance energy off the plot of the phase δ_1.

8.5.4 (a–e) Repeat Exercise 8.5.2 (a–e) for $\ell = 2$. (f) Read the lowest resonance energy off the plot of the phase δ_2. (g) Explain the increase of the lowest resonance energies as ℓ increases from 0 to 2.

8.5.5 For the potential
$$V(r) = \begin{cases} 0 & , \quad 0 \leq r < 2 \\ 4 & , \quad 2 \leq r < 2.5 \\ 0 & , \quad 2.5 \leq r \end{cases}$$
plot for $\ell = 0$ and $0.001 \leq E \leq 20$ (a) the Argand plot $f_0(E)$, (b) Im $f_0(E)$, (c) Re $f_0(E)$, (d) $\delta_0(E)$, (e) the combined plot of (a–d). Start from the descriptor (a) 9, (b) 10, (c) 11, (d) 12 and (e) 8. (f) Read the lowest resonance energies of the plot of the phase δ_0. (g) Give a rough estimate of the resonance energies by calculating the bound-state energies of an infinitely deep square well of width 2.

8.5.6 (a–e) Repeat Exercise 8.5.5 (a–e) for $\ell = 1$.

8.5.7 (a–e) Repeat Exercise 8.5.5 (a–e) for $\ell = 2$.

8.5.8 Plot for the potential
$$V(r) = \begin{cases} 0 & , \quad 0 \leq r < 2 \\ 4 & , \quad 2 \leq r < 2.5 \\ 0 & , \quad 2.5 \leq r \end{cases}$$
in the energy range $0.001 \leq E \leq 8$ the partial cross sections σ_ℓ for (a) $\ell = 0, 1, 2, 3$, (b) $\ell = 4, 5, 6, 7$. Start from descriptor 14.

8.5.9 Plot for the potential
$$V(r) = \begin{cases} 0 & , \quad 0 \leq r < 2 \\ 10 & , \quad 2 \leq r < 2.5 \\ 0 & , \quad 2.5 \leq r \end{cases}$$

in the energy range $0.001 \le E \le 20$ the partial cross sections σ_ℓ for **(a)** $\ell = 0, 1, 2, 3$, **(b)** $\ell = 4, 5, 6, 7$. Start from descriptor 14. **(c)** What is the significance of the small peaks in the plot? **(d)** For zero angular momentum calculate the first few energy eigenvalues of an infinitely deep square well of equal width. **(e)** Why are the energy values of the peaks for $\ell = 0$ in the plot (a) smaller than the energies calculated under (d)?

8.5.10 (a) Plot for the potential

$$V(r) = \begin{cases} 0 & , \quad 0 \le r < 2 \\ 4 & , \quad 2 \le r < 2.5 \\ 0 & , \quad 2.5 \le r \end{cases}$$

in the energy range $0.001 \le E \le 8$ the total cross section obtained by summation of angular momenta up to $L = 15$. Start from descriptor 14. **(b)** Compare the value for $E = 0.001$ with the value of the partial cross section σ_0 at this energy. **(c)** Calculate the total classical cross section for the scattering on a hard sphere of radius 2.5.

8.5.11 Plot for the repulsive square-well potential

$$V(r) = \begin{cases} V_0 & , \quad 0 \le r < 2 \\ 0 & , \quad 2 \le r \end{cases}$$

in the energy range $10^{-5} \le E \le 0.01$ the total cross section σ_{tot} for the summation $L = 3$ for **(a)** $V_0 = 10$, **(b)** $V_0 = 100$, **(c)** $V_0 = 1000$, **(d)** $V_0 = 10\,000$, **(e)** $V_0 = 100\,000$. Start from descriptor 14. Read the values of σ_{tot} at $E = 10^{-5}$ off the plots. **(f)** Compare the values with the limiting formula $\sigma_{\text{tot}}(E = 0) = 4\pi d^2$ for an infinitely high repulsive square-well potential of width d.

9 Special Functions of Mathematical Physics

Contents: Discussion of the most important formulae and construction of plots for some functions of mathematical physics relevant to quantum mechanics. These functions are Hermite polynomials, Legendre polynomials and Legendre functions, spherical harmonics, Bessel functions and spherical Bessel functions and Laguerre polynomials. Directly related to some of these and also discussed are the eigenfunctions of the one-dimensional harmonic oscillator and the radial eigenfunctions of the harmonics oscillator in three dimensions and of the hydrogen atom.

9.1 Basic Formulae

9.1.1 Hermite Polynomials

The *Hermite polynomials* are solutions of the differential equation

$$\frac{d^2 H_n}{dx^2} - 2x \frac{dH_n}{dx} + 2n H_n = 0 \quad , \quad n = 0, 1, \dots \quad . \tag{9.1}$$

They can be computed from the recurrence relation

$$H_0(x) = 1 \quad , \quad H_1(x) = 2x$$

$$H_n(x) = 2x H_{n-1}(x) - 2(n-1) H_{n-2}(x) \quad , \quad n = 2, 3, \dots \tag{9.2}$$

or from *Rodrigues' formula*

$$H_n(x) = (-1)^n e^{x^2} \frac{d^n}{dx^n} e^{-x^2} \tag{9.3}$$

and satisfy the orthogonality relation

$$\int_{-\infty}^{\infty} H_n(x) H_m(x) e^{-x^2} dx = 0 \quad , \quad n \neq m \quad . \tag{9.4}$$

9.1.2 Harmonic-Oscillator Eigenfunctions

The *eigenfunctions of the one-dimensional harmonic oscillator* are

$$\varphi_n(x) = (\sqrt{\pi} 2^n n! \sigma_0)^{-1/2} H_n\left(\frac{x}{\sigma_0}\right) \exp\left(-\frac{x^2}{2\sigma_0^2}\right) \quad , \quad n = 0, 1, 2, \dots \tag{9.5}$$

with

$$\sigma_0 = \sqrt{\hbar/m\omega} \quad ,$$

where m and ω are the mass and angular frequency, respectively. The eigenfunctions are orthonormal

$$\int_{-\infty}^{\infty} \varphi_n(x)\varphi_m(x)dx = \delta_{nm} \quad .$$

(9.6)

9.1.3 Legendre Polynomials and Legendre Functions

The *Legendre polynomials* solve the differential equation

$$(1 - x^2)\frac{d^2 P_\ell(x)}{dx^2} - 2x\frac{dP_\ell(x)}{dx} + \ell(\ell+1)P_\ell(x) = 0 \quad , \quad \ell = 0, 1, 2, \ldots \quad .$$

(9.7)

They follow from the recurrence relation

$$P_0(x) = 1 \quad , \quad P_1(x) = x \quad ,$$

$$(\ell+1)P_{\ell+1}(x) = (2\ell+1)xP_\ell(x) - \ell P_{\ell-1}(x) \quad , \quad \ell = 1, 2, \ldots$$

(9.8)

or from Rodrigues' formula

$$P_\ell(x) = \frac{1}{2^\ell \ell!}\frac{d^\ell}{dx^\ell}\left[(x^2-1)^\ell\right]$$

(9.9)

and satisfy the orthogonality relation

$$\int_{-1}^{1} P_\ell(x)P_n(x)dx = 0 \quad , \quad \ell \neq n \quad .$$

(9.10)

The *associated Legendre functions* are solutions of the differential equation

$$(1 - x^2)\frac{d^2 P_\ell^m}{dx^2} - 2x\frac{dP_\ell^m}{dx} + \left[\ell(\ell+1) - \frac{m^2}{1-x^2}\right]P_\ell^m = 0 \quad ,$$

$$\ell = 0, 1, 2, \ldots \quad , \quad m = 0, 1, \ldots, \ell \quad .$$

(9.11)

With $P_\ell^0(x) = P_\ell(x)$ they can be obtained from the recurrence relation

$$P_m^m(x) = \frac{(2m)!}{2^m m!}(1-x^2)^{m/2} \quad ,$$

$$(2\ell+1)xP_\ell^m(x) = (\ell-m+1)P_{\ell+1}^m(x) + (\ell+m)P_{\ell-1}^m(x)$$

(9.12)

or from

$$P_\ell^m(x) = (1-x^2)^{m/2}\frac{d^m}{dx^m}P_\ell(x) \quad .$$

(9.13)

They are orthogonal:

$$\int_{-1}^{1} P_n^m(x)P_\ell^m(x)dx = 0 \quad , \quad \ell \neq n \quad .$$

(9.14)

9.1.4 Spherical Harmonics

The *spherical harmonics* solve the differential equation

$$\frac{1}{\sin\vartheta}\frac{d}{d\vartheta}\sin\vartheta\frac{dY_{\ell m}}{d\vartheta} + \frac{1}{\sin^2\vartheta}\frac{d^2Y_{\ell m}}{d\varphi^2} + \ell(\ell+1)Y_{\ell m} = 0 \tag{9.15}$$

and can be expressed by the associated Legendre functions

$$Y_{\ell m}(\vartheta,\varphi) = (-1)^m\sqrt{\frac{2\ell+1}{4\pi}\frac{(\ell-m)!}{(\ell+m)!}}P_\ell^m(\cos\vartheta)e^{im\varphi} \quad . \tag{9.16}$$

For negative m values one defines

$$Y_{\ell m} = (-1)^m Y_{\ell,|m|}^* \quad , \quad m < 0 \quad . \tag{9.17}$$

The first few spherical harmonics are

$$Y_{0,0} = \frac{1}{\sqrt{4\pi}} \qquad\qquad Y_{1,0} = \sqrt{\frac{3}{4\pi}}\cos\vartheta$$

$$Y_{1,1} = -\sqrt{\frac{3}{8\pi}}\sin\vartheta e^{i\varphi} \qquad Y_{2,0} = \sqrt{\frac{5}{16\pi}}(3\cos^2\vartheta - 1) \quad .$$

$$Y_{2,1} = -\sqrt{\frac{15}{8\pi}}\sin\vartheta\cos\vartheta e^{i\varphi} \quad Y_{2,2} = \sqrt{\frac{15}{32\pi}}\sin^2\vartheta e^{2i\varphi}$$

The spherical harmonics are orthonormal

$$\int_{-1}^{+1}\int_0^{2\pi} Y_{\ell'm'}^*(\vartheta,\varphi)Y_{\ell m}(\vartheta,\varphi)d\cos\vartheta d\varphi = \delta_{\ell'\ell}\delta_{m'm} \quad . \tag{9.18}$$

The absolute square of a spherical harmonic is directly proportional to the square of the associate Legendre function with the same indices, (9.16):

$$|Y_{\ell m}(\vartheta,\varphi)|^2 = \frac{2\ell+1}{4\pi}\frac{(\ell-m)!}{(\ell+m)!}[P_\ell^m(\cos\vartheta)]^2 \quad . \tag{9.19}$$

9.1.5 Bessel Functions

Bessel's differential equation

$$x^2\frac{d^2Z_\nu(x)}{dx^2} + x\frac{dZ_\nu(x)}{dx} + (x^2 - \nu^2)Z_\nu(x) = 0 \tag{9.20}$$

is solved by the *Bessel functions* of the first kind $J_\nu(x)$, of the second kind $N_\nu(x)$ (also called *Neumann functions*) and of the third kind (also called *Hankel functions*) $H_\nu^{(1)}(x)$ and $H_\nu^{(2)}(x)$ which are complex linear combinations of the former two. The Bessel functions of the first kind are

$$J_\nu(x) = \left(\frac{x}{2}\right)^\nu\sum_{k=0}^\infty\frac{(-1)^k x^{2k}}{4^k k!\Gamma(\nu+k+1)} \quad , \tag{9.21}$$

where $\Gamma(z)$ is Euler's *Gamma function*

$$\Gamma(z+1) = \int_0^\infty t^z e^{-t}dt \quad , \tag{9.22}$$

which for integer and half-integer arguments is

$$\Gamma(1) = 1 \quad , \quad \Gamma(\tfrac{1}{2}) = \sqrt{\pi} \quad , \quad \Gamma(z+1) = z\Gamma(z) \quad . \tag{9.23}$$

For integer arguments $n \geq 1$ this leads to the identification

$$\Gamma(n+1) = n! \quad . \tag{9.24}$$

The Bessel functions of the second kind are

$$N_\nu(x) = \frac{1}{\sin \nu\pi}[J_\nu(x)\cos \nu\pi - J_{-\nu}(x)] \quad . \tag{9.25}$$

For integer $\nu = n$

$$J_{-n}(x) = (-1)^n J_n(x) \tag{9.26}$$

and $N_n(x)$ has to be determined from (9.25) by the limit $\nu \to n$. The *modified Bessel functions* are defined as

$$I_\nu(x) = e^{-i\pi\nu/2} J_\nu(e^{i\pi/2}x) \quad , \quad -\pi < \arg x \leq \pi/2 \quad . \tag{9.27}$$

The Hankel functions are defined by

$$H_\nu^{(1)}(x) = J_\nu(x) + iN_\nu(x) \quad , \tag{9.28}$$

$$H_\nu^{(2)}(x) = J_\nu(x) - iN_\nu(x) \quad . \tag{9.29}$$

9.1.6 Spherical Bessel Functions

The differential equation

$$x^2 \frac{d^2 z_\ell(x)}{dx^2} + 2x\frac{dz_\ell(x)}{dx} + [x^2 - \ell(\ell+1)]z_\ell(x) = 0 \tag{9.30}$$

with integer ℓ is solved by the *spherical Bessel functions* of the first kind

$$j_\ell(x) = \sqrt{\frac{\pi}{2x}} J_{\ell+1/2}(x) \quad , \tag{9.31}$$

the spherical Bessel functions of the second kind (also called *spherical Neumann functions*)

$$n_\ell(x) = -\sqrt{\frac{\pi}{2x}} N_{\ell+1/2}(x) = (-1)^\ell j_{-\ell-1}(x) \quad , \tag{9.32}$$

and the spherical Bessel functions of the third kind (also called *spherical Hankel functions* of the first and second kind)

$$h_\ell^{(+)}(x) = n_\ell(x) + ij_\ell(x) = i[j_\ell(x) - in_\ell(x)] = i\sqrt{\frac{\pi}{2x}} H_{\ell+1/2}^{(1)}(x) \quad , \tag{9.33}$$

$$h_\ell^{(-)}(x) = n_\ell(x) - ij_\ell(x) = -i[j_\ell(x) + in_\ell(x)] = -i\sqrt{\frac{\pi}{2x}} H_{\ell+1/2}^{(2)}(x) \quad . \tag{9.34}$$

The spherical Hankel functions can be written in the form

$$h_\ell^{(\pm)}(x) = C_\ell^\pm \frac{e^{\pm ix}}{x} \quad , \tag{9.35}$$

where

$$C_\ell^\pm = (\mp i)^\ell \sum_{s=0}^{\ell} \frac{1}{2^s s!} \frac{(\ell+s)!}{(\ell-s)!} (\mp ix)^{-s} \tag{9.36}$$

is a polynomial in $1/x$. Explicitly, the first few Hankel functions are

$$h_0^{(\pm)}(x) = \frac{e^{\pm ix}}{x} \quad , \quad h_1^{(\pm)} = \left(\mp i + \frac{1}{x}\right) \frac{e^{\pm ix}}{x} \quad . \tag{9.37}$$

By inversion of (9.33) and (9.34) we obtain

$$j_\ell(x) = \frac{1}{2i} \left[h_\ell^{(+)}(x) - h_\ell^{(-)}(x) \right] \tag{9.38}$$

and

$$n_\ell(x) = \frac{1}{2} \left[h_\ell^{(+)}(x) + h_\ell^{(-)}(x) \right] \quad . \tag{9.39}$$

of which the first few are simply

$$j_0(x) = \frac{\sin x}{x} \quad , \quad j_1(x) = \frac{\sin x}{x^2} - \frac{\cos x}{x} \quad ,$$

$$n_0(x) = \frac{\cos x}{x} \quad , \quad n_1(x) = \frac{\cos x}{x^2} + \frac{\sin x}{x} \quad .$$

The behavior of the spherical Bessel and Neumann functions for small x is

$$j_\ell(x) \sim \frac{x^\ell}{(2\ell+1)!!} \quad , \quad n_\ell(x) \sim (2\ell-1)!! x^{-(\ell+1)} \quad , \quad x \to 0 \quad , \tag{9.40}$$

with

$$(2\ell+1)!! = 1 \times 3 \times 5 \times \ldots \times (2\ell+1) \quad .$$

The $h_\ell^{(\pm)}$ for purely imaginary argument $x = i\eta$ can be expressed as

$$h_\ell^{(\pm)}(i\eta) = (\mp i)^{\ell \pm 1} \sum_{s=0}^{\ell} \frac{1}{2^s s!} \frac{(\ell+s)!}{(\ell+s)!} (\pm \eta)^{-s} \frac{e^{\mp \eta}}{\eta} \quad . \tag{9.41}$$

Thus $i^{\ell+1} h_\ell^{(+)}(i\eta)$ is a real function of η. Its asymptotic behavior for large x is

$$i^{\ell+1} h_\ell^{(+)}(i\eta) \sim \frac{e^{-\eta}}{\eta} \quad , \quad \eta \to \infty \quad . \tag{9.42}$$

Introducing the result (9.41) into (9.33) and (9.39) we get for the spherical Bessel and Neumann functions expressions which can be made explicitly real by appropriate powers of the imaginary unit:

$$(-i)^\ell j_\ell(i\eta) \quad , \quad i^{\ell+1} n_\ell(i\eta) \quad .$$

9.1.7 Laguerre Polynomials

The *Laguerre polynomials* solve the differential equation

$$x\frac{d^2 L_n^\alpha(x)}{dx^2} + (\alpha + 1 - x)\frac{dL_n^\alpha(x)}{dx} + nL_n^\alpha(x) = 0 \quad . \tag{9.43}$$

They are given by the recurrence relation

$$L_0^\alpha(x) = 1 \quad , \quad L_1^\alpha = \alpha + 1 - x \quad ,$$

$$(n + 1)L_{n+1}^\alpha(x) = (2n + \alpha + 1 - x)L_n^\alpha(x) - (n + \alpha)L_{n-1}^\alpha(x) \tag{9.44}$$

or by the explicit formula

$$L_n^\alpha(x) = \sum_{j=0}^{n}(-1)^j\frac{\Gamma(\alpha + n + 1)x^j}{\Gamma(n - j + 1)\Gamma(\alpha + j + 1)j!} \tag{9.45}$$

or by Rodrigues' formula

$$L_n^\alpha(x) = \frac{1}{n!}\frac{e^x}{x^\alpha}\frac{d^n}{dx^n}(x^{n+\alpha}e^{-x}) \quad , \tag{9.46}$$

and satisfy the orthogonality relation

$$\int_0^\infty L_n^\alpha(x)L_m^\alpha(x)x^\alpha e^{-x}dx = 0 \quad , \quad n \neq m \quad . \tag{9.47}$$

9.1.8 Radial Eigenfunctions of the Harmonic Oscillator

The *radial eigenfunctions of the three-dimensional harmonic oscillator* are

$$R_{n\ell}(\varrho) = N_{n\ell}\varrho^\ell \exp\left(-\varrho^2/2\right) L_{n_r}^{\ell+1/2}\left(\varrho^2\right) \tag{9.48}$$

with

$N_{n\ell} = \sqrt{(n_r!2^{n_r+\ell+2})/\{[2(\ell + n_r) + 1]!!\sqrt{\pi}\sigma_0^3\}}$

$(2m + 1)!! = 1 \times 3 \times 5 \times \ldots \times (2m + 1)$

$\varrho = r/\sigma_0$

r: radial distance from origin

$\sigma_0 = \sqrt{\hbar/M\omega}$ ground-state width

$n_r = (n - \ell)/2$

n: principal quantum number

ℓ: angular-momentum quantum number

They are orthonormal:

$$\int_0^\infty R_{n\ell}\left(\frac{r}{\sigma_0}\right) R_{m\ell}\left(\frac{r}{\sigma_0}\right) r^2 dr = \delta_{nm} \quad . \tag{9.49}$$

9.1.9 Radial Eigenfunctions of the Hydrogen Atom

The *radial eigenfunctions of the electron in the hydrogen atom* are

$$R_{n\ell}(r) = N_{n\ell} \left(\frac{2r}{na}\right)^\ell \exp\left(-\frac{r}{na}\right) L_{n-\ell-1}^{2\ell+1}\left(\frac{2r}{na}\right) \tag{9.50}$$

with

$$N_{n\ell} = 2\sqrt{(n-\ell-1)!/(n+\ell)!}/(a^{3/2}n^2)$$

n: principal quantum number

ℓ: angular-momentum quantum number

a: Bohr's radius

r: radial distance from origin

They are orthonormal:

$$\int_0^\infty R_{n\ell}(r)R_{m\ell}(r)r^2 dr = \delta_{nm} \quad . \tag{9.51}$$

Further Reading

Messiah: Vol. 1, Appendix B
Abramowitz, Stegun, Chaps. 6,8,9,10,22

9.2 Hermite Polynomials

Aim of this section: Illustration of the Hermite polynomials (9.2).

Plot type: 2

C3 coordinates: x: argument x y: $H_n(x)$

Input parameters:

```
CH 2 0 40
VO n
```

with

n: index of Hermite polynomial $H_n(x)$; if you ask for a multiple plot the input value n is
taken only for the first plot. The value of n is increased successively by 1 for every
plot

Automatically provided texts: T1, T2, CA

Example plot: Fig. 9.1

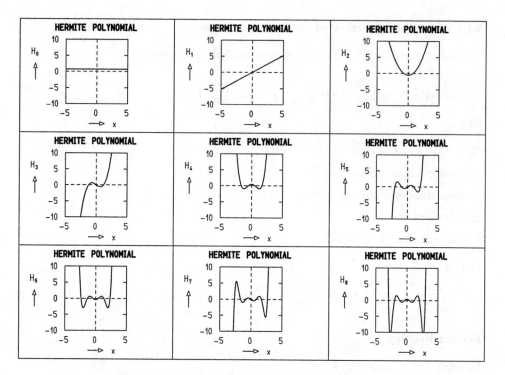

Fig. 9.1. Plot produced with descriptor 1 on file IQ000.DES

9.3 Eigenfunctions of the One-Dimensional Harmonic Oscillator

Aim of this section: Illustration of the eigenfunctions (9.5).

Plot type: 2

C3 coordinates: x: argument x y: $\varphi_n(x)$ or $\varphi_n^2(x)$

Input parameters:

```
CH 2 0 41 fRES
VO n
```

with

$f_{\mathrm{RES}} = 0$: function plotted is $\varphi_n(x)$

$f_{\mathrm{RES}} = 1$: function plotted is $\varphi_n^2(x)$

n: index of eigenfunction $\varphi_n(x)$; if you ask for a multiple plot the input value n is taken
only for the first plot. The value of n is increased successively by 1 for every plot

Automatically provided texts: T1, T2, CA

Example plot: Fig. 9.2

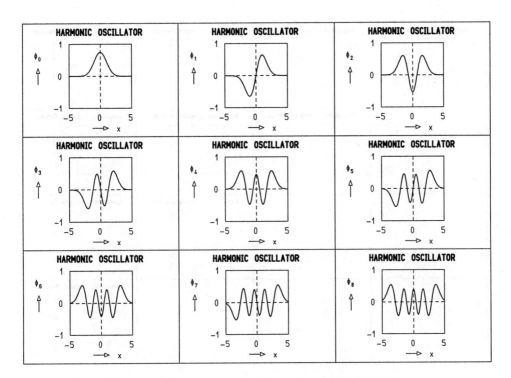

Fig. 9.2. Plot produced with descriptor 2 on file IQ000.DES

9.4 Legendre Polynomials and Associated Legendre Functions

Aim of this section: Illustration of the Legendre polynomials (9.8), the associated Legendre functions (9.12) and the absolute square (9.19) of the spherical harmonics.

9.4.1 Type 2 Plots – Functions of x or $\cos \vartheta$

Plot type: 2

C3 coordinates: x: argument x y: $P_\ell(x)$ or $P_\ell^m(x)$ or $|Y_{\ell m}(x = \cos \vartheta)|^2$

Input parameters:

```
CH 2 0 fCASE
VO ℓ  m
```

with

$f_{CASE} = 30$: function plotted is $P_\ell(x)$

$f_{CASE} = 31$: function plotted is $P_\ell^m(x)$

$f_{CASE} = 32$: function plotted is $|Y_{\ell m}(\cos \vartheta)|^2$

ℓ: index $\ell, \ell \geq 0$

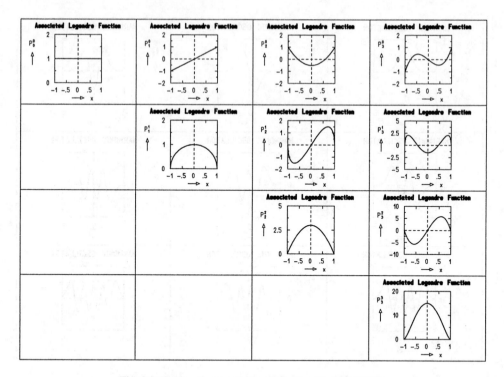

Fig. 9.3. Plot produced with descriptor 4 on file IQ000.DES

m: index m, $0 \le m \le \ell$ (ignored for $f_{CASE} = 30$)

if you ask for a multiple plot the input values of ℓ and m are taken for the first plot. For $f_{CASE} = 30$ the value of ℓ is increased by 1 for each plot. For $f_{CASE} = 31$ or 32 the value of ℓ is increased by 1 for each column and the value of m by 1 for each row

Automatically provided texts: T1, T2, CA

Example plot: Fig. 9.3

9.4.2 Type 2 Plots – Polar Diagrams

Plot type: 2

C3 coordinates: x: variable z y: variable x ϑ: arctan(z/x)

Input parameters:

    CH 2 0 33 f_{CASE}
    VO ℓ m

with

$f_{CASE} = 0$: function plotted is $|Y_{\ell m}(\cos \vartheta)|^2$

$f_{CASE} = 1$: function plotted is $|Y_{\ell m}(\cos \vartheta)|$

ℓ: index ℓ, $\ell \ge 0$

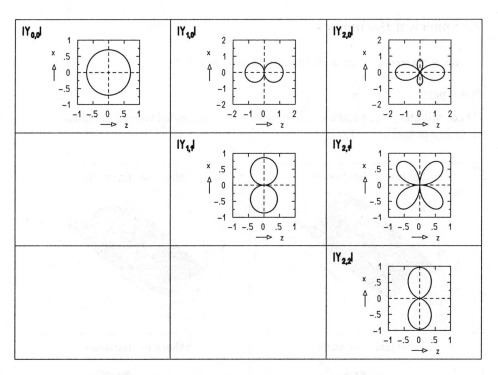

Fig. 9.4. Plot produced with descriptor 20 on file IQ000.DES

m: index m, $0 \le \ell \le m$

if you ask for a multiple plot the input values of ℓ and m are taken for the first plot. The value of ℓ is increased by 1 for each column and the value of m by 1 for each row

Plot recommendation:
Use

    NL 0 2 2

to ensure that a parameter (here the polar angle ϑ) is used as independent variable and that automatic scaling is invoked and the same scale is used in x and y. Use a square as W3 window and, since the parameter is ϑ, use

    RP 0 3.1416

to vary it between 0 and π or – somewhat ambiguously but graphically pleasing – vary it between 0 and 2π. The curve shown is interpreted as follows. A ray from a point on the curve to the origin forms an angle ϑ with the positive z axis. The length of the ray is the function $|Y_{\ell m}(\cos\vartheta)|^2$ or $|Y_{\ell m}(\cos\vartheta)|$ depending on f_{CASE}.

Automatically provided texts: T1, T2, CA

Example plot: Fig. 9.4

9.5 Spherical Harmonics

Aim of this section: Illustration of the spherical harmonics (9.16).

Plot type: 0

C3 coordinates: x: polar angle ϑ in radians y: azimuthal angle ϑ in radians
z: $\operatorname{Re} Y_{\ell m}$ or $\operatorname{Im} Y_{\ell m}$ or $|Y_{\ell m}|^2$

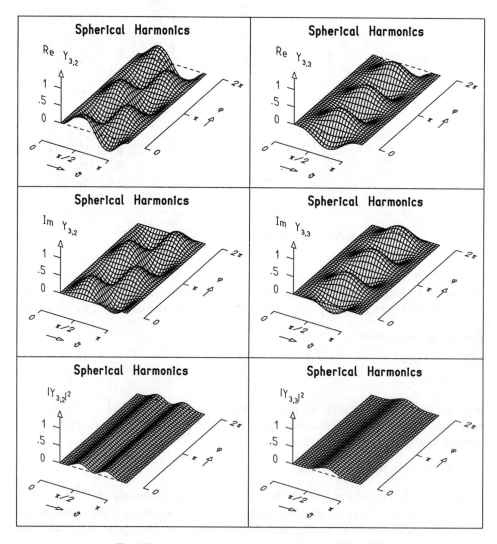

Fig. 9.5. Plot produced with descriptor 6 on file IQ000.DES

Input parameters:

CH 0 0 30

```
VO  fRES
V1  ℓ   m
```

with

$f_{RES} = 0$: function shown in $|Y_{\ell m}|^2$

$f_{RES} = 1$: function shown is $\mathrm{Re}\, Y_{\ell m}$

$f_{RES} = 2$: function shown is $\mathrm{Im}\, Y_{\ell m}$

ℓ, m: indices of $Y_{\ell m}(\ell \geq 0, 0 \leq m \leq \ell)$

if you ask for a multiple plot (Appendix A.5.2) the input values of m and f_{RES} are taken only for the first plot. The value of m is increased by 1 for every column, the value of f_{RES} is increased by one for every row

Automatically provided texts: T1, T2, TF

Example plot: Fig. 9.5

9.6 Bessel Functions

Aim of this section: Illustration of the Bessel function $J_\nu(x)$, (9.21), and of the modified Bessel function $I_\nu(x)$, (9.27), for integer and for real index ν.

9.6.1 Type 2 Plots

Plot type: 2

C3 coordinates: x: argument x y: $J_n(x)$ or $I_n(x)$

Input parameters:

```
CH  2  0  fCASE
VO  n
```

with

$f_{CASE} = 1$: function plotted is $J_n(x)$

$f_{CASE} = 2$: function plotted is $I_n(x)$

n: index of Bessel function, n has to be a non-negative integer; if you ask for a multiple plot (Appendix A.5.2) the index is n for the first plot and increased by one for every successive plot

Automatically provided texts: T1, T2, CA

Example plot: Fig. 9.6

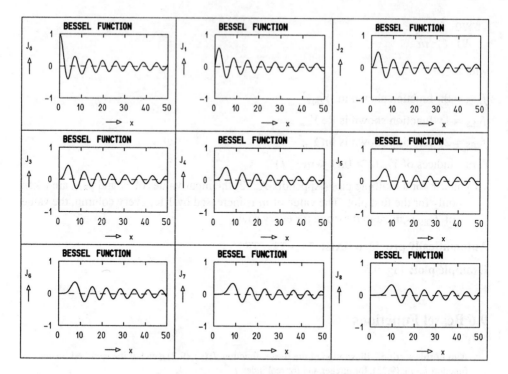

Fig. 9.6. Plot produced with descriptor 7 on file IQ000.DES

9.6.2 Type 0 Plots

Plot type: 0

C3 coordinates: x: argument x y: real index ν z: $J_\nu(x)$ or $I_\nu(x)$

Input parameters:

   CH 0 0 f_{CASE}

with

$f_{CASE} = 1$: function plotted is $J_\nu(x)$
$f_{CASE} = 2$: function plotted is $I_\nu(x)$

Automatically provided texts: T1, T2, TF, CA

Example Plot: Fig. 9.7

Remarks:

1. Restrict the arguments to $x \geq 0$ and $-99 < \nu < 100$.

2. You may cut off infinities by setting NL(3) 1, then the function z plotted is cut off at ZZ(3) and ZZ(4), i.e., for $z > M = \max(ZZ(3), ZZ(4))$ the function is set to $z = M$ and for $z < m = \min(ZZ(3), ZZ(4))$ it is set to $z = m$.

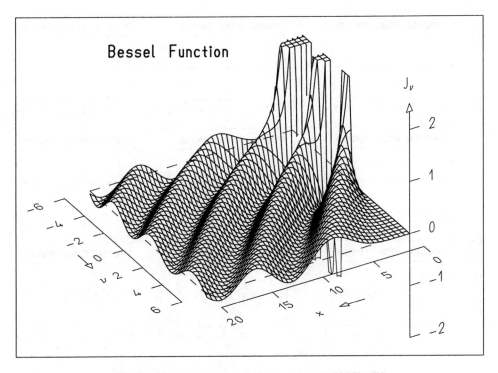

Fig. 9.7. Plot produced with descriptor 17 on file IQ000.DES

9.7 Spherical Bessel Functions

Aim of this section: Illustration of the spherical Bessel functions (9.31) and spherical Neumann functions (9.32) for purely real and purely imaginary arguments and of the spherical Hankel functions of the first kind (9.41) for purely imaginary arguments.

Plot type: 2

C3 coordinates: x: variable ϱ
y: $j_\ell(\varrho)$ or $n_\ell(\varrho)$ or $(-\mathrm{i})^\ell j_\ell(\mathrm{i}\varrho)$ or $\mathrm{i}^{\ell+1} n_\ell(\mathrm{i}\varrho)$ or $\mathrm{i}^{\ell+1} h_\ell^{(+)}(\mathrm{i}\varrho)$

Input parameters:

```
CH 2 0 fCASE
VO ℓ
```

with

$f_{\mathrm{CASE}} = 10$: function plotted is $j_\ell(\varrho)$

$f_{\mathrm{CASE}} = 11$: function plotted is $n_\ell(\varrho)$

$f_{\mathrm{CASE}} = 12$: function plotted is $\mathrm{i}^{\ell+1} h^{(+)}(\mathrm{i}\varrho)$

$f_{\mathrm{CASE}} = 13$: function plotted is $(-\mathrm{i})^\ell j_\ell(\mathrm{i}\varrho)$

$f_{\mathrm{CASE}} = 14$: function plotted is $\mathrm{i}^{\ell+1} n_\ell(\mathrm{i}\varrho)$

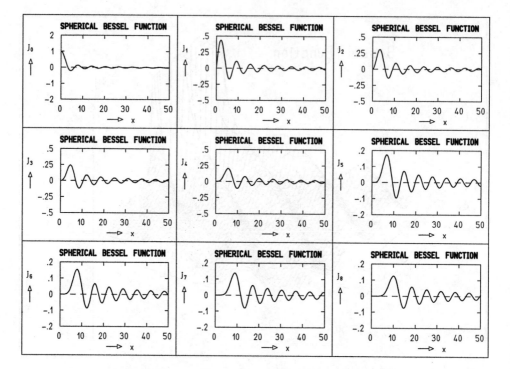

Fig. 9.8. Plot produced with descriptor 9 on file IQ000.DES

if you ask for a multiple plot (Appendix A.5.2) the index is ℓ for the first plot and increased by one for every successive plot

Automatically provided texts: T1, T2, CA

Example plot: Fig. 9.8

9.8 Laguerre Polynomials

Aim of this section: Illustration of the Laguerre polynomials (9.44).

9.8.1 Type 2 Plots

Plot type: 2

C3 coordinates: x: argument x y: $L_n^\alpha(x)$

Input parameters:

```
CH  2  0  20
VO  n   α
```

with

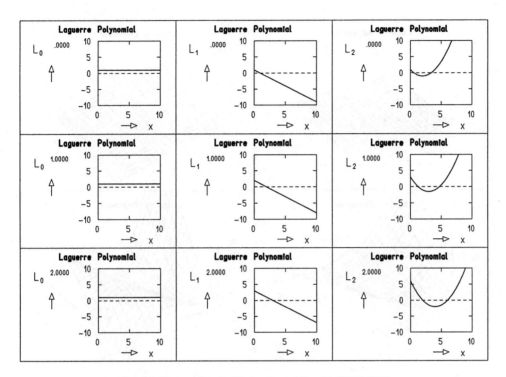

Fig. 9.9. Plot produced with descriptor 14 on file IQ000.DES

n: lower index of Laguerre polynomial

α: upper index of Laguerre polynomial

if you ask for a multiple plot (Appendix A.5.2) the indices n and α are taken for the first plot. The lower index n is increased by one for each column and the upper index α by one for each row in the multiple plot

Automatically provided texts: T1, T2, CA

Example plot: Fig. 9.9

9.8.2 Type 0 Plots

Plot type: 0

C3 coordinates: x: argument x y: upper index α z: $L_n^\alpha(x)$

Input parameters:

    CH 0 0 20
    VO n

with

n: lower index of associated Laguerre polynomial; if you ask for a multiple plot the input value n is taken for the first plot and increased by 1 for each following plot

Automatically provided texts: T1, T2, TF, CA

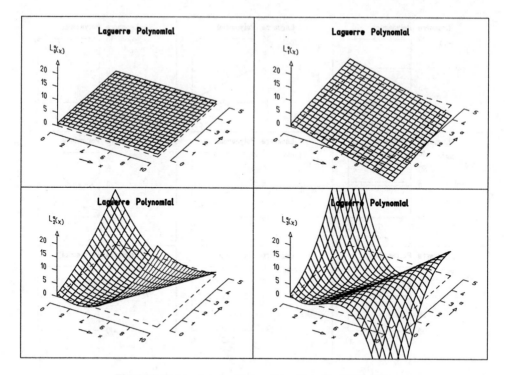

Fig. 9.10. Plot produced with descriptor 19 on file IQ000.DES

Example plot: Fig. 9.10

Remark:
You may cut off very large or very small function values by setting NL(3) 1, then the function z plotted is cut off at ZZ(3) and ZZ(4), i.e., for $z > M = \max(\text{ZZ(3)}, \text{ZZ(4)})$ the function is set to $z = M$ and for $z < m = \min(\text{ZZ(3)}, \text{ZZ(4)})$ it is set to $z = m$.

9.9 Radial Eigenfunctions of the Harmonic Oscillator

Aim of this section: Illustration of the radial eigenfunctions (9.48) of the three-dimensional harmonic oscillator ($\sigma_0 = 1$).

Plot type: 2

C3 coordinates: x: reduced radial distance ϱ y: $R_{n_r\ell}(\varrho)$

Input parameters:

```
CH 2 0 22
VO n_r ℓ
```

with

n_r: radial quantum number

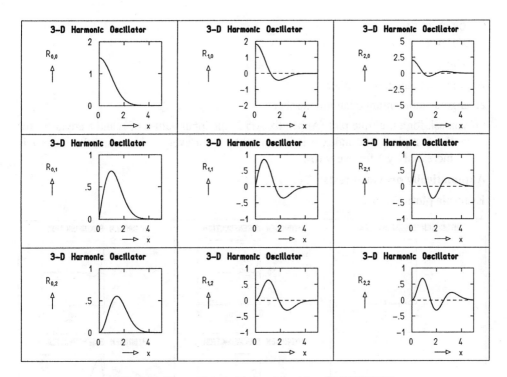

Fig. 9.11. Plot produced with descriptor 16 on file IQ000.DES

ℓ: angular-momentum quantum number

if you ask for a multiple plot (Appendix A.5.2) the input values of n_r and ℓ are taken for the first plot. The value of n_r is increased by 1 for each column the value of ℓ by 1 for each row

Automatically provided texts: T1, T2, CA

Example plot: Fig. 9.11

Remark:
The indices of $R_{n_r\ell}$ appearing in the plot refer to the radial and the angular-momentum quantum number, respectively.

9.10 Radial Eigenfunctions of the Hydrogen Atom

Aim of this section: Illustration of the radial eigenfunctions (9.50) of the electron in the hydrogen atom, where the Bohr radius is set equal to 1.

Plot type: 2

C3 coordinates: x: argument ϱ $\quad$ y: $R_{n\ell}(\varrho)$

Input parameters:

CH 2 0 21
VO n ℓ

with

n: principal quantum number

ℓ: angular-momentum quantum number

if you ask for a multiple plot (Appendix A.5.2) the input values of n and ℓ are taken for the first plot. The index n is increased by 1 for every column and the index ℓ is increased by 1 for every row

Automatically provided texts: T1, T2, CA

Example plot: Fig. 9.12

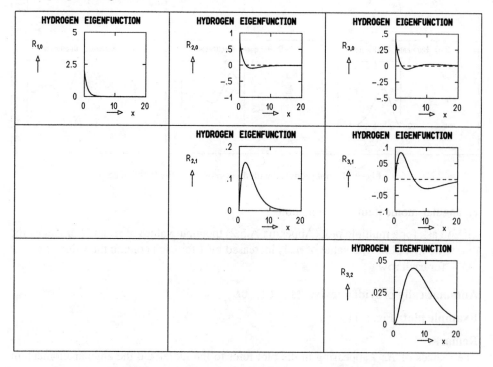

Fig. 9.12. Plot produced with descriptor 15 on file IQ000.DES

9.11 Simple Functions of a Complex Variable

Aim of this section: Illustration of simple complex functions of one complex variable, i.e., $w = w(z)$. The complex variable $z = x + iy = re^{i\varphi}$ has real part $\text{Re } z = x$, imaginary part $\text{Im } z = y$, absolute value $|z| = r$ and argument $\arg z = \varphi$. Presented are plots of $\text{Re } w$, $\text{Im } w$, $|w|$ and $\arg w$ as surfaces over the complex plane spanned by x and y. Plots can be type 0 or type 1. The functions are e^z, $\log z$, $\sin z$, $\cos z$, $\sinh z$, $\cosh z$, z^n and $z^{1/n}$.

Plot type: 0 and 1

C3 coordinates for type 0 plots:

x: Re z y: Im z z: Re w or Im w or $|w|$ or arg w

C3 coordinates for type 1 plots:

r: $|z|$ φ: arg z z: Re w or Im w or $|w|$ or arg w

Input parameters:

    CH f_{PT} 99 f_{FUNC} f_{RES}
    VO n n_{SHEET}

with

$f_{PT} = 0(1)$: plot type is $0(1)$

$f_{FUNC} = 1$: function computed is $w = e^z$

$f_{FUNC} = 2$: function computed is $w = \log z$

$f_{FUNC} = 3$: function computed is $w = \sin z$

$f_{FUNC} = 4$: function computed is $w = \cos z$

$f_{FUNC} = 5$: function computed is $w = \sinh z$

$f_{FUNC} = 6$: function computed is $w = \cosh z$

$f_{FUNC} = 10$: function computed is $w = z^n$, n integer

$f_{FUNC} = 11$: function computed is $w = z^{1/n}$, n positive integer

$f_{RES} = 1$: plot shows Re w

$f_{RES} = 2$: plot shows Im w

$f_{RES} = 3$: plot shows $|w|$

$f_{RES} = 4$: plot shows arg w

n: determines exponent for $f_{FUNC} = 10$ ($w = z^n$, n integer) and $f_{FUNC} = 11$ ($w = z^{1/n}$, n positive integer)

n_{SHEET}: the functions $w = z^{1/n}$ are n valued. The plot shows a surface over the Riemann sheet n_{SHEET} of the complex plane ($1 \leq n_{SHEET} \leq n$)

Example plot: Fig. A.1 and Fig. A.4 in Appendix A

Remark:
You may cut off very large or very small function values by setting NL(3) 1, then the function z plotted is cut off at ZZ(3) and ZZ(4), i.e., for $z > M = \max(\text{ZZ}(3), \text{ZZ}(4))$ the function is set to $z = M$ and for $z < m = \min(\text{ZZ}(3), \text{ZZ}(4))$ it is set to $z = m$.

9.12 Exercises

Please note:

(i) you may watch a semiautomatic demonstration of the material of this chapter by typing
SA DOO

(ii) for the following exercises use descriptor file IQ000.DES

9.2.1 Plot in a 2×2 multiple plot the Hermite polynomials $H_n(x)$ for **(a)** $n = 9, 10, 11, 12$,
(b) $n = 13, 14, 15, 16$, **(c)** $n = 20, 21, 22, 23$. Start from descriptor 1. **(d)** How many zeros
does $H_n(x)$ possess?

9.3.1 Plot in a 2×2 multiple plot the eigenfunctions of the harmonic oscillator for **(a)** $n =$
$9, 10, 11, 12$, **(b)** $n = 13, 14, 15, 16$, **(c)** $n = 20, 21, 22, 23$. Start from descriptor 2.

9.4.1 Plot in a 2×2 multiple plot the Legendre polynomials $P_\ell(x)$ for **(a)** $\ell = 9, 10, 11, 12$,
(b) $\ell = 13, 14, 15, 16$, **(c)** $\ell = 20, 21, 22, 23$. Start from descriptor 3. **(d)** How many zeros
does the polynomial $P_\ell(x)$ possess?

9.4.2 Plot in a 4×4 multiple plot the associated Legendre functions $P_\ell^m(x)$ for **(a)** $\ell = 4$,
$m = 0, \ldots, \ell = 7, m = 3$, **(b)** $\ell = 4, m = 4, \ldots, \ell = 7, m = 7$. Start from descriptor 4.
(c) How many zeros does the function $P_\ell^m(x)$ exhibit?

9.4.3 Plot as polar diagrams in a 3×3 multiple plot the square $|Y_{\ell m}|^2$ of the spherical
harmonic function for **(a)** $\ell = 4, m = 0$ as plot in the left upper field. **(b)** $\ell = 4, m = 4$ as
plot in the left upper field. Start from descriptor 20.

9.5.1 Plot the real and imaginary part of the spherical harmonic function for $\ell = 2$ and its
absolute square in a 3×3 multiple plot. Start from descriptor 6.

9.6.1 Plot the Bessel function $I_n(x)$ in a 3×3 multiple plot for $n = 0, 1, \ldots, 8$. Start from
descriptor 7. Use the autoscale facility NL(3) 1.

9.7.1 Plot in a 2×2 multiple plot the spherical Bessel functions $j_\ell(x)$ in the range $0 \leq$
$x \leq 50$ for **(a)** $\ell = 9, 10, 11, 12$, **(b)** $\ell = 13, 14, 15, 16$. Start from descriptor 9.

9.7.2 Plot in a 2×2 multiple plot the spherical Neumann functions $n_\ell(x)$ in the range
$0 \leq x \leq 50$ for **(a)** $\ell = 9, 10, 11, 12$, **(b)** $\ell = 13, 14, 15, 16$. Start from descriptor 10.

9.7.3 Plot in a 2×2 multiple plot the spherical modified Hankel functions $i^{\ell+1} h^{(+)}(i\varrho)$ in
the range $0 \leq x \leq 20$ for **(a)** $\ell = 9, 10, 11, 12$, **(b)** $\ell = 13, 14, 15, 16$. Start from descriptor
11.

9.7.4 Plot in a 2×2 multiple plot the spherical modified Bessel functions $(-i)^\ell j_\ell(i\varrho)$ in
the range $0 \leq x \leq 20$ for **(a)** $\ell = 9, 10, 11, 12$, **(b)** $\ell = 13, 14, 15, 16$. Start from descriptor
12.

9.7.5 Plot in 2×2 multiple plot the spherical modified Neumann functions $i^{\ell+1} n_\ell(i\varrho)$ in
the range $0 \leq x \leq 20$ for **(a)** $\ell = 9, 10, 11, 12$, **(b)** $\ell = 13, 14, 15, 16$. Start from descriptor
13.

9.8.1 Plot the Laguerre polynomials $L_n^\alpha(x)$ in a 2×2 multiple plot for the index values
(a) $n = 3, \alpha = 0, 1; n = 4, \alpha = 0, 1$, in the range $0 \leq x \leq 11$, **(b)** $n = 3, \alpha = 2, 3; n = 4$,
$\alpha = 2, 3$, in the range $0 \leq x \leq 13$. Start from descriptor 14.

9.8.2 Plot the Laguerre polynomials $L_n^\alpha(x)$ in a 2×2 multiple plot for the range $0 \leq x \leq 10$
with α as a running index in the range $0 \leq \alpha \leq 5$, $n = 4, 5, 6, 7$. Start from descriptor 19.

9.9.1 Plot in a 3×3 multiple plot the radial eigenfunctions of a spherically symmetric harmonic oscillator in the range $0 \leq x \leq 8$ for the angular momenta $\ell = 1, 2, 3$ for the principal quantum numbers **(a)** $n = 3, 4, 5$, **(b)** $n = 6, 7, 8$. Start from descriptor 16.

9.10.1 Plot in a 3×3 multiple plot the radial eigenfunctions of the hydrogen atom in the range $0 \leq x \leq 60$ for the principal quantum numbers $n = 3, 4, 5$ and **(a)** $\ell = 0, 1, 2$, **(b)** $\ell = 3, 4, 5$. Start from descriptor 15.

10 Additional Material and Hints for the Solution of Exercises

10.1 Units and Orders of Magnitude

10.1.1 Definitions

Every physical quantity q can be expressed as a product of a dimensionless *numerical value* q_{Nu} and its *unit* q_u

$$q = q_{\mathrm{Nu}} q_u \quad . \tag{10.1}$$

The index u specifies the particular system of units used. We may factorize the numerical value

$$q_{\mathrm{Nu}} = q_{\mathrm{Mu}} \times 10^{q_{\mathrm{Eu}}} \tag{10.2}$$

into a *mantissa* q_{Mu} and a power of ten with the integer *exponent* q_{Eu}. The factorization (10.2) is by no means unique but it is understood that the mantissa is not too far from one, say $0.001 \leq |q_{\mathrm{Mu}}| \leq 1000$.

It is important that numerical values used in the computer all have similar exponents since otherwise the result of computation may have severe rounding errors. Moreover, numbers with very small or very large exponents, typically $|q_{\mathrm{Eu}}| \geq 38$, cannot be represented at all with simple techniques. Numerical values which we use as input in computer programs should therefore have exponents close to zero, say $|q_{\mathrm{Eu}}| \leq 5$.

In mechanics and quantum mechanics one can choose 3 physical quantities to be *basic quantities*, define units for them as *basic units* and derive the units of all other quantities from these basic units.

10.1.2 SI Units

In the international system of weights and measures (SI) the basic quantities are *length*, *mass* and *time* with the basic units *meter* (m), *kilogram* (kg) and *second* (s), respectively.

The SI units of the more important physical quantities and the numerical values of some constants of nature are given in Table 10.1.

10.1.3 Scaled Units

Unfortunately, the constants typical of quantum phenomena, listed at the bottom of Table 10.1 all have very large or small exponents in SI units. A simple way out of this

Table 10.1. SI units

SI Units (Based on m, kg, s)	
Quantity	Unit
Energy	$1E_{SI} = 1m^2\,kg\,s^{-2}$
Length	$1\ell_{SI} = 1m$
Time	$1t_{SI} = 1s$
Angular Frequency	$1\omega_{SI} = 1t_{SI}^{-1} = 1s^{-1}$
Action	$1A_{SI} = 1E_{SI}t_{SI} = 1m^2\,kg\,s^{-1}$
Velocity	$1v_{SI} = 1\ell_{SI}t_{SI}^{-1} = 1m\,s^{-1}$
Mass	$1M_{SI} = 1kg$
Momentum	$1p_{SI} = 1M_{SI}v_{SI} = 1m\,kg\,s^{-1}$
Constants	
Planck's Constant	$\hbar = \hbar_{NSI}A_{SI}, \quad \hbar_{NSI} = 1.055 \times 10^{-34}$
Speed of Light	$c = c_{NSI}v_{SI}, \quad c_{NSI} = 2.998 \times 10^{8}$
Electron Mass	$M_e = M_{eNSI}M_{SI}, \quad M_{eNSI} = 9.110 \times 10^{-31}$
Proton Mass	$M_p = M_{pNSI}M_{SI}, \quad M_{pNSI} = 1.673 \times 10^{-27}$
Bohr radius	$a = a_{NSI}\ell_{SI}, \quad a_{NSI} = 0.5292 \times 10^{-10}$
Energy of Hydrogen Ground State	$E_1 = E_{1NSI}E_{SI}, \quad E_{1NSI} = 2.180 \times 10^{-18}$

problem is the use of *scaled units*, i.e., units multiplied by suitable powers of ten. We reformulate the factorization (10.1) and (10.2)

$$
\begin{aligned}
q &= q_{Nu}q_u = q_{Mu} \times 10^{q_{Eu}}q_u = q_{Mu} \times 10^{q'_{Eu}} \times 10^{\bar{q}_{Eu}}q_u \\
&= q_{Mu} \times 10^{q'_{Eu}}q'_u = q'_{Nu}q'_u
\end{aligned}
\tag{10.3}
$$

by writing the exponent q_{Eu} as a sum of two integers,

$$
q_{Eu} = q'_{Eu} + \bar{q}_{Eu} \quad ,
\tag{10.4}
$$

where q'_{Eu} is chosen close to zero, and thus defining a scaled unit

$$
q'_u = 10^{\bar{q}_{Eu}}q_u
\tag{10.5}
$$

and the corresponding numerical value

$$
q'_{Nu} = q_{Mu} \times 10^{q'_{Eu}} \quad .
\tag{10.6}
$$

The decomposition (10.4) of the exponent is not unique. It has to be chosen in such a way that the numerical values (10.6) expressed in scaled units are not too far away from unity, i.e., that exponents q'_{Eu} are close to zero, say $|q'_{Eu}| \leq 5$. Since three basic units can be chosen, three scaling exponents q'_{Eu} may also be chosen. All other scaling exponents are fixed by this choice.

As examples in Table 10.2 we give two sets of scaled SI units. The set in the left column is based on the choice of scale factors for action, mass and velocity,

Table 10.2. Scaled SI units

Examples of Scaled SI Units	
with a choice of scale factors for action, mass, velocity	with a choice of scale factors for action, mass, energy
$E'_{SI} = M'_{SI}v'^2_{SI} = 10^{-14}m^2kg\,s^{-2}$	$E'_{SI} = 10^{-18}E_{SI} = 10^{-18}m^2kg\,s^{-2}$
$\ell'_{SI} = A'_{SI}/M'_{SI}v'_{SI} = 10^{-12}m$	$\ell'_{SI} = A'_{SI}/\sqrt{E'_{SI}M'_{SI}} = 10^{-10}m$
$t'_{SI} = A'_{SI}/M'_{SI}v'^2_{SI} = 10^{-20}s$	$t'_{SI} = A'_{SI}/E'_{SI} = 10^{-16}s$
$\omega'_{SI} = 1/t'_{SI} = 10^{20}s^{-1}$	$\omega'_{SI} = 1/t'_{SI} = 10^{16}s^{-1}$
$A'_{SI} = 10^{-34}A_{SI} = 10^{-34}m^2kg\,s^{-1}$	$A'_{SI} = 10^{-34}A_{SI} = 10^{-34}m^2kg\,s^{-1}$
$v'_{SI} = 10^8v_{SI} = 10^8m\,s^{-1}$	$v'_{SI} = \sqrt{E'_{SI}/M'_{SI}} = 10^6m\,s^{-1}$
$M'_{SI} = 10^{-30}M_{SI} = 10^{-30}kg$	$M'_{SI} = 10^{-30}M_{SI} = 10^{-30}kg$
$p'_{SI} = M'_{SI}v'_{SI} = 10^{-22}m\,kg\,s^{-1}$	$p'_{SI} = \sqrt{E'_{SI}M'_{SI}} = 10^{-24}m\,kg\,s^{-1}$
Constants	
$\hbar = 1.055\,A'_{SI}$	$\hbar = 1.055\,A'_{SI}$
$c = 2.998\,v'_{SI}$	$c = 299.8\,v'_{SI}$
$M_e = 0.9110\,M'_{SI}$	$M_e = 0.9110\,M'_{SI}$
$M_p = 1673\,M'_{SI}$	$M_p = 1673\,M'_{SI}$
$a = 52.92\,\ell'_{SI}$	$a = 0.5292\,\ell'_{SI}$
$E_1 = 0.000218\,E'_{SI}$	$E_1 = 2.180\,E'_{SI}$

$$A'_{SI} = 10^{-34}A_{SI} \quad , \quad v'_{SI} = 10^8v_{SI} \quad , \quad M'_{SI} = 10^{-30}M_{SI} \quad ,$$

which ensures that $\hbar, c$ and M_e have numerical values close to unity. For the set in the right column,

$$A'_{SI} = 10^{-34}A_{SI} \quad , \quad E'_{SI} = 10^{-18}E_{SI} \quad , \quad M'_{SI} = 10^{-30}M_{SI} \quad ,$$

were chosen. Note that in this case the powers of ten of the scale factors for E'_{SI} and M'_{SI} have to be chosen either even for both or odd for both to ensure that the square roots appearing will again be integer powers of ten.

10.1.4 Atomic and Subatomic Units

Scaling factors, or at least scaling factors with large absolute powers of ten, are unnecessary if one chooses units which are "natural" to the system studied. One selects three quantities typical for the system and sets their numerical values equal to one. For questions of atomic physics it is most natural to choose Planck's constant, the velocity of light and the electron mass as these quantities. The so-defined *atomic units* are listed in Table 10.3. All actions are expressed in multiples of $\hbar$, all velocities in multiples of c and all masses in multiples of M_e. The unit of length is

$$1\ell_a = \hbar/M_e c = \lambdabar_e = 3.863 \times 10^{-13}m \quad ,$$

which is called the *Compton wavelength* of the electron. The Bohr radius is $137\lambdabar_e$. The unit of time is the time it takes for a light pulse to traverse one unit of length.

Table 10.3. Atomic units

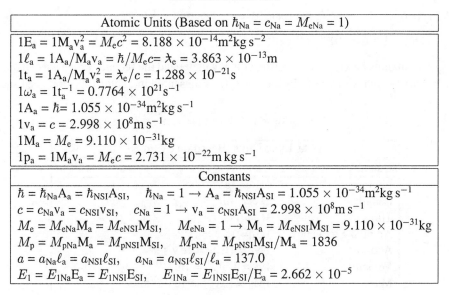

Atomic Units (Based on $\hbar_{Na} = c_{Na} = M_{eNa} = 1$)
$1E_a = 1M_a v_a^2 = M_e c^2 = 8.188 \times 10^{-14} m^2 kg\, s^{-2}$
$1\ell_a = 1A_a/M_a v_a = \hbar/M_e c = \lambdabar_e = 3.863 \times 10^{-13} m$
$1t_a = 1A_a/M_a v_a^2 = \lambdabar_e/c = 1.288 \times 10^{-21} s$
$1\omega_a = 1t_a^{-1} = 0.7764 \times 10^{21} s^{-1}$
$1A_a = \hbar = 1.055 \times 10^{-34} m^2 kg\, s^{-1}$
$1v_a = c = 2.998 \times 10^8 m\, s^{-1}$
$1M_a = M_e = 9.110 \times 10^{-31} kg$
$1p_a = 1M_a v_a = M_e c = 2.731 \times 10^{-22} m\, kg\, s^{-1}$

Constants
$\hbar = \hbar_{Na} A_a = \hbar_{NSI} A_{SI}, \quad \hbar_{Na} = 1 \rightarrow A_a = \hbar_{NSI} A_{SI} = 1.055 \times 10^{-34} m^2 kg\, s^{-1}$
$c = c_{Na} v_a = c_{NSI} v_{SI}, \quad c_{Na} = 1 \rightarrow v_a = c_{NSI} A_{SI} = 2.998 \times 10^8 m\, s^{-1}$
$M_e = M_{eNa} M_a = M_{eNSI} M_{SI}, \quad M_{eNa} = 1 \rightarrow M_a = M_{eNSI} M_{SI} = 9.110 \times 10^{-31} kg$
$M_p = M_{pNa} M_a = M_{pNSI} M_{SI}, \quad M_{pNa} = M_{pNSI} M_{SI}/M_a = 1836$
$a = a_{Na} \ell_a = a_{NSI} \ell_{SI}, \quad a_{Na} = a_{NSI} \ell_{SI}/\ell_a = 137.0$
$E_1 = E_{1Na} E_a = E_{1NSI} E_{SI}, \quad E_{1Na} = E_{1NSI} E_{SI}/E_a = 2.662 \times 10^{-5}$

Many phenomena in nuclear physics are best treated in *subatomic units* which are obtained by using $\hbar$, c and the proton mass M_p as basic units, Table 10.4.

Table 10.4. Subatomic units

Subatomic Units (Based on $\hbar_{Ns} = c_{Ns} = M_{pNs} = 1$)
$1E_s = 1M_s v_s^2 = M_p c^2 = 1.504 \times 10^{-10} m^2 kg\, s^{-2}$
$1\ell_s = 1A_a/M_a v_a = \hbar/M_p c = \lambdabar_p = 2.103 \times 10^{-16} m$
$1t_s = 1A_a/M_a v_a^2 = \lambdabar_p/c = 7.015 \times 10^{-25} s$
$1\omega_s = 1t_a^{-1} = 1.426 \times 10^{24} s^{-1}$
$1A_s = \hbar = 1.055 \times 10^{-34} m^2 kg\, s^{-1}$
$1v_s = c = 2.998 \times 10^8 m\, s^{-1}$
$1M_s = M_p = 1.673 \times 10^{-27} kg$
$1p_s = 1M_s v_s = M_p c = 5.016 \times^{-19} m\, kg\, s^{-1}$

Constants
$\hbar = \hbar_{Ns} A_s = \hbar_{NSI} A_{SI}, \quad \hbar_{Ns} = 1 \rightarrow A_s = \hbar_{NSI} A_{SI} = 1.055 \times 10^{-34} m^2 kg\, s^{-1}$
$c = c_{Ns} v_s = c_{NSI} v_{SI}, \quad c_{Ns} = 1 \rightarrow v_s = c_{NSI} A_{SI} = 2.998 \times 10^8 m\, s^{-1}$
$M_e = M_{eNs} M_s = M_{eNSI} M_{SI}, \quad M_{eNs} = M_{eNSI} M_{SI}/M_s = 5.445 \times 10^{-4}$
$M_p = M_{pNs} M_s = M_{pNSI} M_{SI}, \quad M_{pNs} = 1 \rightarrow M_s = M_{pNSI} M_{SI} = 1.673 \times 10^{-27} kg$
$a = a_{Ns} \ell_s = a_{NSI} \ell_{SI}, \quad a_{Ns} = a_{NSI} \ell_{SI}/\ell_s = 2.516 \times 10^5$
$E_1 = E_{1Ns} E_s = E_{1NSI} E_{SI}, \quad E_{1Ns} = E_{1NSI} E_{SI}/E_s = 1.449 \times 10^{-8}$

10.1.5 Data-Table Units

Often energies of atomic systems are given in electron volts

$$1\text{eV} = 1.602 \times 10^{-19}\text{m}^2\text{kg s}^{-2} \quad .$$

A system of units based on the electron volt, the meter and the second, which we call *data-table units* (which we identify by an index d), is presented in Table 10.5. You need scale

Table 10.5. Data-table units

Data Table Units (Based on eV, m, s)	
Quantity	Unit
Energy	$1E_d = 1\text{eV}$
Length	$1\ell_d = 1\text{m}$
Time	$1t_d = 1\text{s}$
Angular Frequency	$1\omega_d = 1t_d^{-1} = 1\text{s}^{-1}$
Action	$1A_d = 1E_d t_d = 1\text{eV s}$
Velocity	$1v_d = 1\ell_d t_d^{-1} = 1\text{m s}^{-1}$
Mass	$1M_d = E_d v_d^{-2} = 1\text{eV s}^2\,\text{m}^{-2}$
Momentum	$1p_d = 1M_d v_d = 1\text{eV s m}^{-1}$
Constants	
Planck's Constant	$\hbar = \hbar_{\text{Nd}} A_d, \quad \hbar_{\text{Nd}} = 0.6582 \times 10^{-15}$
Speed of Light	$c = c_{\text{Nd}} v_d, \quad c_{\text{Nd}} = 2.998 \times 10^8$
Electron Mass	$M_e = M_{\text{eNd}} M_d, \quad M_{\text{eNd}} = 5.685 \times 10^{-12}$
Proton Mass	$M_p = M_{\text{pNd}} M_d, \quad M_{\text{pNd}} = 1.044 \times 10^{-8}$
Bohr radius	$a = a_{\text{Nd}} \ell_d, \quad a_{\text{Nd}} = 0.5292 \times 10^{-10}$
Energy of Hydrogen Ground State	$E_1 = E_{1\text{Nd}} E_d, \quad E_{1\text{Nd}} = -13.61$

factors in order to use this system. Table 10.6 contains two useful sets of scaled data-table units.

A different system of units still (which we identify by an index c) measures masses in units $1M_c = 1\text{eV}/c^2$. It is obtained through the following set of equations:

$$\begin{aligned} M &= M_{\text{Nd}} M_d = M_{\text{Nd}} \text{eV s}^2\text{m}^{-2} = M_{\text{Nd}} \frac{\text{eV}}{c^2}(c^2\text{s}^{-2}\text{m}^{-2}) \\ &= M_{\text{Nd}}(2.998 \times 10^8)^2 \text{eV}/c^2 = M_{\text{Nc}} \text{eV}/c^2 \end{aligned}$$

$$M_{\text{Nc}} = M_{\text{Nd}} \times 8.998 \times 10^{16}$$

or

$$1\text{eV}/c^2 = 8.998 \times 10^{16}\text{eV s}^2\text{m}^{-2} = 1.783 \times 10^{-36}\text{kg} \quad .$$

In units eV/c^2 the electron and proton mass are

$$M_e = 0.5110 \times 10^6 \text{eV}/c^2, \quad M_p = 938.3 \times 10^6 \text{eV}/c^2$$

Table 10.6. Scaled data-table units

Examples of Scaled Data-Table Units	
with a choice of scale factors for energy, length and mass suitable for problems of atomic physics	with a choice of scale factors for energy, length and mass suitable for problems of nuclear physics
$E_d' = 1\text{eV}$	$E_d' = 10^6 E_d = 1\text{MeV}$
$\ell_d' = 10^{-9}\ell_d = 10^{-9}\text{m}$	$\ell_d' = 10^{-14}\ell_d = 10^{-14}\text{m}$
$t_d' = \ell_d'\sqrt{M_d'/E_d'} = 10^{-15}\text{s}$	$t_d' = \ell_d'\sqrt{M_d'/E_d'} = 10^{-21}\text{s}$
$\omega_d' = 1/t_d' = 10^{15}\text{s}^{-1}$	$\omega_d' = 1/t_d' = 10^{21}\text{s}^{-1}$
$A_d' = \ell_d'\sqrt{M_d'E_d'} = 10^{-15}\text{eVs}$	$A_d' = \ell_d'\sqrt{M_d'E_d'} = 10^{-15}\text{eVs}$
$v_d' = \sqrt{E_d'/M_d'} = 10^6\text{m s}^{-1}$	$v_d' = \sqrt{E_d'/M_d'} = 10^7\text{m s}^{-1}$
$M_d' = 10^{-12}M_d = 10^{-12}\text{eV m}^{-2}\text{s}^2$	$M_d' = 10^{-8}M_d = 10^{-8}\text{eV m}^{-2}\text{s}^2$
$p_d' = \sqrt{E_d'M_d'} = 10^{-6}\text{eV m}^{-1}\text{s}$	$p_d' = \sqrt{E_d'M_d'} = 10^{-7}\text{eV m}^{-1}\text{s}$
Constants	
$\hbar = 0.6582\,A_d'$	$\hbar = 0.6582\,A_d'$
$c = 299.8\,v_d'$	$c = 29.98\,v_d'$
$M_e = 5.685\,M_d'$	$M_e = 0.0005685\,M_d'$
$M_p = 10440\,M_d'$	$M_p = 1.044\,M_d'$
$a = 0.05292\,\ell_d'$	$a = 0.5292 \times 10^{-4}\ell_d'$
$E_1 = -13.61\,E_d'$	$E_1 = -13.61 \times 10^{-6}E_d'$

10.1.6 Special Scales

The Hydrogen Atom

The energy spectrum in the hydrogen atom is given by (7.51). In Sect. 7.2 we use atomic units but we allow a change of input value to $\sqrt{\alpha}$. What, then, is the meaning of the energy scale in Fig. 7.4 in the default case $\alpha = 1$? If the right-hand side of (7.51) is written with α^2 replaced by 1 it means that a factor α^2 is missing in the equation. We absorb this factor into the energy unit by defining a new unit

$$E_h = \alpha^2 E_a = 5.328 \times 10^{-5}E_a$$

where E_a is the energy in atomic units. For the lowest eigenvalue in Fig. 7.4 (which has $\ell = 2$ and then $n = 3$) we read off $E_3 = 0.0555\,E_h = 2.957 \times 10^{-6}E_a$ as expected since $E_3 = \frac{1}{9}E_1$ and $E_1 = 2.662 \times 10^{-5}E_a$, see Table 10.3

The Harmonic oscillator

Similarly the energy scale of the harmonic-oscillator eigenvalue spectrum described by (7.44) and shown in Fig. 7.3 for the default value of $\omega = 1\omega_a$ in atomic units can be interpreted as the spectrum of an oscillator with arbitrary angular frequency $\omega = \omega_{Na}\omega_a$ if the numerical values on the energy scale are given the units

$$E_o = E_a\omega_{Na} \quad .$$

10.2 Argand Diagrams and Unitarity for One-Dimensional Problems

10.2.1 Probability Conservation and the Unitarity of the Scattering Matrix

Scattering processes in one spatial dimension offer a simple study of the properties of the S matrix. In stepwise constant potentials of the kind (3.16) the wave function is of complex exponential form in the various regions. We represent the wave functions in the region 1 of vanishing potential far to the left as

$$\psi_1(x) = A'_1 e^{ik_1 x} + B'_1 e^{-ik_1 x} \tag{10.7}$$

and in the region N far to the right as

$$\psi_N(x) = A'_N e^{ik_N x} + B'_N e^{-ik_N x} \quad . \tag{10.8}$$

Here, we have also included the term $B'_N e^{-ik_N x}$ representing a wave coming in from large values of x propagating to the left. Obviously, there are two physical scattering situations:

i) incoming right-moving wave at negative x values represented by the term $A'_1 e^{ik_1 x}$: the outgoing waves are the transmitted wave $A'_N e^{ik_N x}$ and the reflected wave $B'_1 e^{-ik_1 x}$;

ii) incoming left-moving wave at large positive x values, represented by the term $B'_N e^{-ik_N x}$: the outgoing waves are the transmitted wave $B_1 e^{-ik_1 x}$ and the reflected wave $A_N e^{ik_N x}$.

For real potentials $V(x)$ every solution of the time-dependent Schrödinger equation keeps the normalization at all times. Thus the integral of the probability density over the whole x axis does not change with time. This is interpreted as probability conservation. It can also be expressed as the conservation of the *probability current density*

$$j(x, t) = \frac{\hbar}{2mi} \left(\psi^* \frac{\partial}{\partial x} \psi - \psi \frac{\partial}{\partial x} \psi^* \right) \quad . \tag{10.9}$$

through a *continuity equation*

$$\frac{\partial \varrho}{\partial t} + \frac{\partial j}{\partial x} = 0 \quad , \tag{10.10}$$

where

$$\varrho(x, t) = \psi^*(x, t)\psi(x, t) \tag{10.11}$$

is the probability density. For stationary states

$$\psi(x, t) = e^{-\frac{i}{\hbar} Et} \varphi(x) \tag{10.12}$$

the probability density ϱ and current density j are time independent and probability conservation is tantamount to

$$\frac{d}{dx} j(x) = 0 \quad , \tag{10.13}$$

i.e., the probability current density is constant in x.

For the wave functions (10.7), (10.8) this means

$$k_1(|A_1'|^2 - |B_1'|^2) = k_N(|A_N'|^2 - |B_N'|^2) \tag{10.14}$$

or

$$\left| \sqrt{\frac{k_N}{k_1}} A_N' \right|^2 + |B_1'|^2 = |A_1'|^2 + \left| \sqrt{\frac{k_N}{k_1}} B_N' \right|^2 \tag{10.15}$$

for arbitrary values of A_1' and B_N'. We associate the quantities on either side of the above equation with the absolute squares of the components of two-dimensional complex vectors

$$\begin{pmatrix} A_N \\ B_1 \end{pmatrix} = \begin{pmatrix} \sqrt{\frac{k_N}{k_1}} A_N' \\ B_1' \end{pmatrix} \quad , \quad \begin{pmatrix} A_1 \\ B_N \end{pmatrix} = \begin{pmatrix} A_1' \\ \sqrt{\frac{k_N}{k_1}} B_N' \end{pmatrix} \quad . \tag{10.16}$$

The equation (10.15) derived from current conservation then states the equality of the length of these two vectors. Thus they may be related by a complex 2×2 matrix

$$S = \begin{pmatrix} S_{11} & S_{12} \\ S_{21} & S_{22} \end{pmatrix}$$

defined by

$$\begin{pmatrix} A_N \\ B_1 \end{pmatrix} = S \begin{pmatrix} A_1 \\ B_N \end{pmatrix} \quad , \tag{10.17}$$

which is unitary, i.e., S and its adjoint

$$S^+ = \begin{pmatrix} S_{11}^* & S_{21}^* \\ S_{12}^* & S_{22}^* \end{pmatrix} \tag{10.18}$$

fulfill the relation

$$S S^+ = 1 \quad , \tag{10.19}$$

or

$$\begin{pmatrix} S_{11}S_{11}^* + S_{12}S_{12}^* & S_{11}S_{21}^* + S_{12}S_{22}^* \\ S_{21}S_{11}^* + S_{22}S_{12}^* & S_{21}S_{21}^* + S_{22}S_{22}^* \end{pmatrix} = \begin{pmatrix} 1 & 0 \\ 0 & 1 \end{pmatrix} \quad . \tag{10.20}$$

The unitary matrix S is called the *scattering matrix* or *S matrix*. If we consider the amplitudes A_1 and B_N given to be one or zero, we have two cases:

i) wave coming in from the left:

$$A_1 = 1 \quad , \quad B_N = 0 \quad , \quad A_N = S_{11} \quad , \quad B_1 = S_{21} \quad , \tag{10.21}$$

i.e., S_{11} is the transmission coefficient and S_{21} the reflection coefficient for a right-moving incoming wave;

ii) wave coming in from the right:

$$A_1 = 0 \quad , \quad B_N = 1 \quad , \quad A_N = S_{12} \quad , \quad B_1 = S_{22} \quad , \tag{10.22}$$

i.e., S_{12} is the reflection coefficient and S_{22} the transmission coefficient for a left-moving wave.

10.2.2 Time Reversal and the Scattering Matrix

Invariance under *time reversal* implies that $\varphi^*(x)$ is also a solution of the stationary Schrödinger equation with the real potential $V(x)$. Because of the change of the sign of the exponents in the wave functions, incoming and outgoing waves are interchanged and we find that the scattering matrix also relates the vectors

$$\begin{pmatrix} B_N^* \\ A_1^* \end{pmatrix} = S \begin{pmatrix} B_1^* \\ A_N^* \end{pmatrix} = \begin{pmatrix} S_{11} & S_{12} \\ S_{21} & S_{22} \end{pmatrix} \begin{pmatrix} B_1^* \\ A_N^* \end{pmatrix} \quad . \tag{10.23}$$

By complex conjugation we find

$$\begin{pmatrix} B_N \\ A_1 \end{pmatrix} = S^* \begin{pmatrix} B_1 \\ A_N \end{pmatrix} = \begin{pmatrix} S_{11}^* & S_{12}^* \\ S_{21}^* & S_{22}^* \end{pmatrix} \begin{pmatrix} B_1 \\ A_N \end{pmatrix} \tag{10.24}$$

or, by rearranging,

$$\begin{pmatrix} A_1 \\ B_N \end{pmatrix} = \begin{pmatrix} S_{22}^* & S_{21}^* \\ S_{12}^* & S_{11}^* \end{pmatrix} \begin{pmatrix} A_N \\ B_1 \end{pmatrix} \quad . \tag{10.25}$$

Putting this into the form of (10.17) we have

$$\begin{pmatrix} A_N \\ B_1 \end{pmatrix} = \begin{pmatrix} S_{22}^* & S_{21}^* \\ S_{12}^* & S_{11}^* \end{pmatrix}^{-1} \begin{pmatrix} A_1 \\ B_N \end{pmatrix} \quad . \tag{10.26}$$

So, by comparison, we find the relation

$$\begin{pmatrix} S_{11} & S_{12} \\ S_{21} & S_{22} \end{pmatrix}^{-1} = \begin{pmatrix} S_{22}^* & S_{21}^* \\ S_{12}^* & S_{11}^* \end{pmatrix} \quad . \tag{10.27}$$

Because of the unitarity of the S matrix we have $S^{-1} = S^+$ and thus

$$\begin{pmatrix} S_{11}^* & S_{21}^* \\ S_{12}^* & S_{22}^* \end{pmatrix} = \begin{pmatrix} S_{22}^* & S_{21}^* \\ S_{12}^* & S_{11}^* \end{pmatrix} \quad , \tag{10.28}$$

so that time reversal is equivalent to

$$S_{11} = S_{22} \quad . \tag{10.29}$$

The time-reversal invariant S matrix has the particular form

$$S = \begin{pmatrix} S_{11} & S_{12} \\ S_{21} & S_{11} \end{pmatrix} \quad . \tag{10.30}$$

For real potentials, time reversal invariances holds true. Thus (10.21, 10.22, 10.29) show that the transmission coefficients for right-moving and left-moving incoming waves are equal in this case. Just as a side remark we note that *space-reflection* symmetry reduces S further to

$$S_{12} = S_{21} \tag{10.31}$$

so that the spatial-reflection invariance leads to

$$S = \begin{pmatrix} S_{11} & S_{12} \\ S_{12} & S_{11} \end{pmatrix} \quad . \tag{10.32}$$

10.2.3 Diagonalization of the Scattering Matrix

We return to the time-reversal invariant form (10.30) and investigate the restrictions of unitarity (10.20):

$$\begin{matrix} S_{11}S_{11}^* + S_{12}S_{12}^* = 1 & S_{11}S_{21}^* + S_{12}S_{11}^* = 0 \\ S_{21}S_{11}^* + S_{11}S_{12}^* = 0 & S_{21}S_{21}^* + S_{11}S_{11}^* = 1 \end{matrix} \quad , \tag{10.33}$$

which represent only two independent relations. The off-diagonal relations yield

$$S_{21} = -\frac{S_{11}}{S_{11}^*}S_{12}^* \quad , \tag{10.34}$$

i.e., the second off-diagonal element is determined by the first. Thus the only relation remaining besides (10.34) is

$$S_{11}S_{11}^* + S_{12}S_{12}^* = 1 \quad , \tag{10.35}$$

which can be solved by

$$S_{11} = e^{2is_{11}}\cos\sigma \quad , \quad S_{12} = e^{2is_{12}}\sin\sigma \quad , \tag{10.36}$$

yielding

$$S_{21} = e^{2is_{21}}\sin\sigma \quad , \quad s_{21} = 2s_{11} - s_{12} + \pi/2 \quad , \tag{10.37}$$

with real phases s_{11}, s_{12}, s_{21} and the real angle σ. Since S is a unitary matrix it can be diagonalized by a unitary transformation

$$S^D = U^+ S U \quad , \tag{10.38}$$

or, equivalently

$$S = U S^D U^+ \quad . \tag{10.39}$$

The *diagonalized S matrix* turns out to be

$$S^D = \begin{pmatrix} S_1^D & 0 \\ 0 & S_2^D \end{pmatrix} = \begin{pmatrix} e^{2i\delta_1} & 0 \\ 0 & e^{2i\delta_2} \end{pmatrix} \tag{10.40}$$

with the scattering phases δ_1 and δ_2 determined by

$$\delta_1 = s_{11} + \frac{1}{2}\sigma \quad , \quad \delta_2 = s_{11} - \frac{1}{2}\sigma \quad . \tag{10.41}$$

It is unitary itself, i.e.,

$$S_1^{D*}S_1^D = 1 \quad , \quad S_2^{D*}S_2^D = 1 \quad . \tag{10.42}$$

The unitary matrix U diagonalizing S has the form

$$U = \frac{1}{\sqrt{2}}\begin{pmatrix} -ie^{-2is_{11}} & ie^{-2is_{11}} \\ e^{-2is_{12}} & e^{-2is_{12}} \end{pmatrix} \quad . \tag{10.43}$$

The matrix elements of the S matrix can be expressed in terms of S_1^D and S_2^D by

$$S_{11} = \tfrac{1}{2}(S_1^D + S_2^D) \quad , \qquad S_{12} = \tfrac{1}{2i}e^{-2i(s_{11}-s_{12})}(S_1^D - S_2^D) \quad ,$$

$$S_{21} = -\tfrac{1}{2i}e^{2i(s_{11}-s_{12})}(S_1^D - S_2^D) \quad , \qquad S_{22} = S_{11} \quad . \tag{10.44}$$

10.2.4 Argand Diagrams

We decompose the S matrix elements into real and imaginary parts

$$S_{ik} = \varrho_{ik} + i\,\sigma_{ik} \quad \text{and} \quad S_i^D = \varrho_i^D + i\,\sigma_i^D \quad . \tag{10.45}$$

From

$$\frac{1}{2}S_1^D = S_{11} - \frac{1}{2}S_2^D \quad \text{and} \quad S_1^D S_1^{D*} = 1 \tag{10.46}$$

we have

$$\left(\varrho_{11} - \frac{1}{2}\varrho_2^D\right)^2 + \left(\sigma_{11} - \frac{1}{2}\sigma_2^D\right)^2 = \frac{1}{4} \quad . \tag{10.47}$$

This represents the equation of the curve of $S_{11}(k) = \varrho_{11}(k) + i\,\sigma_{11}(k)$ in an *Argand plot* in the complex plane. Here, the wave number k of the incident wave plays the role of the curve parameter.

Whenever the transmission amplitude S_{11} has the absolute square $|S_{11}|^2 = 1$, because of (10.35) the reflection amplitude S_{12} vanishes so that the wave numbers k_i of the intersections of the curve $S_{12}(k_i)$ with the origin of the Argand plot are the transmission resonances.

If $\varrho_2^D(k)$ and $\sigma_2^D(k)$ are slowly varying with k in a range where $\varrho_{11}(k)$ and $\sigma_{11}(k)$ are quickly varying functions[1] the above equation describes a circle of radius $\frac{1}{2}$ about a center with the coordinates $\frac{1}{2}\varrho_2^D, \frac{1}{2}\sigma_2^D$. These coordinates define a point on the circle

$$\frac{1}{4}\left(\varrho_2^D\right)^2 + \frac{1}{4}\left(\sigma_2^D\right)^2 = \frac{1}{4}S_2^{D*}S_2^D = \frac{1}{4} \tag{10.48}$$

of radius $\frac{1}{2}$ about the origin of the complex plane. By the same argument in a range in which $\varrho_1^D(k)$, $\sigma_1^D(k)$ are slowly changing whereas $\varrho_{11}(k)$, $\sigma_{11}(k)$ are quickly varying we have instead

$$\left(\varrho_{11} - \frac{1}{2}\varrho_1^D\right)^2 + \left(\sigma_{11} - \frac{1}{2}\sigma_1^D\right)^2 = \frac{1}{4} \quad , \tag{10.49}$$

i.e., a circle of radius $\frac{1}{2}$ about the center $\frac{1}{2}\varrho_1^D, \frac{1}{2}\sigma_1^D$. The center itself moves slowly on a circle of radius $\frac{1}{2}$ about the origin.

The absolute square of the off-diagonal element S_{21} is given by the unitarity relation, see (10.33), (10.36),

$$S_{21}S_{21}^* = 1 - S_{11}S_{11}^* = 1 - \cos^2\sigma = \sin^2\sigma \quad . \tag{10.50}$$

In terms of the real and imaginary parts

$$S_{21} = \varrho_{21} + i\,\sigma_{21} \quad ; \tag{10.51}$$

this is equivalent to

$$\varrho_{21}^2 + \sigma_{21}^2 = \sin^2\sigma \quad . \tag{10.52}$$

---

[1] In this range $\varrho_1^D(k)$, $\sigma_1^D(k)$ are quickly varying because of $S_{11} = (1/2)(S_1^D + S_2^D)$.

The behavior of the reflection coefficient S_{21} is read off the equation (10.37), (10.44)

$$S_{21} - \frac{1}{2} e^{2i(s_{21}-s_{11}+\pi/4)} S_2^D = \frac{1}{2} e^{2i(s_{21}-s_{11}-\pi/4)} S_1^D \quad , \tag{10.53}$$

which can be rewritten using the explicit form of S_1^D and S_2^D, (10.40, 10.41),

$$S_{21} - \frac{1}{2} e^{i(2s_{21}-\sigma+\pi/2)} = \frac{1}{2} e^{i(2s_{21}+\sigma-\pi/2)} \tag{10.54}$$

or in terms of the real and imaginary parts (10.51) as ($r_{21} = 2s_{21} - \sigma + \pi/2$)

$$\left[\varrho_{21} - \frac{1}{2} \left(\varrho_2^D \cos 2r_{21} - \sigma_2^D \sin 2r_{21} \right) \right]^2$$
$$+ \left[\sigma_{21} - \frac{1}{2} \left(\varrho_2^D \sin 2r_{21} + \sigma_2^D \cos 2r_{21} \right) \right]^2 = \frac{1}{4} \quad . \tag{10.55}$$

Again, this is the equation of a circle in the complex $\varrho_{21}\sigma_{21}$ plane about a center at $\exp[i(2s_{21} - \sigma + \pi/2)]$ in the complex plane if this center only slowly varies with k.

10.2.5 Resonances

The nonvanishing matrix elements (10.40), (10.41)

$$S_1^D = e^{i(2s_{11}+\sigma)} \quad \text{and} \quad S_2^D = e^{i(2s_{11}-\sigma)} \tag{10.56}$$

of the diagonalized S matrix lie on the unit circle in the complex plane. For varying wave number k the pointers represented by $S_1^D(k)$ or $S_2^D(k)$ move on the unit circle.

A *resonance* phenomenon occurs whenever one of the two matrix elements $S_1^D(k)$ or $S_2^D(k)$ moves through a large part of the unit circle in a small interval to both sides of the wave number k_r at resonance. For the scattering phases δ_1, δ_2 of the diagonalized S matrix this means a fast increase by an angle close to π.

For definiteness we assume that the element $S_1^D(k)$ exhibits this fast variation in the interval

$$k_r - \kappa < k < k_r + \kappa \tag{10.57}$$

surrounding the resonance at k_r, whereas the other element $S_2^D(k)$ remains practically unchanged in this region.

$$S_2^D(k) \approx S_2^D(k_r) = e^{2i\delta_2(k_r)} \quad , \tag{10.58}$$

with

$$\delta_2(k_r) = s_{11}(k_r) - \frac{1}{2}\sigma(k_r) = \delta_{2r} \quad . \tag{10.59}$$

Here δ_{2r} is the phase at the wave number k_r.

Under these assumptions we find for

$$S_{11}(k) = \frac{1}{2} S_1^D(k) + \frac{1}{2} S_2^D(k_r) \tag{10.60}$$

the behavior

$$S_{11} = \frac{1}{2}e^{2i\delta_1(k)} + \frac{1}{2}e^{2i\delta_{2r}} = e^{2is_{11}}\cos\sigma \quad . \tag{10.61}$$

This represents a circle of radius $\frac{1}{2}$ about the center $\frac{1}{2}e^{2i\delta_{2r}}$, see Fig. 4.5. From the condition of a constant phase δ_{2r} we conclude that

$$\sigma(k_r) = 2[s_{11}(k_r) - \delta_{2r}] \tag{10.62}$$

and find that

$$S_{11} = e^{2is_{11}}\cos 2[s_{11}(k_r) - \delta_{2r}] \quad . \tag{10.63}$$

This shows that for $s_{11}(k_r) = \delta_{2r} = \delta_2(k_r)$ the circle of radius $\frac{1}{2}$ about the center $\exp(2i\delta_{2r})/2$ touches the unitarity circle of radius 1 about the origin of the complex plane.

Thus the wave number k_r of the resonance of the diagonal element S_1^D also determines the resonance of the element S_{11}.

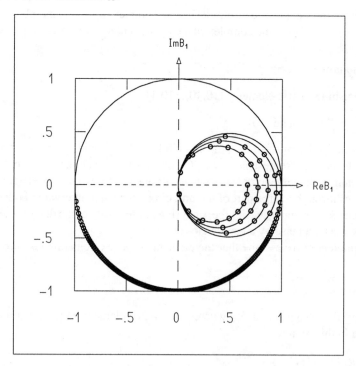

Fig. 10.1. Argand plot of the scattering-matrix element $S_{21} = B_1$ for the physical situation of Fig. 4.5.

The fast variation of the phase δ_1 over the range $0 < \delta_1 < \pi$ in the neighborhood of a resonance leads to a change of the phase s_{11} of the matrix element S_{11} passing quickly through the value $\delta_{2r} = \delta_2(k_r)$. The size of the jump at k_r in the phase δ_1 depends on how completely the curve $S_1^D(k)$ overlaps with the unit circle in the neighborhood of k_r. For a full circle of $S_1^D(k)$ the phase $\delta_1(k)$ increases by $\pi/2$ from $\delta_2(k_r) - \frac{\pi}{4}$ to $\delta_2(k_r) + \frac{\pi}{4}$. For a half circle of $S_1^D(k)$ the phase $s_{11}(k)$ of $S_{11}(k)$ changes by $\frac{\pi}{4}$ from $\delta_{2r} - \frac{\pi}{8}$ to $\delta_{2r} + \frac{\pi}{8}$.

The reflection coefficient exhibits a resonance for minimal transmission. According to (10.36)

$$|S_{21}|^2 = \sin^2 \sigma \quad ;$$

this means $\sigma \simeq \pi/2$ and $|S_{21}|^2 \leq 1$. For simple square-well potentials the approximate size of the phase s_{21} of S_{21} can be determined by coarse arguments. As an example we take the single repulsive square well, already discussed in (4.49). For a kinetic energy low compared to the height V_0 of the repulsive potential, no transmission ($A_2 = 0$), i.e., only reflection with a phase shift π occurs. Thus, for $A_1 = 1$, this means that the reflection coefficient $B_1 = S_{21}$ starts at $B_1 = -1$ in the Argand plot. As long as transmission remains negligible, i.e., for $E < V_0$, the coeffcient B_1 keeps the absolute value $|B_1| = 1$ and thus moves on the unit circle in the Argand plot. For $E = V_0$ reflection at a "thinner medium" (see Sect. 4.1) occurs causing no phase shift, (4.42), (4.43) so that $S_{21} = B_1 = 1$. For growing energies $E > V_0$, the first transmission resonance at $k_2 = \ell\pi/d$ (4.52) is soon reached, so that there is no reflection, i.e., $S_{21} = B_1 = 0$. As soon as the energy is further increased, transmission goes through a minimum before it reaches the next resonance. This minimum occurs for reflection at a denser medium, so that no phase shift occurs, $S_{21} = 0$ and $S_{21} = B_1 = \sin\sigma$, $\sigma \approx \pi/2$, so that $S_{21} \approx 1$. Further increasing the energy leads to a spiral about a point close to $1/2$ in the Argand plot, Fig. 10.1.

The determination of the Argand plot of more complicated potentials requires explicit calculation of the S-matrix elements from the solutions of the Schrödinger equation.

Further Reading

Brandt, Dahmen: Chaps. 5,11
Gasiorowicz: Chaps. 24
Merzbacher: Chap. 6
Messiah: Vol. 1, Chap. 3

10.3 Hints and Answers to the Exercises

Hints and Answers to the Exercises in Chapter 2

For those exercises of Chap. 2, in which physical quantities are given with units use the Data-Table Units, see Sect. 10.1.5.

2.3.2 The quantum-mechanical wave packet widens as time passes, whereas the optical wave packet keep its initial width.

2.3.3 The group velocity of this wave packet is twice as high as in Exercise 2.3.1.

2.3.4 The group velocity of the optical wave packet remains unchanged. The vacuum speed of light does not depend on the wave number.

2.3.5 (c) The square of the wave function is a probability distribution of the position of the particle. Its expectation values $\langle x \rangle = x_0 + p_0\Delta t/m$ follow the trajectory of free motion with the velocity $v_0 = p_0/m$. If the particle position after the time interval Δt is measured to be larger than its expectation value $\langle x \rangle = x_0 + p_0\Delta t/m$ at time Δt, its momentum p must have been larger than the momentum expectation value p_0. Thus the wavelength $\lambda = h/p$ in the spatial region in front of $\langle x \rangle$ must be smaller than $\lambda_0 = h/p_0$. For the spatial region $x < \langle x \rangle$ the analogous arguments result in $\lambda > \lambda_0$.

2.3.7 The wavelength is halved because of the doubling of the momentum.

2.3.9 The wave packet widens as time passes. The phase velocity of the de Broglie waves is $p/2m$. Thus the waves superimposed to form the wave packet move with different phase velocities so that the width of the wave packet spreads in time.

2.3.10 As time passes the shape of the wave packet remains unaltered. The phase velocity of electromagnetic waves in the vacuum is independent of their frequencies. Therefore all waves move with the same speed as light in vacuum and the wave packet does not change its shape.

2.3.11 (d) $p_0 = 5.685 \times 10^{-12}\,\mathrm{meVsm}^{-1}$. (e) The momentum width $\sigma_p = 0.5 \times 10^{-12}\,\mathrm{eVsm}^{-1}$ is relatively small compared to the momentum p_0. Thus the spreading of the phase velocities is small. Therefore the dispersion of the wave packet with time is small. (f) The unit at the x axis is mm.

2.3.12 (d) $p_0 = 17.06 \times 10^{-12}\,\mathrm{eVsm}^{-1}$.

2.3.13 (d) $p_0 = 12\,\mathrm{eVsm}^{-1}$. (e) See Exercise 2.3.11 (f). (f) The momentum expectation value is $p_0 = 12$, the momentum width of the wave packet is $\sigma_p = 9$. Thus, in contrast to descriptor 7, the Gaussian spectral function extends also into the range of negative momenta. Therefore the wave packet contains contributions propagating to the left. (g) The small wavelengths to the right belong to high momenta of a particle moving right. The large wavelengths to the left of the center indicate that a particle found at these x values is at rest. The small wavelengths to the left indicate a particle with some momentum pointing to the left.

2.4.4 (b) $p_0 = 5.685 \times 10^{-12}\,\mathrm{eVsm}^{-1}$ and $p_0 = 17.06 \times 10^{-12}\,\mathrm{eVsm}^{-1}$. (c) The physical unit at the abscissa is $10^{-12}\,\mathrm{eVsm}^{-1}$. (d) $E_0 = 2.843 \times 10^{-12}\,\mathrm{eV}$, $E_0 = 25.58 \times 10^{-12}\,\mathrm{eV}$.

2.4.5 (b) $p_0 = 3.372 \times 10^{-6}\,\mathrm{eVsm}^{-1}$ and $p_0 = 5.84 \times 10^{-6}\,\mathrm{eVsm}^{-1}$. (c) $10^{-6}\,\mathrm{eVsm}^{-1}$. (d) $v_0 = 0.5932 \times 10^6\,\mathrm{ms}^{-1}$ and $v_0 = 1.027 \times 10^6\,\mathrm{ms}^{-1}$. (e) The order of the quotient v^2/c^2 is equal to 10^{-4}.

2.4.6 (b) $p_0 = 10.44 \times 10^{-9}\,\mathrm{eVsm}^{-1}$ and $p_0 = 31.32 \times 10^{-9}\,\mathrm{eVsm}^{-1}$. (c) $10^{-9}\,\mathrm{eVsm}^{-1}$. (d) $E_0 = 5.22 \times 10^{-9}\,\mathrm{eV}$ and $E_0 = 93.96 \times 10^{-9}\,\mathrm{eV}$.

2.4.7 (b) $p_0 = 1.370 \times 10^{-3}\,\mathrm{eVsm}^{-1}$ and $p_0 = 2.373 \times 10^{-3}\,\mathrm{eVsm}^{-1}$. (c) $10^{-3}\,\mathrm{eVsm}^{-1}$. (d) $v_0 = 1.460 \times 10^6\,\mathrm{ms}^{-1}$ and $v_0 = 2.529 \times 10^6\,\mathrm{ms}^{-1}$. (e) The order of magnitude of the quotient v^2/c^2 is 10^{-4}.

2.5.4 The sum (2.3) over the contributions of the harmonic waves $\psi_{p_n}(x,t)$, (2.1), with the momentum vectors $p_n = p_0 - 3\sigma_p + (n-1)\Delta p$, see Sect. 2.5, is a Fourier series. It is periodic in space. The wavelength of this periodicity is $\Lambda = 2\pi\hbar/(\Delta p)$. The wave patterns constituting the wave packet reoccur along the x axis at distances Λ. For more densely spaced p_n, i.e., smaller Δp the reoccurrences are spaced more widely. Only the Fourier integral (2.4) yields the Gaussian wave packet without periodic repetitions.

Hints and Answers to the Exercises in Chapter 3

For those exercises of Chap. 3 in which physical quantities are given in terms of physical units use the Data-Table Units, see Sect. 10.1.5.

3.2.2 For an infinitely deep square-well potential symmetrically placed with respect to the origin of the coordinate frame, the expectation values of position and momentum vanish.

3.2.3 (a) The values of the wave functions $\varphi_n^{(\pm)}(-d/2)$ and $\varphi_n^{\pm}(d/2)$ do not vanish. Thus they violate the boundary conditions to be satisfied by solutions of the Schrödinger equation with an infinitely deep square-well potential. (b) The eigenfunctions $\varphi_n(x)$ are standing waves resulting as even and odd superpositions of the right-moving and left-moving waves $\varphi_n^{(\pm)}(x)$.

3.2.4 (c) Using $\Delta p \Delta x \geq \hbar/2$ and $E = (\Delta p)^2/2m$ for the kinetic energy of the ground state we get $E \geq 1.06 \times 10^{-3}$ eV.

3.2.6 (a) The energy of the ground state is $E_1 = 14.9$ eV. (b) The value of the Coulomb potential at the Bohr radius $a = \hbar c/\alpha c^2$ is $V(a) = -\alpha^2 c^2 = -27.2$ eV. The sum of this value and E_1 is -12.3 eV. The binding energy of the hydrogen ground state is -13.65 eV. The order of magnitude of the true ground-state energy in hydrogen and the rough estimate is the same.

3.2.7 (g) The unit of the energy scale is eV. (h) The spring constants are for (a) 5.685, (b) 12.97, (c) 22.74 eV(nm)^2. (i) The unit of the x scale is nanometers (nm). (j) Close to the wall of the harmonic-oscillator potential the speed of the particle is lowest; thus its probability density is largest here.

3.2.8 (c) The angular frequencies for the two spring constants are $\omega = 0.3095 \times 10^4$s^{-1} and $\omega = 0.3095 \times 10^6$s^{-1}. (f) $\sigma_0 = 0.4513 \times 10^{-5}$m and $\sigma_0 = 0.4513 \times 10^{-6}$m.

3.3.1 (c) The energy eigenvalues are $E_1 = -5.7$, $E_2 = -5.1$, $E_3 = -3.9$, $E_4 = -2.3$, $E_5 = -0.6$. The differences $\Delta_i = E_i - V_0$ are $\Delta_1 = 0.3$, $\Delta_2 = 0.9$, $\Delta_3 = 2.1$, $\Delta_4 = 3.7$, $\Delta_5 = 5.4$.

3.3.2 (c) The energy eigenvalues for the width $d = 2$ are $E_1 = -5.3$, $E_2 = -3.2$, $E_3 = -0.4$; the differences Δ_i are $\Delta_1 = 0.7$, $\Delta_2 = 2.8$, $\Delta_3 = 5.6$. For the width $d = 6$ we find $E_1 = -5.9$, $E_2 = -5.6$, $E_3 = -5$, $E_4 = -4.2$, $E_5 = -3.2$, $E_6 = -2$, $E_7 = -0.7$; the differences Δ_i are $\Delta_1 = 0.1$, $\Delta_2 = 0.4$, $\Delta_3 = 1$, $\Delta_4 = 1.8$, $\Delta_5 = 2.8$, $\Delta_6 = 4$, $\Delta_7 = 5.3$.

3.3.3 (a) The energy eigenvalues of the stationary wave functions in an infinitely deep square well are given by (3.12). For $d = 2$ the eigenvalues are $E_1' = 1.23$, $E_2' = 4.93$, $E_3' = 11.10$, etc.; for $d = 4$ they are $E_1' = 0.31$, $E_2' = 1.23$, $E_3' = 2.78$, $E_4' = 4.93$, $E_5' = 7.71$ etc.; for $d = 6$ they are $E_1' = 0.14$, $E_2' = 0.55$, $E_3' = 1.23$, $E_4' = 2.19$, $E_5' = 3.43$, $E_6' = 4.93$, $E_7' = 6.72$. (b) The differences Δ_i of the energy eigenvalues to the depth V_0 of the potential are smaller than the eigenvalues of the stationary states in the corresponding potential well of the same width. The separation of the energy eigenvalues are also smaller than for the infinitely deep square well. (c) The eigenfunctions in the infinitely deep square well vanish at the edges of the potential well whereas the eigenfunctions in the well of finite depth extend also into regions outside the well. Their curvature $d^2\varphi/dx^2$ is smaller than in the infinitely deep square well, and thus the contribution of the kinetic energy term $T = -(\hbar^2/2m)d^2\varphi/dx^2$ is smaller for the well of finite depth. This difference between the two potentials becomes the more prominent the larger the eigenvalue E_i is.

3.3.4 (c) In the central region of the potential the eigenvalue is equal to the value of the potential. Therefore the contribution of the kinetic-energy term T in the Schrödinger equation vanishes. Thus the second derivative of the wave function vanishes in the central region of the potential so that it has to be a straight line. (d) The two lowest eigenfunctions in the double potential well can be closely approximated by a symmetric and an antisymmetric superposition of the two ground-state eigenfunctions of the two single potential wells constituting the double well. The two ground-state wave functions of the single wells have the same energy eigenvalues so that the difference in energy originates from the minor differences between the two wave functions. This explains the narrow spacing of their energy eigenvalues. (e) The double-well potential is symmetric about the point $x = 1.75$. Thus the Hamiltonian $H = T + V$ is symmetric under a reflection with respect to this point. Therefore the corresponding parity operator P commutes with H. As a consequence, the eigenfunctions of H have to be eigenfunctions of P as well, so that they have to be symmetric or antisymmetric.

3.3.5 (c) The potential of this exercise is no longer symmetric with respect to a point x.

3.3.6 (c) The energy eigenvalue of the ground state is the same as the value of the potential in the second well. The contribution of the kinetic energy T vanishes in the second well so that the second derivative $d^2\varphi/dx^2$ vanishes in this region. Therefore the wave function in this region is a straight line.

3.3.7 (e) The behavior of the eigenstates of a double-well potential with different wells is most easily discussed in terms of the eigenstates of the single wells. The lowest eigenvalues in the two single wells appear in the wide right well. The ground-state eigenvalue in the narrower left well is much higher. Thus the ground state of the double well is given by a wave function closely resembling the ground state in the single right well and is very small in the left well. The same argument is valid for the highest state. However, the ground state in the narrow well to the left has an eigenvalue not very much different from the first excited state with one node of the wide well to the right. Two linear combinations can be formed out of them. One with a number of nodes equal to the sum of the number of nodes of the two wave functions (in this case altogether one) and one with an additional node located in the region between the two wells (in this case altogether two nodes). The combination with the lower number of nodes possesses a lower (average) curvature, i.e., at average a lower second spatial derivative. Thus its kinetic energy is lower and so is the corresponding eigenvalue of the energy. The linear combination with one more node has a higher kinetic energy and thus its energy eigenvalue is slightly higher. In the process of increasing the width of the second well the distance between the two eigenvalues of the single wells decreases. The eigenvalues of the two eigenstates of the single wells cross each other. For a narrower right well the nodeless eigenstate of the left well has a lower eigenvalue. During the widening of the second well the eigenvalue of its first excited state with one node decreases so that, eventually, it becomes lower than the eigenvalue of the nodeless ground state in the left well. This process is called *level crossing*. However, since the eigenstate for the double well with the lower number of nodes stays related to the lower energy eigenvalue, there is no level crossing for this energy eigenvalue with the one of the state with one more node. During the widening of the second well the eigenvalues of the two states of the double well at first approach each other up to a minimum distance. With further broadening of the right well the eigenvalues of the two states depart again from each other.

3.3.9 (k) The single potential well (a) contains two bound states. The assembly of the nodeless degenerate ground states in the r single wells splits into a band of the r nondegenerate states with no node within the single potential wells. The assembly of the first excited states with one node within the single potential wells analogously splits into a band of states with one node within the single wells. The energy separation of the states within

the bands is small compared to the energy separation between the ground state and the
first excited state in a single well.

3.3.10 (e) The states in the r single wells constituting the quasiperiodic potential are degenerate. They are linearly independent and can be superimposed into linear combinations which approximate the eigenstates of the quasiperiodic potential. The superpositions follow the node structure of the eigenstates of the wide single-well potential which remains after all the inner walls have been removed from the quasiperiodic potential. The eigenstates of the wide potential well are narrowly spaced with energy eigenvalues becoming larger as the number of nodes of the eigenstates grows. The nodeless ground states of the single wells are combined to wave functions with $0, 1, \ldots, r-1$ nodes. These states, if they have an even number of nodes, are symmetric with respect to the center of the quasiperiodic potential. The ones with an odd number of nodes are antisymmetric. They form the lowest band of states in the quasiperiodic potential. Their energy separation corresponds to the small energy difference between the states in the wide single-well potential that contains all the narrow wells of the quasiperiodic potential. The eigenstates of the quasiperiodic potential with $r, r+1, \ldots, 2r-1$ nodes are composed of the r one-node first excited states of the single wells and from the second band. The separation in energy of the two bands corresponds to the energy separation of the ground and first excited state of the single well and thus is much larger than the separation of the states within the lowest band.

3.3.11 (e) The lowest band of 10 states splits into a set of 8 states narrowly spaced in energy and another of two states closely spaced. The reason for the separation into two sets is that the states in the outermost wells possess a somewhat larger energy since their curvature is larger because of the higher walls at the right and left boundary of the quasiperiodic potential. Since the states in the outer wells have a somewhat larger energy, their superposition with the ground states of the inner wells is less pronounced. Thus the two states locate the particle mainly close to the right and left boundary of the potential. The two states are narrowly spaced because they are again superpositions of wave functions that are mainly different from zero close to the left and right boundary. The lower one is the symmetric, the higher of the two the antisymmetric, superposition of the two wave functions that are mainly different from zero either at the left or at the right boundary. The two eigenstates correspond to the *surface states* in a crystal of finite extension.

3.4.1 (d) The period of the wave function is $T = 4$. (e) The period of the absolute square is $T = 2$. (f) $\psi(x, t+T) = e^{i\alpha}\psi(x, t)$.

3.4.2 (d) The wave packets plotted are Gaussian so that for the initial state we have $\Delta x \Delta p = \hbar/2$. A wave packet initially wide in x is thus narrow in momentum. The period of a classical particle in a harmonic-oscillator potential is independent of its amplitude. Independent of its actual position in the spatially wide initial state the particle will reach the point of the potential minimum after the same time interval, since its momentum spread about the initial value $p_0 = 0$ is small. Thus the width of the spatial distribution will shrink and assume its minimal width at the center position in the oscillator. While moving towards the other turning point the distribution widens in space to the same shape as in the initial state. An initially narrow Gaussian wave packet has a wide momentum distribution. A classical particle the initial position of which is described by a probability distribution will reach the center position of the oscillator after a time interval in general dependent on the initial position and momentum. Since the original momentum probability distribution is wide, the spatial probability distribution is wide at the center of the harmonic oscillator. The wave packet shrinks to its original shape at the other turning point. The wave function observed in (c) describes the coherent state with an initial width equal to that of the ground state.

3.4.3 (d) The Gaussian wave packet, which is initially wide in space, possesses a narrow momentum distribution about $p_0 = 0$. If the particle is at an off-center position, the harmonic force drives it back towards the center. Independent of its initial position, the particle reaches the center after the same time interval if its initial momentum is zero. Since there is a small initial momentum spread, the probability distribution does not oscillate between its initial width and zero width but between a maximum and a minimum value.

3.4.4 (d) The initial state at the center position of the oscillator has relative width $f_\sigma = 1$. It is therefore the ground state of the oscillator and is stationary. Its probability distribution is time independent and so is its width.

3.5.1 (d) The phase velocity of a right-moving or left-moving wave in the deep square well is $v_{pi} = \hbar k_i/2m$. It depends on the wave number k_i. A wave packet being a superposition of stationary wave functions of different k_i is thus made up of waves of different phase velocity. Therefore it widens as time elapses. (e) The wiggly patterns occurring when the wave packet is close to the wall are an interference effect between the incident and the reflected wave. As soon as the wave packet becomes wide enough the wiggly patterns occur everywhere in the deep well.

3.5.2 (n) The wave number of the lowest state in the infinitely deep square well is $k = \pi/d = 0.31416$ for $d = 10$. The corresponding temporal period is $T_1 = 2\pi/\omega$ with $\omega = E/\hbar$ and $E = \hbar^2 k^2/(2m)$. Thus we find $T_1 = 4\pi m/(\hbar k^2) = 127.32$. (o) At the time $T = T_1/2$ the wave packet looks the same as the initial wave packet. (p) At the beginning of the motion the wave packet is so narrow that it does not touch the walls of the deep square well. Therefore the classical particle position and the expectation value of the wave packet coincide as in a free wave packet. As soon as the wave packet hits the wall it consists of incident and reflected parts. Its position expectation value cannot reach the wall because of the width of the packet. Thus it stays behind the classical expectation value. (q) Because of the dispersion, the wave packet becomes so wide that both ends are reflected at the left-hand and right-hand wall. It consists of right-moving and left-moving parts with equal probability. Thus its momentum expectation value vanishes. It also spreads over the whole width of the square well with – at average – equal probability density. Thus the position expectation value equals zero.

3.5.3 (d) The wave function possesses the period $T = T_1$ as calculated in Exercise 3.5.1 (n). Thus the wave packet has reassumed its original shape at the initial time. It represents a wave packet narrow compared to the width of the well. The classical particle follows a periodic motion between the walls of the well with a classical period $2d/v = 2dm/p = 3.858$. After 33 classical periods the time T_1 has elapsed and the classical position and the expectation value are at the same point. Since the absolute square of the wave packet reassumes its original well-localized shape after half the time T_1, the coincidence of the classical position and the expectation value occurs at $T_1/2$.

3.5.4 (b) The system of a point particle in an infinitely deep square well is conservative. Thus the expectation value of the energy is constant and thus all the time equal to its initial value. (c) The momentum expectation value of the particle is practically zero in periods where the wave packet fills the whole well and its position expectation value is at rest.

3.5.5 (h) The momentum of the particle, as well as its mass, have been increased by a factor of ten leaving the velocity unchanged. However, because of the larger mass, the wave packet disperses much more slowly than in Exercise 3.5.4. Thus it oscillates more periods as a localized wave packet between the two walls of the square well.

Hints and Answers to the Exercises in Chapter 4

4.2.1 (d) The kinetic energy of the third wave function is zero in the region of the repulsive square-well potential. Since the kinetic energy T is proportional to $d^2\psi/dx^2$, only the first derivative of the wave function can be different from zero in the region of the square well. (e) The trend of the transmission coefficient A_3 is most easily read off the plot of the absolute square (c). The transmission coefficient increases with increasing energy. For kinetic energies $E < 2$ the tunnel effect increases as the difference between the barrier height and the kinetic energy decreases. For energies above the barrier height the reflection decreases.

4.2.2 (d) The tunnel probability increases exponentially as the barrier width decreases. (e) There is only an outgoing wave of the type $A_N \exp(ik_N x)$, $N = 3$, in the region N beyond the potential, see (4.18). Its absolute square is just the constant $|A_N|^2$. (f) In the region 1 the incident and reflected wave are superimposed, see (4.22). We use the representation $A_1' = |A_1'|\exp(i\alpha_1)$, $B_1' = |B_1'|\exp(i\beta_1)$. Then the absolute square of the superposition yields $|A_1'|^2 + |B_1'|^2 + 2|A_1'||B_1'|\cos(2k_1 x + \alpha_1 - \beta_1)$. The third member is the interference term between incident and reflected waves. It exhibits twice the original wave number k_1, i.e., half the wavelength of the incident wave.

4.2.3 (d) The energy at which the wave pattern in the absolute square is smallest is $E = 1.18$. The wavelength of the incident wave is $\lambda = 2\pi/k = 2\pi/\sqrt{2mE} = 1.54$. Thus half a wavelength fits roughly into the interval of the two walls forming the double barrier. Therefore the incident and reflected wave interfere destructively and the interference pattern of the two vanishes almost.

4.2.4 (d) The width of the region of the repulsive double-well potential is $d = 2$. A nodeless wave function fitting into the well exactly has a wave number $k = \pi/d = \pi/2$. The corresponding kinetic energy is $E = (\hbar k)^2/2m$. With the values $\hbar = 1$, $m = 1$, we get $E = 1.2$, a value very close to the observed $E = 1.18$ of the resonance wave function without a node. (e) Any other wave function fitting into the region of the double-well potential has one or more nodes and thus a higher wave number and a higher kinetic energy. Thus there is no resonance at an energy lower than the value $E = 1.12$.

4.2.5 (a) Resonances occur at $E = 0.6$ and $E = 2.0$. (f) The first resonance at $E = 0.53$ has no node whereas the second resonance at $E = 2.28$ possesses one node. Since the wave function with the node has a second derivative much larger than the one of the lowest resonance, its kinetic energy – being proportional to the second derivative – is larger. Thus, the total energy of the second resonance (with one node) is larger than the one of the lowest resonance. As a general rule one may say that the resonance energy increases with the number of nodes in the resonance wave function between the two barriers.

4.2.6 (a) First resonance $E_1 = 0.73$; large bump in the wave function in the left potential well. (b) Second resonance at $E_2 = 1.2$; large bump in the wave function in the right potential well. (c) Third resonance at $E_3 = 2.9$; wave function with two bumps, i.e., one node, in the left potential well. (d) Fourth resonance at $E_4 = 4.41$; wave function with two bumps, i.e., one node, in the right potential well. (e) Fifth resonance at $E_5 = 6.51$, wave function with three large bumps, i.e., two nodes, in the left potential well.

4.2.7 The maxima of the transmission coefficient occur at $E_1 = 0.73$, $E_2 = 1.2$, $E_3 = 2.95$, $E_4 = 4.5$, $E_5 = 6.52$.

4.2.8 (a) First resonance at $E_1 = 0.72$; no node. (c) Second resonance at $E_2 = 2.82$; one node. (g) Third resonance at $E_3 = 6.42$; two nodes.

4.2.9 The maxima of the transmission coefficient occur at $E_1 = 0.75$, $E_2 = 2.92$, $E_3 = 6.35$.

4.2.10 (b) First resonance at $E_1 = 1.1$. (e) Second resonance at $E_2 = 4.3$.

4.2.11 The maxima of the transmission coefficient occur at $E_1 = 1.15$, $E_2 = 4.4$.

4.2.12 (f) The trend of the differential resistance is understood as follows: (i) For a voltage $U \ll U_{res}$ far below the resonance voltage $U_{res} = V_1 - V_5 = 1$ the tunnel effect is very small; thus the square of the coefficient $|A_5|^2$ is small. Therefore the current through the quantum-well device is small. Also the variation $\Delta|A_5|^2$ for the given increase $\Delta U = 0.5$ from (a) to (b) is small. Thus the variation of the current ΔI associated with ΔU is small. Therefore the differential resistance $R = \Delta U/\Delta I = (1/\alpha)\Delta U/\Delta|A_5|^2$ is large. (ii) In the region close below the resonance voltage the absolute square $|A_5|^2$ grows quickly from (b) to (c) with the increase of the voltage $\Delta U = 0.5$ and thus the current ΔI; the differential resistance in the region close below the resonance region $R = \Delta U/\Delta I$ decreases quickly. (iii) In the region close above the resonance voltage (c,d) U_{res} the square $|A_5|^2$ of the coefficient A_5 decreases quickly from (c) to (d) for the given increase in voltage $\Delta U = 0.5$. Thus the corresponding change in current ΔI is negative, so that the differential resistance $R = \Delta U/\Delta I$ becomes negative above the resonance voltage. At the resonance voltage U_{res} the differential resistance goes through zero.

4.2.13 (f) The potential walls ($V_2 = 60$) are much higher in this case than in Exercise 4.2.12 ($V_2 = 7$). Therefore the resonance region is much more narrow than in Exercise 4.2.12. The variation of $|A_5|^2$ is much larger for the potential of Exercise 4.2.12. Therefore the differential resistance varies much faster with increasing voltage.

4.2.14 (f) The second transmission resonance is much wider than the first resonance. Thus the variation of the differential resistance is much slower.

4.2.15 (f) The discussion of the differential resistance follows the same lines of argument as in Exercise 4.2.12 (f).

4.2.17 (f) Because of the triple-wall structure of the potential, the transmission resonance is much more narrow than in the case of a double-wall potential as in Exercise 4.2.16 (f). Thus the variation of the differential resistance $R = \Delta U/\Delta I = -\Delta V_3/\Delta I$ is much faster.

4.3.1 (d) The answer is obvious from (4.48) and the accompanying text of Sect. 4.1.

4.3.2 (d) The answer is obvious from (4.45) and the accompanying text of Sect. 4.1.

4.3.3 (d) The answer is obvious from (4.42) and the accompanying text of Sect. 4.1.

4.3.4 (d) Upon taking the absolute square of a harmonic wave (2.1), see Sect. 2.1, the time-dependent factor $\exp(-iEt/\hbar)$ yields 1. Thus the absolute square of a harmonic wave looks the same as the absolute square of a stationary scattering wave function at the same momentum.

4.3.5 (d) To the right of the potential barrier there is an outgoing wave propagating to the right. It possesses a time-independent amplitude A_3. (e) To the left of the potential barrier the time-dependent wave function is a superposition of the incident and reflected waves. The real part of the wave function thus exhibits a time-dependent amplitude.

4.4.1 (d) The wave packet is reflected completely for $E = 2$, an energy smaller than the height of the step potential. For $E = 6.5$ the reflection has not ceased to occur even though the kinetic energy is larger than the height of the step potential so that a classical particle would only be transmitted. For $E = 8$ the transmission probability has grown. Upon increasing the energy even further, the reflection disappears. (e) The transmitted part of the wave packet is faster than the classical particle since its average energy corresponds to a value higher than the energy expectation value $E = 6.5$ of the incident wave packet. The potential step acts like a discriminator threshold.

4.4.3 (d) At the potential step at $x = 0$ the continuity conditions (3.34) are $A_1 + B_1 = A_2$, $k_1(A_1 - B_1) = k_2 A_2$. The solution is $A_2 = 2A_1 k_1/(k_1 + k_2)$, $B_1 = A_1(k_1 - k_2)/(k_1 + k_2)$, with $k_1 = \sqrt{2mE}/\hbar$ and $k_2 = \sqrt{2m(E - V_2)}/\hbar = \sqrt{2m(E + |V_2|)}/\hbar$. The modulus $|B_1|$ of the reflection coefficient is a monotonously increasing function of $|V_2|$.

4.4.4 $E_1 = 16.3$, $E_2 = 17.25$, $E_3 = 18.8$, $E_4 = 20.95$.

4.4.5 (c) In an analogy to optics the resonances occur because the reflection at a step increase of the potential corresponds to a reflection at a thinner medium whereas the subsequent reflection at a step decrease of the potential corresponds to a reflection on a denser medium. Thus, according to (4.52), resonances occur if the wavelength λ_2 of the wave function in region 2 is approximately equal to $\lambda_{2\ell} = 2d/\ell$, $\ell = 1, 2, 3, \ldots$. The corresponding energies are $E_\ell = V_2 + \ell^2\pi^2/(2d^2)$, see (4.54). The numerical values are $E_1 = 16.308$, $E_2 = 17.23$, $E_3 = 18.78$, $E_4 = 20.93$. (d) At resonance energy a large percentage of probability is stored in the region $0 \le x \le 4$. The resonance wave function decays exponentially in time after some time of the resonance formation. The exponential time dependence of the amplitude is equivalent to constant ratios of the amplitudes of the wave functions in the barrier region for equidistant instants in time.

4.4.7 (c) According to the formulae of Exercise 4.4.3 the reflection probability at the two down steps of the potential at $x = 0$ and $x = 4$ decreases with increasing energy $E = \hbar^2 k_1^2/2m$. Thus the exponential decay is faster for higher resonance energy. The uncertainty relation relates the lifetime τ and the width Γ of the resonance energy by $\tau \times \Gamma \approx h$. Thus for the shorter-lived higher resonance the energy width Γ increases. This is obvious from the Argand diagram of Exercise 4.4.4.

4.4.8 (f) The low-lying transmission resonances occur as narrow maxima in the absolute square $|A_N|^2$. They correspond to fast sweeps of the pointer A_N in the complex Argand plot through the regions at which the pointer A_N has its maximal length $|A_N| = 1$.

4.4.9 (f) The relation of the prominent features of A_N to the T-matrix element T_T is most easily understood using (4.34).

4.4.10 (f) Compare (4.32).

4.4.11 (f) Compare (4.34).

4.4.12 $E = 0.08$.

4.4.13 There is one bound state at $E = -3.2$.

4.4.14 (d) The wave function inside the potential region possesses one node. The energy $E = 0.1$ corresponds to the lowest resonance energy (see Exercise 4.4.12). There is no resonance without a node because of the existence of one nodeless bound state in this potential, see Exercise 4.4.13.

4.4.15 (f) $E_1 = 0.9$, $E_2 = 3.4$, $E_3 = 7.8$, $E_4 = 14.1$.

4.5.1 (a) The resonance energies read off $|A_N|^2$ are $E_1 = 21$, $E_2 = 35.0$. (f) See Exercise 4.4.8 (f) and Sect. 10.2. (g) $E_1 = 20.93$, $E_2 = 35, 74$.

4.5.2 (f) See Exercise 4.4.9 (f) and Sect. 10.2.

4.5.3 (f) See (4.32) and Sect. 10.2.

4.5.4 (f) See (4.34) and Sect. 10.2.

4.5.5 (a) The resonance energies read off $|A_N|^2$ are $E_1 = 17$, $E_2 = 21$, $E_3 = 27$, $E_4 = 36$. (f) See Exercise 4.4.8 (f) and Sect. 10.2. (g) $E_1 = 17.23$, $E_2 = 20.93$, $E_3 = 27.10$, $E_4 = 35.74$.

4.5.6 (f) See Exercise 4.4.9 (f) and Sect. 10.2.

4.5.7 (f) See (4.32) and Sect. 10.2.

4.5.8 (f) See (4.34) and Sect. 10.2.

4.5.9 The potential barrier of Exercise 4.5.5 is twice as wide as the one of Exercise 4.5.1. Thus in Exercise 4.5.5 the low-lying resonances according to the relation (4.54) occurring at the energy values $E_\ell = V_2 + \ell^2\pi^2\hbar^2/2md^2$ show up at values $(E_\ell - V_2)$ which are a quarter of the values of Exercise 4.5.1. If we increase d again by a factor of two we expect the values $(E_\ell - V_2)$ at energies which are approximately another factor of four lower than the ones of Exercise 4.5.4. The transmission resonances show up as minima in the reflection coefficient B_1. The behavior of the coefficients T_T and T_R related to the transmission resonances is easily understood with the help of the relations (4.34).

4.5.10 (a) $E_1 = 16.2$, $E_2 = 17.2$, $E_3 = 18.8$, $E_4 = 20.9$, $E_5 = 23.7$, $E_6 = 27.1$, $E_7 = 31.1$, $E_8 = 35.9$. (f) See Exercise 4.4.8 (f) and Sect. 10.2. (g) $E_1 = 16.3$, $E_2 = 17.23$, $E_3 = 17.78$, $E_4 = 20.93$, $E_5 = 23.71$, $E_6 = 27.10$, $E_7 = 31.11$, $E_8 = 35.74$.

4.5.11 (f) See Exercise 4.4.9 and Sect. 10.2.

4.5.12 (f) See (4.32) and Sect. 10.2.

4.5.13 (f) See (4.34) and Sect. 10.2.

4.8.1 (d) For a refractive index n_2 of glass the wavelength of light in the glass is given by $\lambda_2 = \lambda/n_2$, where λ is the wavelength in vacuum. Thus the wavelength in region 2 is $\lambda/2$. (e) In region 2 there is only an outgoing transmitted wave $E_2 = A_2 \exp(ik_2x)$. Its absolute square is constant: $|E_2|^2 = |A_2|^2$. (f) The superposition of incident and reflected waves in region 1 causes the wiggly pattern. It indicates the contribution of the interference term of incident and reflected wave, see Exercise 4.2.2 (f).

4.8.2 (d) For the boundary between regions 1 and 2 at $x = 0$ the continuity equations (4.66) require $A_1' + B_1' = A_2'$ and $k_1(A_1' - B_1') = k_2 A_2'$. With $k_2/k_1 = n_2$ we find $A_2' = 2A_1'/(1+n_2)$ and $B_1' = (1 - n_2)A_1'/(1+n_2)$. Thus the transmission coefficient A_2' is small for refractive index $n_2 \gg 1$ of a much denser medium in region 2.

4.8.3 (d) The expression for $A_2' = 2A_1'(1 + n_2)$ obtained above in Exercise 4.8.1 (d) shows that for $n_2 \ll 1$ the amplitude A_2' becomes larger than 1.

4.8.4 (a) The values of the physical electric field strength are the real parts of the complex field strength. The real part Re $E_{1+}(0)$ at the surface at $x = 0$ of the glass has the opposite sign to Re E_{1-} at $x = 0$. Thus the phase shift upon reflection at a denser medium is π, since $\cos(\alpha + \pi) = -\cos\alpha$. (b) The real part Re E_{1+} at $x = 0$ is equal to the real part Re E_{1-} at $x = 0$. Thus there is no phase shift upon reflection on a thinner medium.

4.8.5 (d) The absence of wiggles in the absolute square of the complex electric field strength in region 1 signals the vanishing or smallness of the interference term and thus the absence or smallness of the reflected wave at the wave numbers $k = \pi/2, \pi, 3\pi/2$. (e) At these wave numbers we observe transmission resonances upon transmission through a denser medium, see (4.76) of Sect. 4.7.

4.8.6 (d) The wavelengths of the resonances occurring now upon two-fold transmission into denser media follow from (4.72), Sect. 4.7. (e) The speed of light in region 3 is only $1/4$ of the speed in region 1. Since there is no reflection at a resonant wavelength the energy current density of the transmitted wave must be the same as the one of the incident wave. Therefore the energy density, and thus the absolute square of the complex electric field strength, must change, see (4.67), (4.68) of Sect. 4.7.

4.9.1 (d) Only a transmitted outgoing harmonic wave propagates to the right in region 3. Its complex form is given by (4.56), see Sect. 4.7. The amplitudes of real and imaginary part are time independent. (f) In region 1 the superposition of incident and reflected waves leads to a time-dependent amplitude.

4.9.2 (g) Only the transmitted wave E_{3+} moving to the right propagates in region 3.

4.9.4 (g) The incident wave number $k = 3\pi/2$ leads to $k_2 = n_2 k = 3\pi$. The corresponding wavelength in region 2 is $\lambda_2 = 2\pi/k_2 = 2/3$. Since the thickness d of the region 2 fulfills $\lambda_2 = 2d/m$ for $m = 3$. Thus at the incident wave number $k = 3\pi/2$ we find a transmission resonance, so that there is no reflected wave moving to the left in region 1. Therefore there is no constituent wave E_{1-} in region 1.

4.10.1 (c) The speed of light in the glass of refractive index $n = 4$ is $1/4$ of the vacuum speed of light. (d) See Exercise 4.8.6 (d). (e) The spatial extension of the light wave packet shrinks upon entering the glass sheet since the speed of light in glass is only $1/4$ of the vacuum speed of light. All wavelengths superimposed in the wave packet shrink according to $\lambda_2 = \lambda/n = \lambda/4$.

4.10.2 (c) The incident wave number $k_0 = 7.854$ leads to a wave number $k_2 = 15.708$ in the layer with $n = 2$. This wave number fulfills the resonance condition $2k_2 d_2 = (2m+1)\pi$ for the thickness $d_2 = 0.1$ of the layer with $n = 2$ and for $m = 0$, see (4.72) and accompanying text of Sect. 4.7. Thus for light of wave number $k_0 = 7.854$ there is maximal destructive interference in region 1 between the waves reflected at the front and rear boundaries of the layer with $n = 2$. The refractive indices $n = 1$, $n_2 = 2$ and $n_3 = 4$ satisfy the condition (4.73) so that there is no reflection at the front surfaces of the three glass layers. The corresponding arguments hold true at the rear surfaces of the glass sheets.

4.10.3 (b) The wavelength in the region of the coating is given by $\lambda_1 = \lambda/n_1 = 449$ nm. The thickness of the coating is $d = \lambda_1/4 = 112$ nm.

4.10.4 (b) The modulus of the reflection coefficient B_1' of light is larger for larger refractive index n of the material, see Exercise 4.8.1 (d). We have $n = 1.5$, in Exercise 4.10.1 we had $n = 4$. The reflection is smaller for $n = 1.5$ than for $n = 4$.

4.10.5 (b) The simple coating for a lens is adjusted to an average wavelength within the spectrum of visible light. For larger and shorter wavelengths in the spectrum some reflection at the surfaced of the lens remain. This is the reason for the bluish reflection of coated lenses.

4.11.1 (e) The resonant wave numbers read off $|A_N|^2$ are $k_1 = 7.85$, $k_2 = 23.6$. The resonant wave numbers calculated using (4.72) are $k_1 = 7.854$, $k_2 = 23.56$.

4.11.2 (e) The resonant wave number read off $|A_N|^2$ is $k_1 = 12.8$. The resonant wave number calculated using (4.72) is $k_1 = 12.83$.

Hints and Answers to the Exercises in Chapter 5

5.2.2 (e) The number of nodes in the center-of-mass coordinate in $R = (x_1 + x_2)/2$ is equal to N, in the relative coordinate $r = x_2 - x_1$ it is equal to n. The nodes in R lead to node lines in the $x_1 x_2$ plane parallel to the diagonal in the first and third quadrant. The nodes in r show up as node lines parallel to the diagonal in the second and fourth quadrant of the $x_1 x_2$ plane.

5.2.3 (e) The doubling of the mass reduces the spatial extension of the wave function. The ground-state width $\sigma_0 = \sqrt{\hbar/m\omega}$ is reduced by a factor of $\sqrt{2}$ upon the doubling of the mass. The same factor $\sqrt{2}$ reduces the spatial extension in x and y also for the wave function of the higher states. The classical frequencies $\omega_R = \sqrt{k/m} = \omega_r$ for uncoupled oscillators are also reduced by a factor $\sqrt{2}$ upon doubling the mass. Thus the energy eigenvalues are reduced by $\sqrt{2}$. Therefore for a particle with twice the mass the eigenfunctions belonging to the same quantum numbers N, n are deeper down in the harmonic-oscillator potential and thus their spatial extension has shrunk.

5.2.4 (e) The coupling κ is positive. The coupling spring pulls the two particles towards each other. The oscillator potential in the relative variable r becomes steeper, see (5.12). Thus the spatial extension of the stationary wave function in the direction of the diagonal $r = x_2 - x_1$ in the $x_1 x_2$ coordinate system shrinks. This effect also shows in the dashed curve in the $x_1 x_2$ plane which marks the classically allowed region for the positions of the particles. In contrast to Exercises 5.2.2 and 5.2.3 it is no longer a circle but an ellipse with the shorter principle axis in r direction.

5.2.5 (e) The shrinking in the diagonal direction $r = x_2 - x_1$ becomes more prominent. (f) The probability of particle 1 being close to particle 2 is larger than the probability of it being far from particle 2. The particle coordinates exhibit a positive correlation.

5.2.6 (e) The coupling κ is negative, i.e., the coupling spring pushes the two particles apart. The potential in the relative variable becomes shallower, see (5.12). Thus the spatial extension of the wave function in the diagonal direction $r = x_2 - x_1$ widens. This feature is also obvious from the dashed ellipse in the $x_1 x_2$ plane. It is more likely to find particle 2 at a position distant from that of particle 1 than close by. The particle coordinates exhibit a negative correlation.

5.2.7 (b) At the outer maxima of the probability density the potential energy of the particles is larger than in the inner ones. Thus the kinetic energy of the particles is lower in the outer maxima than in the inner ones. Hence, their speed is lower and therefore the time they need to pass the region of the outer maxima is larger. The probability density of the particle being in the outer maxima is larger than the probability density of it being in the inner ones. (c) The particle speed and thus the particle momentum close to the origin in the $x_1 x_2$ plane is larger; thus the wavelength of the wave function is smaller than in the outer regions of the plot. Hence, the widths of the inner maxima of the plot are smaller. (d) The system of two one-dimensional coupled oscillators has two uncoupled degrees of freedom: the center of mass motion and the relative motion. The energy remains constant in either degree of freedom. Its value is given by E_N and E_n, see (5.17). For the case $N = 0$, $n = 2$ the motion of the center of mass has the ground-state energy only. Its amplitude, or equivalently the spatial extension of the ground-state wave function in the center of mass coordinate R, is small. Thus in this direction the classically allowed region is not completely exhausted by the probability density and the wave function.

5.2.8 (b) The lines of the rectangular grid confining the areas of the maxima in the plots of the probability density are the node lines in the center-of-mass coordinate R and in the relative coordinate r.

5.2.9 (e) $\omega_R = 1$, $\omega_r = 3$. The energy values of the states with the quantum number N, n are $E_{N,n} = \left(N + \frac{1}{2}\right) + 3\left(n + \frac{1}{2}\right) = N + 3n + 2$. (f) The graphs for $N \gg n$ represent stationary probability densities of the coupled oscillators where a large part of the total energy is in the degree of freedom R of the center-of-mass motion. This causes the wide extension of the wave function in the diagonal in the first and third quadrant of the $x_1 x_2$ plane.

5.3.1 (b) Since the two oscillators are uncoupled, the expectation value $\langle x_{20} \rangle$ being zero initially remains zero at all times. (c) The widths of the two coupled oscillators vary with time, since neither of the two initial widths σ_{10}, σ_{20} is equal to the ground-state width σ_0 divided by $\sqrt{2}$, see Exercise 3.4.4 and (3.31). (d) The initial correlation c_0 is equal to zero. Since the oscillators are uncoupled, no correlation is produced during the motion of the wave packet.

5.3.2 (b) The time dependence of the correlation reflects the periodic change of the widths in the two uncoupled oscillators, (3.31).

5.3.3 (a) $\sigma_0/\sqrt{2} = \sqrt{\hbar/2m\omega} = \sqrt{\hbar\sqrt{m}/2\sqrt{k}} = k^{-1/4}/\sqrt{2} = 0.5946$ for $\hbar = 1$, $m = 1$. (c) The ground-state width of the two uncoupled oscillators is σ_0. The initial widths have been chosen equal to $\sigma_0/\sqrt{2}$, the initial state is a coherent state of the two uncoupled oscillators, see Exercise 3.4.4 and (3.31). (d) For nonvanishing correlation the initial state is no longer a product of two Gaussian wave packets in the variables x_1, x_2. Thus, it is not a product of coherent states. The widths will no longer remain time independent.

5.3.4 (b) See Exercise 5.3.3. (d). (c) See Exercise 5.3.2 (b).

5.3.5 (b) Because of the positive initial correlation, the time dependence of the correlation of Exercise 5.3.4 starts with a positive value and oscillates. In this exercise the correlation is negative and thus the function $c(t)$ starts with the negative initial value. The time dependences of the two correlations differ only by a phase shift.

5.3.6 (b) The initial conditions for the position expectation values do not influence the correlation or the widths. Thus they exhibit no time dependence as in Exercise 5.3.3.

5.3.7 (b) The oscillation of the expectation values $\langle x_1(t) \rangle$, $\langle x_2(t) \rangle$ is the same as for the position of classical particles. The two position expectation values exhibit a beat because of the coupling of the two oscillators. (c) Even though the initial correlation vanishes, a correlation emerges which is positive most of the time because of the attractive coupling.

5.3.9 (b) The motion of the expectation value of one particle is a mirror image of the motion of the other one. Thus the center of mass of the two particles is at rest, the oscillation takes place in the relative coordinate. The coupled oscillator system is in a normal mode. Only one of the two uncoupled degrees of freedom – the center of mass and the relative motion – oscillates.

5.3.10 (b) The system of the two coupled oscillators is still in the normal mode of the relative motion. (c) Even though the initial correlation is negative the time average of the correlation is positive because of the positive coupling.

5.3.12 (b) The motion of the two particles is the same in x_1 and x_2. The relative coordinate is zero all the time. The oscillation takes place in the center-of-mass coordinate R. The system of coupled oscillators is in the normal mode which refers to the center-of-mass motion.

5.4.1 (c) The initial correlation $c_0 = 0$ vanishes. Thus the initial Gaussian wave packet has axes parallel to the coordinate axes. Since the coupling of the two oscillators vanishes no correlation is introduced during the motion.

5.4.2 (c) There is a positive initial correlation $c_0 = 0.8$. It shows in the first graph of the multiple plot in which the orientation of the covariance ellipse is not parallel to the axes. (d) The covariance ellipse in the plot of the initial wave packet has its large principal axis oriented under a small angle with the diagonal direction $x_1 = x_2$. Thus it is more likely than not that x_1 is close to x_2.

5.4.3 (c) The large principal axis is oriented under a small angle to the diagonal direction $x_1 = -x_2$.

5.4.4 (c) The initial expectation value $\langle x_2 \rangle$ at $t = t_0$ is zero. Without coupling between the two oscillators, $\langle x_2 \rangle$ would remain zero while $\langle x_1 \rangle$ moves. The attractive coupling pulls the particle 2 towards the position of particle 1, which is at positive x_1 values in the beginning. Thus the expectation value $\langle x_2 \rangle$ takes on positive values for $t > t_0$.

5.4.5 (c) The coupling κ is larger, so the force exerted on particle 2 by particle 1 is larger. Thus the acceleration of particle 2 out of its original position is bigger.

5.4.6 (c) The attractive coupling κ causes the appearance of a correlation in the initially uncorrelated wave packet.

5.4.7 (c) The coupling κ is now larger than the coupling constant in the former exercises. Thus the frequency of the oscillation in the relative coordinate $r = x_2 - x_1$ is larger than before and there are more oscillations in r in the same time interval.

5.4.8 (c) The initial wave packet is wide compared to the ground state in the coupled oscillator. With the same arguments as in Exercise 3.4.2 we expect the width of the wave packet to decrease as it moves close to the center.

5.4.9 (c) The oscillation observed is the normal oscillation taking place in the relative coordinate $r = x_2 - x_1$.

5.4.10 (c) The oscillation observed is the normal oscillation taking place in the center-of-mass coordinate $R = (x_1 + x_2)/2$. (d) The normal mode in the relative motion in Exercise 5.4.9 is fast because of the large coupling $\kappa = 20$, which means that the oscillator of relative motion has a high eigenfrequency. The normal mode of the center-of-mass motion in this exercise has a spring constant of only $k = 2$. The oscillator of the center-of-mass motion has a much lower frequency.

5.4.11 (c) The repulsive coupling $\kappa < 0$ of the two oscillators pushes the particle 2 away from particle 1. Thus particle 2 is accelerated into the negative x_2 direction instead of being pulled towards positive x_2 values, as in the case of an attractive coupling $\kappa > 0$.

5.4.12 (c) The effective spring constant in the oscillator of relative motion is $(k/2+\kappa)$, see (5.12). For the values $k = 2$, $\kappa = -0.95$ the effective spring constant of relative motion is 0.1. Thus the oscillator in the relative coordinate is very shallow. Therefore the wave packet spreads into the direction of the relative coordinate. This way it gets anticorrelated.

5.4.13 (c) The wave function for bosons is symmetric, see (5.25). The symmetrization of the initial wave function of Exercise 5.4.1 with one hump at a nonsymmetric expectation value $\langle x_1 \rangle \neq \langle x_2 \rangle$ leads to an $x_1 x_2$-symmetric two-hump probability density. (d) The very narrow peak in plots where the two humps almost completely overlap is due to the symmetrization of the wave function.

5.4.14 (c) The initial Gaussian wave function is centered around a symmetric position $x_1 = x_2 = 3$ in the $x_1 x_2$ plane. Thus symmetrization does not create a second hump.

5.4.15 (c) The direction of maximal width of the two humps in the initial probability distribution forms a small angle with the diagonal $x_1 = -x_2$ in the $x_1 x_2$ plane. Thus the wave packet is anticorrelated.

5.4.17 (c) The oscillations in the relative coordinate $r = x_2 - x_1$ exhibit a much higher frequency than the oscillation in the center-of-mass motion; thus the spring constant in the oscillator of the relative motion is larger, i.e., the coupling constant κ is large.

5.4.18 (c) The two-hump structure of the initial probability distribution says that it is just as likely that particle 1 is at $x_1 = 3$ and particle 2 at $x_2 = -3$ (hump to the left) as it is that particle 1 is at $x_1 = -3$ and particle 2 is at $x_2 = 3$ (hump to the right).

5.4.19 (c) The width in $r = x_2 - x_1$ is the width in the oscillator of relative motion. It has a strong attractive spring constant since the coupling $\kappa = 20$ is large. Thus the frequency of this oscillator is high. Therefore the width in the oscillator of relative motion changes very quickly. Even though we are looking at the normal mode of the center-of-mass oscillator, which keeps the expectation value of the relative coordinate time independent, the strong coupling shows in the oscillation of the width in the relative coordinate.

5.4.20 (c) See Exercise 5.4.11 (c).

5.4.21 (c) The two-particle wave function for fermions is antisymmetric. Therefore it has to vanish at $x_1 = x_2$. Hence, its absolute square, the probability distribution vanishes.

5.4.22 (c) See Exercise 5.4.4.

5.4.25 (c) Because of vanishing correlation, equal widths and equal initial positions, the initial wave function is obtained through the antisymmetrization of a product of two identical Gaussian wave packets, in x_1 and x_2, see (5.18). Thus the result of the antisymmetrization vanishes. The two fermions cannot be in the same state, as the Pauli principle requires.

5.4.26 (c) Also for nonvanishing correlation the initial wave function for distinguishable particles (5.18) remains symmetric in x_1 and x_2.

5.4.27 (c) The Pauli principle states that two fermions cannot occupy the same state. The different widths in the two coordinates allow for different states in the wave packet to be occupied by the two particles.

5.5.3 (d) The dips between the humps are slightly more pronounced for the fermions. (e) The marginal distributions are obtained as integrals in one variable x_1 (or x_2) over the joint probability distributions. Since the zero line $x_1 = x_2$ of the fermion probability distribution forms an angle of 45 degrees with either the x_1 axis or x_2 axis, the effect of the zero line gets washed out upon integration over x_1 or x_2.

Hints and Answers to the Exercises in Chapter 6

6.3.1 (n) The minimal impact parameter b of the particle not entering the region $0 \leq r \leq 2\pi$ is $b = 2\pi$. The minimal classical angular momentum of the particle is $L = kb\hbar = 4\pi\hbar = 12.57\hbar$. (o) The partial wave with $\ell = 13$ starts becoming different from values close to zero only in the outer region adjacent to $r = 2\pi$. (p) The radial dependence of the ℓ-th partial wave is given by the spherical Bessel functions $j_\ell(kr)$. Their behavior for $kr \ll 1$ is proportional to $(kr)^\ell$, see (9.40). Hence, the region in (kr) close to zero in which the partial wave is very small grows with ℓ.

6.3.2 (n) A partial sum up to N of the partial-wave decomposition (6.53) of the plane wave approximates the plane wave in a radial region close to $r = 0$. This region extends as far as the values of the partial wave of index $N+1$ remain small compared to one. This is approximately the region $r < b$, where b is the classical impact parameter determined by $b = N/k$. A particle of momentum $p = \hbar k$ and impact parameter b has the classical angular momentum $L = bk\hbar = N\hbar$.

6.4.1 (d) The Gaussian wave packet of the form (6.24) widens independently in all three space coordinates. The time dependence of the width in every coordinate follows (2.10). The initial width is $\sigma_{x0} = 0.5$ so that with $\hbar = 1$, $M = 1$ we get
$$\sigma_x(t) = \sigma_{x0}\left(1 + t^2/4\sigma_{x0}^4\right)^{1/2} = 0.5\left(1 + 4t^2\right)^{1/2}.$$

6.4.2 (e) The vector of the classical angular momentum is $\boldsymbol{L} = (0, 0, 4)$.

6.4.3 (d) The initial width in the x direction, σ_{x0}, is smaller than σ_{y0} in the y direction. The formula given in 6.4.1 (d) above shows that the coefficient of t^2 in $\sigma_x(t)$ is much larger than the corresponding coefficient in $\sigma_y(t)$. (e) The ripples occur along the direction of the momentum.

6.4.4 (d) The wave packet is at rest, thus there is no special direction besides the radial direction.

6.5.1 (b) The classical angular momentum vector is $\boldsymbol{L} = (-0.5, -2, 5.5)$.

6.5.3 (b) The initial widths in all three coordinates are larger than in Exercise 6.5.2. Thus the coefficients of t^2 in the formulae (2.10) for the three time-dependent widths $\sigma_x, \sigma_y, \sigma_z$ are smaller than for the situation in Exercise 6.5.2.

6.5.4 (b) Formula (2.10) can be rewritten to look like $\sigma_x = \hbar t \left(1 + 4\sigma_{x0}^4 m^2/\hbar^2 t^2\right)^{1/2} / 2\sigma_{x0}m$, i.e., for large t the width σ_x grows approximately linearly in t.

6.6.1 (b) The classical angular momentum is $L = bp_0 = 1$, it points along the z direction. (c) The angular momentum for a particle with impact parameter $b' = b + \sigma_0$ is $L' = 2.5$; for $b'' = b - \sigma_0$ it is $L'' = -0.5$. (d) The maximum of the probabilities W_ℓ to find the angular momentum ℓ in the wave packet at 1 corresponds to the classical angular momentum of the wave packet. Because of the spatial extension of the Gaussian wave packet there is a finite probability density for particles with larger and smaller impact parameters. As typical values for the width of the distribution of impact parameters we choose $b' = b + \sigma_0 = 1.6667$ and $b'' = b - \sigma_0 = 0.3333$. The corresponding angular momenta – as calculated above – are $L' = 2.5$ and $L'' = -0.5$. Actually, the distribution of m values in the plot shows that the values $\ell = 3$, $m = 3$ and $\ell = 1$, $m = -1$ have about half of the probability of the value $\ell = 1$, $m = 1$. This explains the features of the distribution of the $W_{\ell m}$.

6.6.2 (b) The average angular momentum of the distribution is the same as in Exercise 6.6.1. The halving of the impact parameter b is compensated by the doubling of the momentum p_0. The distribution of the angular momenta in the wave packet is the same as in Exercise 6.6.2 since the initial width σ_0 is halved in this exercise.

6.6.3 The Gaussian wave packet moves in the plane $z = 0$. It is an even function in the z coordinate. In polar coordinates it is thus an even function with respect to $\vartheta = \pi/2$. The spherical harmonics of $m = \ell - (2n + 1)$, $n = 0, 1, \ldots, \ell - 1$ are odd with respect to $\vartheta = \pi/2$. Thus they do not contribute to the partial-wave decomposition. The values $W_{\ell m}$ for $m = \ell - (2n + 1)$ vanish.

Hints and Answers to the Exercises in Chapter 7

7.2.2 (e) The centrifugal barrier $\hbar^2 \ell(\ell + 1)/(2Mr^2)$ is a steeply increasing function as r goes to zero. It keeps the probability density low at small values of r. Since the barrier height at a given r grows with $\ell(\ell + 1)$, the suppression becomes the more apparent the higher the angular momentum ℓ.

7.2.3 (e) Heisenberg's uncertainty principle $\Delta x \Delta p \geq \hbar/2$ can be adapted to the ground-state energy E_0 in a potential well of radius R to yield $R\sqrt{2ME} \geq h$. Thus the ground-state energy grows proportional to $1/R^2$, i.e., $E \geq h^2/2MR^2$.

7.2.5 (e) The particle in a potential well of finite depth is not strictly confined to the inside of the well as is the case in the infinitely deep square well. Therefore the curvature $d^2\varphi/dx^2$ of the wave function $\varphi(x)$ of stationary states with the same number of nodes is smaller for the potential of finite depth than for one with infinite depth. Thus the contributions of the radial energy and of the effective potential energy in a state with a given number of nodes are smaller for the potential of finite depth.

7.2.6 (e) For spherical square-well potentials the Schrödinger equation is the same as for an infinitely deep square-well potential except for the constant value of the potential depth. This value can be absorbed into the energy eigenvalue of the stationary Schrödinger equation. Only the boundary condition at the value of the radius of the potential well for bound states in the infinitely deep well is different from the one for a potential well of finite depth. At the origin of the coordinate frame the boundary conditions are the same. Therefore the wave functions look alike for small values of the radial coordinate r.

7.2.7 (e) Under the action of a three-dimensional harmonic force a particle of high angular momentum moves on an orbit far away from the center of the force.

7.2.8 (e) In a three-dimensional harmonic oscillator the relation between energy and angular momentum is given by $E_{n_r \ell} = \left(2n_r + \ell + \frac{3}{2}\right)\hbar\omega$, see (7.40), (7.41). Thus the lowest energy for fixed angular momentum ℓ is $E_{0\ell} = \left(\ell + \frac{3}{2}\right)\hbar\omega$. It is assumed for vanishing radial quantum number n_r. Except for the zero-point energy $3\hbar\omega/2$, the energy is solely rotational and potential energy $E = L\omega$, $L = \ell\hbar$.

7.2.10 (e) The effective potential is given by $V^{\text{eff}}(r) = M\omega^2 r^2/2 + \ell(\ell + 1)\hbar^2/(2Mr^2)$. Its minimum is the value $r_0 = \left[\ell(\ell + 1)\hbar^2/M^2\omega^2\right]^{1/4}$ of the radial variable. The potential at this value is $V^{\text{eff}}(r_0) = \sqrt{\ell(\ell + 1)}\hbar\omega$. Its curvature at r_0 is $d^2 V^{\text{eff}}(r)/dr^2\big|_{r=r_0} = 4M\omega^2$. Thus the oscillator potential approximating the effective potential close to its minimum is $V_\ell^{\text{app}}(r) = \sqrt{\ell(\ell + 1)}\hbar\omega + \frac{M}{2}(2\omega)^2(r - r_0)^2$. (f) The wave functions in the approximating potential $V_\ell^{\text{app}}(r)$ are harmonic-oscillator eigenfunctions centered about r. (g) The angular frequency of the approximating oscillator is 2ω, twice the frequency ω of the three-dimensional oscillator. It is independent of the angular momentum. (h) For fixed ℓ the spacing of the energy levels in the three-dimensional oscillator is $2\hbar\omega$. The eigenvalues are $E_{n_r\ell} = \left(2n_r + \ell + \frac{3}{2}\right)\hbar\omega$. Since the approximating potential has the frequency 2ω, its level spacing is also $2\hbar\omega$. The eigenvalues in the approximating oscillators are $E_{\text{app}} = V(r_0) + 2(n + 1/2)\hbar\omega = \left(2n + \sqrt{\ell(\ell + 1)} + 1\right)\hbar\omega$. The two expressions differ very little for large angular momentum $\ell \gg 1$.

7.2.11 (e) For vanishing angular momentum $\ell = 0$ the three lowest energy values are $E = -1/2, -1/8, -1/18$; for $\ell = 1$ the lowest three eigenvalues are $E = -1/8, -1/18, -1/32$.

7.2.12 (e) Eigenvalues for $\ell = 2$ are $E = -1/18, -1/32, -1/50$; eigenvalues for $\ell = 3$ are $E = -1/32, -1/50, -1/72$.

7.2.13 (e) The effective potential for the Coulomb interaction is given by $V^{\text{eff}}(r) = -\alpha\hbar c/r + \ell(\ell + 1)\hbar^2/2Mr^2$. The minimum occurs at the value $r_0 = \ell(\ell + 1)\hbar c/\alpha M c^2$. Here the effective potential has the value $V^{\text{eff}}(r_0) = -\alpha^2 M c^2/2\ell(\ell + 1)$. The curvature at r_0 has the value $d^2 V^{\text{eff}}(r)/dr^2\big|_{r=r_0} = \alpha^4 (Mc^2)^3/\ell^3(\ell + 1)^3\hbar^2 c^2$. The approximating oscillator potential centered about r_0 is $V^{\text{app}}(r) = -\alpha^2 M c^2/2\ell(\ell+1) + (1/2)M\omega_\ell^2(r - r_0)^2$ with the square of the effective angular frequency $\omega_\ell^2 = \alpha^4(Mc^2)^2/\ell^3(\ell + 1)^3\hbar^2$. (f) The energy eigenvalues of the approximating oscillator are $E_{\ell n} = -\alpha^2 M c^2/2\ell(\ell+1) + (n+1/2)\hbar\omega_\ell = -\alpha^2 M c^2 \left[1 - 2(n + 1/2)/\sqrt{\ell(\ell + 1)}\right]/2\ell(\ell + 1)$. The exact formula $E = -\alpha^2 M c^2/2n^2$ can be expanded for large angular momentum ℓ if we use $n = n_r + \ell + 1$, the decomposition of the principal quantum number n into ℓ and the radial quantum number n_r. For large ℓ we arrive at the approximation $E = -\alpha^2 M c^2[1 - 2(n + 1)/(\ell + 1/2)]/2(\ell + 1/2)^2$. Again for large ℓ the two expressions for E converge to the same value. (g) The exact Bohr radii are $a_n = n^2 a_0$, where $a_0 = \hbar c/(\alpha M c^2)$ is the innermost Bohr radius. For small $n_r \ll \ell$ the above value for r_0 approaches the Bohr radius.

7.3.1 (e) In the neighborhood of the energy of the state there are no eigenvalues of states in the first well. Therefore there is no strong contribution of an eigenstate of the first well to the third eigenstate of the double well. Thus the wave function is different from zero chiefly in the second well.

7.3.4 (c) Because of the strong repulsive potential in the innermost region $0 \leq r < 0.5$ the potential is a hard-core potential. It keeps the values of the wave function in this region small as long as the energy of the state is small compared to the height of the hard core. Therefore the influence of the repulsive centrifugal barrier on the wave function and the energy eigenvalues is much smaller than in a potential without a hard core.

7.4.1 (b) The number of node lines at fixed polar angles is equal to the quantum number ℓ of angular momentum.

7.4.3 (b) The node half circles are the nodes of the radial wave functions. With increasing number of node lines the energy eigenvalues of the wave function increases.

7.4.4 The number of ϑ-node lines is equal to the quantum number ℓ of angular momentum.

7.4.5 (b) The node lines in ϑ are given by the zeros of the spherical harmonics in the polar angle ϑ, in this case $\vartheta = 0$, $\vartheta = \pi/2$.

7.4.7 (b) Exercise 7.4.6 shows the probability density of the eigenstate with principal quantum number $n = 1$ and the angular-momentum quantum numbers $\ell = 15$, $m = 15$. This corresponds to a classical angular-momentum vector parallel to the quantization axis of angular momentum, which is the z axis in this case. Exercise 7.4.7 (a) exhibits the probability density for the angular-momentum quantum numbers $\ell = 15$, $m = 0$. A state with these quantum number is classically interpreted as representing an angular momentum in a direction perpendicular to the z axis. Since L_x and L_y do not commute with L_z no particular direction in the xy plane can be assigned to the classical vector of angular momentum in this case. In fact, the probability distribution is cylindrically symmetric about the z axis. The probability density is largest at the large values of z close to the wall of the spherically symmetric potential well. This can be interpreted as an assembly of classical orbits in the planes which contain the z axis. No particular one of these planes is distinguished, so that no special direction of angular momentum in the xy plane can be assigned to the state with $m = 0$.

7.4.8 (b) The wall of the infinitely deep square well confines the wave function strictly to the range $r \le a$. In the harmonic-oscillator potential the wave function falls off with $\exp(-r^2/2\sigma_0^2)$. Thus the decrease in the infinitely deep square-well potential is much faster.

7.4.12 (b) For eigenstates in the Coulomb potential the quantum number ℓ of angular momentum satisfies the relation $\ell \le n - 1$. Thus no eigenstate exists with the quantum number $n = 1$, $\ell = 1$.

7.4.13 (b) The spherical harmonics $Y_{\ell m}(\vartheta, \varphi)$ possess ℓ nodes in the polar angle ϑ. The number of nodes in the radial variable r is $(n - 1 - \ell)$.

7.4.14 (b) The normalization of the radial wave function is given by an intgral containing the measure $r^2 dr$. The wave function with the quantum numbers $m = \ell$ has its large values at large values of the radial variable. Here r^2 is large; thus the normalization of the wave function to one suppresses the height of the wave function. (c) For $m = \ell$ the wave function represents a particle with the vector of angular momentum in the direction of the axis of quantization, i.e., the z axis. This forces the particle to a far-out Bohr orbit.

7.5.1 (b) The classical angular-momentum vector has the components $L_x = -1$, $L_y = -7$, $L_z = 5$.

7.5.3 (b) The classical angular-momentum vector has the components $L_x = 0$, $L_y = 0$, $L_z = 0$. The motion of the wave packet is an oscillation through the center of the spherically symmetric potential.

7.5.5 (b) The ground-state width of the three-dimensional oscillator is $\sigma_0 = \sqrt{\hbar/M\omega}$. For $T = 1$ the angular frequency equals $\omega = 2\pi/T = 2\pi$. Thus $\sigma_0 = (2\pi)^{-1/2} = 0.56$. The initial widths σ_{x0} and σ_{y0} are smaller than $\sigma_0/\sqrt{2} = 0.4$. In these directions the Gaussian wave packet is initially narrower than $\sigma_0/\sqrt{2}$. In the z direction the wave packet is initially wider than $\sigma_0/\sqrt{2}$. The probability ellipsoid of the initial state has prolate shape with the large principal axis in the z direction. After a quarter period the widths σ_x and σ_y have reached their maximum values, which are now larger than $\sigma_0/\sqrt{2}$, whereas σ_z is at its minimum, which is smaller than $\sigma_0/\sqrt{2}$, see Sect. 3.1, (3.31). After a quarter period $T/4$ the Gaussian wave packet possesses a probability ellipsoid of oblate shape with the smallest principal axis in z direction.

Hints and Answers to the Exercises in Chapter 8

8.2.1 (d) The centrifugal barrier $V_\ell(r) = \hbar\ell(\ell+1)/2Mr^2$ grows quadratically with ℓ. For fixed energy of the incoming particle the wave function becomes more and more suppressed in a region of small r because of the r^ℓ behavior of the spherical Bessel functions $j_\ell(kr)$ for $kr \ll 1$. The range in which the wave function is smaller than a given value widens with increasing ℓ. This way the contribution of the rotational energy represented by the centrifugal barrier remains low enough to keep the total energy constant.

8.2.3 (d) The total energy contains three contributions, the radial, the rotational and the potential energy. For growing total energy and fixed angular momentum the contribution of the rotational energy grows. This requires larger values of the wave function at low values of r where the centrifugal barrier is large.

8.2.5 (b) The energies of the first two resonances are $E_1 = 10.1$ and $E_2 = 10.5$.

8.2.9 For increasing ℓ the centrifugal barrier pushes the resonance wave function out of the region of small r, see Exercise 8.2.1. This increases the curvature of the wave function in the region close to the range of the potential. Thus the kinetic-energy contribution grows and therefore so does the total energy.

8.3.4 For distances $r \gg r_N$ large compared to the range r_N of the potential the radial scattering wave function has the form $R_\ell(kr) \sim \exp(i\delta_\ell)\sin(kr - \ell\pi/2 + \delta_\ell)/kr$. For the scattering at a repulsive infinitely high potential of radius d the phase shift δ_ℓ is simply given by $\delta_\ell = -kd$, since the wave function has to fulfill the condition $R_\ell(ka) = 0$ at the wall of the potential at $r = a$. For the scattering at a repulsive potential of finite height V_0 the wave function R_ℓ has the same form as above if $r \gg d$. However the phase shift δ_ℓ is given by $\delta_\ell = -ka$ where the scattering length a replaces the radius of the repulsive potential. We have $a < d$ since the wave function penetrates somewhat into the repulsive square well of finite height. The free wave functions ($V = 0$) have the same form for the $R_\ell(kr)$ as above, however for $\delta_\ell = 0$. This phase shift determines the partial scattering amplitude f_ℓ, (8.47), and because of (8.38) and (8.42) both the differential cross section and the partial cross section as well.

8.4.1 (j) Summation of the partial scattering amplitudes f_ℓ to the scattering amplitude $f(k,\vartheta)$, (8.31) up to $L = 0$ includes the angular momentum $\ell = 0$ only. Thus only the zeroth Legendre polynomial $P_0(\cos\vartheta) = 1$ contributes. Hence, no ϑ dependence shows up for $L = 0$ in (a).

8.4.2 (c) For $E = 50$ the wavelength is $\lambda = 0.6282$, for $E = 5000$, $\lambda = 0.06282$. (d) The decrease of the differential cross section in the forward direction for energies increasing from the value $E = 0$ follows a $1/E$ behavior, see (8.38), (8.31).

8.4.4 (c) The two differential cross sections for the high energy $E = 5000$ look alike. This can be easily understood by using the lowest-order Born approximation, which is valid at high energies $E \gg |V_0|$. The scattering amplitude is given by a volume integral over a product of the free incoming and outgoing waves and the potential. The differential cross section, being proportional to the absolute square of the scattering amplitude, is in this high-energy approximation independent of the sign of the potential.

8.4.5 (i) For low energies of the incident plane wave the zeroth partial wave yields the largest contribution to the differential cross section. The phase shift δ_0 for very low energies is determined by $\tan \delta_0 = -ka$, where a is the scattering length of the potential. For repulsive potentials we have $0 < a \leq d$ at low energies, see Exercise 8.3.4. For an attractive square-well potential the wave number κ within the range of the potential is $\kappa = \sqrt{2M|V_0| + k^2}$, where $|V_0|$ is the depth of the square-well potential. For $\ell = 0$ the radial wave function in region 1 has the form $R_{01} = (1/\kappa r) \sin \kappa r$. Outside the range of the potential the radial wave function can be written as $R_{02} = (1/kr) \sin(kr + \delta_0)$, where δ_0 is the scattering phase for angular momentum zero. The continuity conditions yield $(1/k) \tan \delta_0 = (1/\kappa) \tan \kappa d$. For $k \to 0$ the scattering phase approaches zero, $\delta_0 \to 0$. Thus for $k \to 0$ we get for the scattering length $a = -\delta_0/k \simeq -(1/k) \tan \delta_0$ the equation $a = -(1/\kappa) \tan \kappa d$. The numerical value of κ for $M = 1$ and $|V_0| = 3$ is $\kappa = 2.45$. For this value and $d = 2$ the scattering length for the attractive potential is obtained to be $a = 2.16$. For the repulsive potential we have $a \leq 2$. The differential cross section for low energies is given by $d\sigma/d\Omega = (1/k^2) \sin^2 \delta_0 = a^2$. Thus for low energy the differential cross section for the repulsive potential of Exercise 8.4.3 is smaller than for the attractive potential of Exercise 8.4.1.

8.5.1 (f) The energy of the lowest resonance with vanishing angular momentum is $E_{01} = 3.9$.

8.5.2 (f) The energies of the two lowest resonances with angular momentum zero appear at the energies $E_{01} = 3.9$, $E_{02} = 8.2$. The values of the scattering phase at these resonances are $\delta_{01} = -\pi$, $\delta_{02} = -\pi/2$. (g) The partial scattering amplitudes at the two lowest resonances with vanishing angular momentum are calculated using (8.47). We get $f_{01} = 0$, $f_{02} = i$.

8.5.3 (f) The energy of the lowest resonance with angular momentum $\ell = 1$ is $E_{11} = 5.5$.

8.5.4 (f) The energy of the lowest resonance with angular momentum $\ell = 2$ is $E_{21} = 7$. (g) The energy of the lowest resonance increases with angular momentum ℓ because of the positive energy contribution of the repulsive centrifugal barrier $V_\ell(r) = \hbar^2 \ell(\ell + 1)/(2Mr^2)$ and the increase of the radial energy with ℓ.

8.5.5 (f) The energies of the lowest resonances of angular momentum zero are $E_{01} = 1$, $E_{02} = 3.6$, $E_{03} = 7.7$. (g) In an infinitely deep square-well potential of radius 2 the energies of the lowest bound states of vanishing angular momentum are $E_1 = 1.23$, $E_2 = 4.93$, $E_3 = 11.10$.

8.5.9 (c) The small peaks in the plots of the partial cross sections σ_ℓ indicate the existence of resonances of angular momentum ℓ at the energies at which the peaks occur. (d) For vanishing angular momentum the energy eigenvalues of an infinitely deep square-well potential of width 2 are $E_{01} = 1.23$, $E_{02} = 4.93$, $E_{03} = 11.10$. The resonance peaks showing up in σ_0 are at $E_{01} = 1.1$, $E_{02} = 4.8$, $E_{03} = 8.6$. (e) The eigenfunctions in the infinitely deep square-well potential of radius d have to vanish at $r = d$. For the resonance wave functions in the barrier potential of this exercise this is not so. Thus the curvature of the eigenfunctions in the infinitely deep square well is larger than the corresponding resonance energies in the barrier potential.

8.5.10 (b) For the energy $E = 0.001$ of the incoming particle the partial cross section σ_0 has practically the same value as the total cross section. This has to be so, since there are no contributions of higher angular momenta $\ell = 1, 2, \ldots$ to the incoming wave function of the low energy $E = 0.001$. (c) The classical total cross section for the elastic scattering of particles on a hard sphere of radius d is equal to its geometrical area $\sigma_{cl} = \pi d^2$. For $d = 2.5$ we get $\sigma_{cl} = 19.6$.

8.5.11 (a) $\sigma_{tot} = 39.5$, (b) $\sigma_{tot} = 46.8$, (c) $\sigma_{tot} = 49.3$, (d) $\sigma_{tot} = 49.8$, (e) $\sigma_{tot} = 50.2$. (f) The total cross section for quantum-mechanical scattering on a hard sphere is $\sigma_{tot} = 4\pi d^2$. For $d = 2$ the numerical value comes out to be $\sigma_{tot} = 50.26$. It represents the upper limit for the series of numbers for σ_{tot} obtained in (a–e).

Hints and Answers to the Exercises in Chapter 9

9.2.1 (d) The Hermite polynomial H_n possesses n zeros.

9.4.1 (d) The Legendre polynomial P_ℓ possesses ℓ zeros.

9.4.2 (c) The associated Legendre function P_ℓ^m possesses $(\ell - m)$ zeros.

Appendix A: A Systematic Guide to IQ

A.1 Dialog Between the User and IQ

A.1.1 A Simple Example

Let us assume that you are working in a directory in which the program **IQ** and the other necessary files exist. (To bring them there use the Installation Guide in Appendix B).

You may start the program by typing

    IQ <RET>

(You have to press the RETURN key after each line of input as indicated by the symbol <RET>. We shall in most cases omit writing <RET> where the use of the RETURN key is obvious.) After first showing a few lines of welcome the program answers with

    IQ>

This is the *prompt* you will usually get when the program asks you for input. Answer by giving the command

    GD 21

(**g**et **d**escriptor 21). You will get the prompt IQ> again. With the command GD 21 you have loaded the *descriptor* number 21 from disk into memory. A descriptor is a record containing all the information **IQ** needs to solve a problem and to present the solution in graphical form.

You may list the descriptor with the command

    LD

(**l**ist **d**escriptor). **IQ** answers by showing, on the display screen, the contents of Table A.1. (The list is very long and it may not fit on your screen in one piece, but you can list the top and bottom parts separately by using the commands LP or LB, see Sect. A.1.4.5.) It also prompts you for further input with IQ>.

Using the command

    PL

(**pl**ot) you may now ask **IQ** to solve the problem posed in descriptor 1 and plot the result. The display screen will be switched to graphic mode and the plot of Fig. A.1 will appear (in color if you are working with a color display). Our example shows a plot of the imaginary part of the complex function $w = e^z$ of the complex variable $z = x + iy$,

Table A.1. Output obtained with the LD command

```
DESCRIPTOR NB.   21, 27-Jun-89 14:12:49, DESCRIPTOR FILE = IQ000.DES
TI:Example of type 0 plot in user's guide
CA:Demonstration of Cartesian 3D Plot
TX:w=e^z
```

CH	.000	99.000	1.000	2.000	NG	.000	.000	.000	.000
XX	-3.000	3.000	-1.000	1.000	YY	-3.142	3.142	-3.142	3.142
ZZ	-1.000	1.000	-1.000	1.000	RP	.000	.000	.000	.000
AC	.000	.000	.000	.000	NL	11.000	9.000	.000	.000
PJ	60.000	-30.000	3.000	.000	BO	2.000	.000	2.000	.000
FO	5.000	.000	.000	.000	SI	.000	.000	.000	15.000
VO	.000	.000	.000	.000	V1	.000	.000	.000	.000
V2	.000	.000	.000	.000	V3	.000	.000	.000	.000
V4	.000	.000	.000	.000	V5	.000	.000	.000	.000
V6	.000	.000	.000	.000	V7	.000	.000	.000	.000
V8	.000	.000	.000	.000	V9	.000	.000	.000	.000
A1	5.000	.000	1.000	4.000	A2	5.000	1.000	1.000	6.000
AF	1.000	-3.000	-3.142	1.000	AP	.000	.000	.000	.000
TS	.000	.000	.000	.000	TL	.000	90.000	50.000	80.000
X1	1.000	2.500	.000	.000	X2	.000	.500	4.000	.200
Y1	1.000	4.500	.000	.000	Y2	.000	1.571	.000	.200
Z1	1.000	-3.000	-3.142	.000	Z2	.000	1.000	2.000	.200
P1	.000	.000	.000	.000	P2	.000	.000	.000	.000
C1	.000	.000	.000	.000	C2	.000	.000	.000	.000
R1	.000	.000	.000	.000	R2	.000	.000	.000	.000

```
T1:x
T2:y
TF:Im w
TP:
F1:
F2:
F3:
F4:
```

$$f(x, y) = \mathrm{Im}\, w = \mathrm{Im}\, e^{x+iy} = e^x \sin y \quad .$$

The function defines a surface in a Cartesian coordinate system spanned by the variables x, y, f. The surface is indicated by two sets of lines

$$f(x, y_i) \quad , \quad y_i = \mathrm{const} \quad ,$$
$$f(x_i, y) \quad , \quad x_i = \mathrm{const} \quad .$$

The surface in three-dimensional x, y, f space is projected onto the two-dimensional screen for plotting.

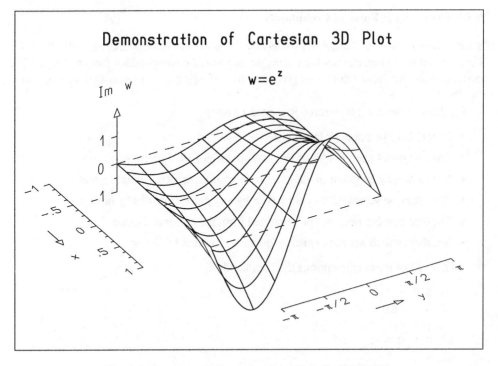

Fig. A.1. Plot corresponding to descriptor 21 on file IQ000.DES

You may want to see more lines in the surface. Switch the display screen back to alphanumeric mode, by pressing the RETURN key

    `<RET>`

Using the command LD again or simply looking back at Table A.1 you can find the line

    `NL 11.000 9.000 0.000 0.000`

in the descriptor, which indicates that 11 lines x = const and 9 lines y = const are to be drawn. You can change the descriptor with

    `NL 21 17`

(**number of lines**). The command

    `PL`

now produces a plot with the required number of lines.

As before `<RET>` brings you back to alphanumeric mode. The command

    `ST`

(**stop**) causes the program to terminate.

A.1.2 The General Form of Commands

In the example dialog of Sect. A.1.1 we have encountered commands like PL or NL 21 17.
They consist of 2 characters, which form the *command name* possibly followed by up to
four numbers. All input which you give directly following the *command prompt* IQ> has
this form.

Commands need not be written in a rigid format:

- Letters may be upper or lower case.
- The command need not be on the first cursor position after the prompt.
- Numbers may be given in integer, floating point, or exponential format.
- They may be separated by commas or by any number of blanks or tabs.
- The first number need not be separated from the command name.
- Numbers which are not explicitly given are assumed to be zero.

As a result of these conventions the commands

```
nl 0 5
NL 0.,.5E1
NL0,5
NL 0,5,0,0
NL,5,,
NL,5
```

are equivalent.

Many commands (like NL) are used to change numerical parameters in the descriptor,
see Sect. A.1.4. There are 40 such commands. Their names can be read from Table A.1,
in which each command name is followed by 4 numbers. If you do not want to change all
4 parameters you can use commands of the following form:

NL(2)5   will replace only the second parameter,
NL(1:2)0,5 or NL(:2)0,5   will replace the first two parameters,
NL(2:4)5,0,0 or NL(2:)5,0,0   will replace parameters number 2, 3 and 4.

After some commands **IQ** asks for additional input. You will be asked for this input by
special prompts, not the standard prompt IQ>. The prompt and the format of the additional
input is discussed in the sections where these commands are explained.

A.1.3 The Descriptor File

We have seen in the example of Sect. A.1.1 that a plot is completely described by a *descriptor (record)*. A number of descriptor records (usually describing related plots) are
grouped together in a *descriptor file* which resides on disk. The set of all descriptor files
forms the *descriptor library*, Fig. A.2. Less technically we also call it the *example library*.

Each descriptor file is a *direct access file*. This means that each descriptor can be read
or overwritten without the need to handle the other descriptors in the file.

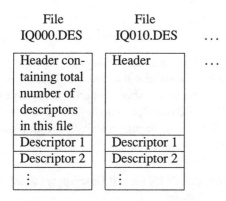

Fig. A.2. The descriptor library is composed of several descriptor files. Each descriptor file contains a certain number of descriptors

A.1.3.1 The Default Descriptor File

When you start **IQ** you automatically open the *default descriptor file* with the *file name* IQ000.DES. You may work with that file or choose another.

A.1.3.2 Choosing a Descriptor File

When you type the command

    CD

(**c**hoose **d**escriptor file), **IQ** prompts you with the line

    ENTER 3 CHARACTERS TO COMPLETE FILE NAME IQ???.DES >

to type another 3 characters. After you have typed, say, XYZ <RET>, **IQ** opens the descriptor file IQXYZ.DES, with which you may then work. You can abbreviate this piece of dialog by providing the required 3 characters directly together with the CD command in the form

    CD XYZ

A.1.3.3 Showing the Current Descriptor File Name

You may ask **IQ** to show you the name of the descriptor file you are currently working with using the command

    SD

(**s**how **d**escriptor file name).

A.1.3.4 Creating a New Descriptor File

If, using the command CD, you have asked for a descriptor file that does not yet exist, **IQ** writes the message

```
FILE IQXYZ.DES DOES NOT EXIST
DO YOU WANT TO CREATE THIS FILE? (Y)
```

The (Y) indicates that the default answer is *yes*. If you type Y <RET> or simply <RET> a descriptor file with this name will be created. If you type N <RET> no new file will be created and you will continue working with the original descriptor file.

The new descriptor file will contain just one descriptor in which all numerical parameters are set to zero and all character strings are blank except for the string TITLE which reads BLANK DESCRIPTOR.

A.1.3.5 Listing or Printing the Titles for All Descriptors in a File

Each descriptor has a *title* which describes its contents in one line of alphanumeric text. With the command

LT

(list titles) you get, for the current descriptor file, a list of all descriptor numbers, each followed by the descriptor title. The command has two optional arguments. In the general form

LT n_1, n_2

it lists the titles of descriptors $n_1, n_1 + 1, \ldots, n_2$. If n_1 is zero or not given, listing starts with descriptor 1. If n_2 is zero or not given or larger than the total number of descriptors in the file, listing proceeds up to the last descriptor in the file.

The command

PT n_1, n_2

(print titles) will list to the *output file* IQOUT.OUT, for later printing.

A.1.3.6 Printing All Descriptors in a File

The command

PA

(print all descriptors) writes not only the titles, but the full contents of all descriptors in the output file IQOUT.OUT for later printing. Like the command LT it can be used with two optional arguments in the form

PA n_1, n_2

A.1.3.7 Copying (Part of) a Descriptor File

You may copy a complete descriptor file or some descriptors from that file into another file. It is useful to prepare a file with selected descriptors for a particular session with **IQ**. After giving the command

CC

(choose copy file) you are prompted, as in Sect. A.1.3.2, to enter 3 characters to complete the name of the file into which you want to copy. You may also provide the 3 characters directly with the CC command, e.g.,

    CC XYZ

Again, as in Sect. A.1.3.4, you can create a new file. With the command

    SC

(show copy file name) you may ask **IQ** to show you the name of the copy file you are currently using.

The command

    CO n_1, n_2, n_3

(copy descriptors n_1 through n_2) copies the descriptors $n_1, n_1 + 1, \ldots, n_2$ from the current descriptor file to the current copy file. The copied descriptors are appended to the copy file, if n_3 is zero (or not given). Otherwise n_3 is the number of the first descriptor to be replaced in the copy file. The arguments n_1, n_2 have the same default values as for the command LT, see Sect. A.1.3.5. If, for example, the current descriptor file has 20 descriptors, the commands

    CO ,3
    CO 5,8
    CO 15,15
    CO 19

will copy descriptors $1, 2, 3, 5, 6, 7, 8, 15, 19, 20$. You can copy descriptors from different descriptor files into the same copy file by changing the current descriptor file with the CD command.

A.1.4 The Descriptor (Record)

A.1.4.1 General Structure of a Descriptor

An example of a descriptor was given in Table A.1. In the first line we find:

- the number of the descriptor in the descriptor file,
- the date and time of creation or last change of the descriptor,
- the descriptor file name.

You can only change these three items indirectly. The remaining descriptor contents can be modified by simple commands.

In Table A.1 there are lines beginning with a two-letter code and a colon, followed by alphanumeric text. The two-letter codes are

    TI, CA, TX, T1, T2, TF, TP, F1, F2, F3, F4

To change and use this information refer to Sect. A.4.4.1.

The rest of the descriptor consists of numerical information. There are 40 groups of 4 numbers, each group headed by a two-letter name. The meaning of these groups is discussed in Sects. A.2–A.5. You can change a group of numbers by giving the name, followed by the new numbers, see Sect. A.1.2.

A.1.4.2 Getting a Descriptor

You always work with exactly one descriptor. This descriptor is loaded from the current descriptor file into the computer memory by the command

>GD n

(**g**et **d**escriptor), where n is the number of the descriptor in the file. We call the descriptor which is in memory the *current descriptor*. You may now ask for the plot corresponding to this descriptor, change the current descriptor, plot again, etc. You may store the changed descriptor in two ways which are discussed in the next two sections.

A.1.4.3 Replacing a Descriptor in a File

With the command

>RD n

(**r**eplace **d**escriptor) you can overwrite the descriptor n in the current descriptor file with the current descriptor. The date and time recorded for the new descriptor is that of the moment of overwriting.

A.1.4.4 Appending a New Descriptor to a File

In a similar manner, with the command

>ND

(**n**ew **d**escriptor) you can append the current descriptor to the current descriptor file. Before you do that, you will usually wish to give the descriptor a new title using the TI command, see Sect. A.4.4.

A.1.4.5 Listing or Printing a Descriptor

In the example of Sect. A.1.1 we used the command

>LD

(**l**ist **d**escriptor) to list the current descriptor on the display screen in the form of Table A.1. Similarly the command

>PD

(**p**rint **d**escriptor) writes the current descriptor to the output file IQOUT.OUT for later printing.

Your display screen is probably too small to hold the full descriptor listing. You may get partial lists with the two following commands. Typing

>LP

(**l**ist **p**arameter part of descriptor) you get a list of the first part of the descriptor ending with the parameter groups V8 and V9. It is this part of the descriptor you will mostly be working with. The command

LB

(list background part of descriptor) gets you a listing of the second half preceded by the heading and the title line of the full descriptor. This half of the descriptor contains information on the "background" of the plots, see Sect. A.4.

Occasionally you may want to change the format in which the descriptor appears. With the command

OF 1

(output format) you change the way in which the floating point numbers in a descriptor are listed or printed from the fractional format of Table A.1 to exponential format. This format will be used until you change it back again with

OF 0

If you want more decimal digits given than available in the format of Table A.1 use the PD and LD command in the form

LD 1 and PD 1

Only one set of four numbers will then be placed in one line. Therefore in this format even the restricted output caused by the LP and LB commands will not fit on your screen. With the commands

LP 1 and LP 2

you will get a listing of the first half and the second half of the parameter part of the current descriptor with many decimal digits. For the background part use

LB 1 and LB 2

A.1.5 The PLOT Command

Practically all commands serve "administrative" purposes such as listing or changing descriptors and are executed almost instantaneously. The command

PL

(plot), however, starts the process of numerical calculation and graphical presentation. The time it takes depends on the problem posed and ranges between several seconds and several minutes. The display screen is automatically switched into graphic mode and you can watch how the plot is built up on the screen.

When the plot is complete you may study it at leisure, but before you can resume the dialog with **IQ** you must *switch the display back into alphanumeric mode* by just pressing

<RET>

The PL command has two optional arguments. In the general form

PL n_1 n_2

it performs consecutive plotting for the descriptors $n_1, n_1+1, \ldots, n_2$. If you do not specify n_2 only one plot is produced which corresponds to descriptor n_1.

A.1.6 The STOP Command

To end an **IQ** session you simply type the command

    ST

(**stop**). **IQ** acknowledges with the line

    IQ TERMINATED

and returns control to the operating system of your machine.

A.1.7 HELP: The Commands HE and PH

You may get on-line help on the explanation of the various commands available by typing

    HELP

or just

    HE

You will be prompted with the line

    ENTER KEYWORD FOR HELP>

to enter the command for which you want help. You will then get a short explanation on your display, followed by the usual prompt IQ>. If your input does not correspond to a command used by **IQ** you will get a list of all valid commands. You may then repeat the HELP command using that information.

A little faster than the procedure above is the HELP command in the form

    HELP *cm*

or

    HE *cm*

where *cm* stands for the command you want explained.

The parameters which determine in detail the physics problems you want to work on are given by the variables V0 through V9 in the descriptor. The association of physical parameters to these variables depends on the physics problem you have chosen with the CH command (Sect. A.3.1). With the command

    PH

(**physics help**) you get help on the physics problem chosen by the CH command in your current descriptor. If you want help on a different physics problem type

    PH c_1 c_2 c_3 c_4

with c_1, c_2, c_3, c_4 being the contents of CH(1), ..., CH(4) which would specify that particular physics problem. As answer to the PH command you get a menu of topics each preceded by a letter in parentheses. After typing that letter you get help on that topic. The most extensive topic is "Input parameters". Here subtopics are available. You get help on a subtopic by typing V0, V1, ..., V9. You can leave the subtopic level by typing <RET>. In the main topic level you can get a fresh menu of topics (which you may want if the old one has been scrolled off your screen) by typing S. You leave the help menu with <RET>.

A.2 Coordinate Systems and Transformations

A.2.1 The Different Coordinate Systems

A.2.1.1 3D World Coordinates (W3 Coordinates)

Figure A.1 shows a structure in three-dimensional space (3D space). Let us consider the whole structure to be built up of thin wires. We call the Cartesian coordinate system, in which this wire structure is described, the *3D world coordinate system* or *W3 coordinates* and denote a point in W3 coordinates by (X, Y, Z).

A.2.1.2 3D Computing Coordinates (C3 Coordinates)

Looking at the scales in Fig. A.1 we observe that, although the x scale and the y scale have approximately equal lengths in W3 coordinates (i.e., if regarded as two pieces of wire suspended in space), their lengths are quite different if expressed by the numbers written next to the scales. These latter lengths are $\Delta x = 2$ and $\Delta y = 2\pi$.

The plot illustrates the function

$$z = f(x, y)$$

where each point (x, y, z) is placed at the position (X, Y, Z) in W3 space. The coordinates x, y, z are called *3D computing coordinates* or simply *C3 coordinates*. They are connected to the W3 coordinates by a simple linear transformation given in Sect. A.2.2.

A.2.1.3 2D World Coordinates (W2 Coordinates)

Our three-dimensional structure given in W3 coordinates must of course be projected onto the two-dimensional display screen or on the paper in a plotter. This is done in two steps. We first project onto a plane placed in W3 space, and then (see Sect. A.2.1.4) from that plane onto the display screen or plotting paper.

Let us consider an observer situated at some point in W3 space looking at the origin $(X = 0, Y = 0, Z = 0)$. The unit vector $\hat{n}$ pointing from the origin to the observer is characterized by the *polar angle* ϑ (the angle between $\hat{n}$ and the Z direction) and the *azimuthal angle* φ (the angle between the projection of $\hat{n}$ onto the XY plane and the X axis). We now construct a plane somewhere in space perpendicular to $\hat{n}$ and a two-dimensional ξ, η coordinate system in that plane. The ξ axis is chosen parallel to the XY plane.

We call the coordinates ξ, η the system of *2D world coordinates* or simply *W2 coordinates*. We can now perform a projection parallel to $\hat{n}$ from W3 to W2 coordinates. The projection transformation is given in detail in Sect. A.2.2.

A.2.1.4 Device Coordinates (D Coordinates)

A final transformation leads us to the sensitive plane of the plotting device which can be the display screen or the paper in a plotter. We call the system of uv coordinates, used by the plotting device to address a point in the sensitive plane, the system of *device coordinates* or *D coordinates*. For the transformation from W2 to D coordinates see Sect. A.2.2.

A.2.2 Defining the Transformations

A.2.2.1 The Window–Viewport Concept

A linear transformation from a variable x to another variable X can be uniquely defined by specifying just 2 points x_a, x_b and the corresponding points X_a, X_b, see Fig. A.3. A general point x is transformed to

$$X = X_a + (x - x_a)\frac{X_b - X_a}{x_b - x_a} \quad .$$

If the range of the variable x is bounded it is useful to choose the pairs $(x_a, \ x_b)$ and (X_a, X_b) as the bounds of the variables. The interval

$$x_a \leq x \leq x_b$$

is called the *window* in x, whereas the interval

$$X_a \leq X \leq X_b$$

is called the *viewport* in X.

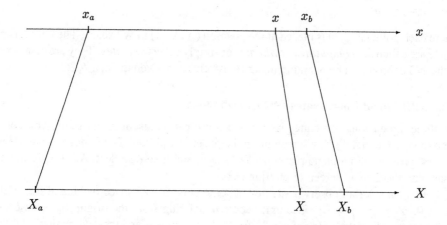

Fig. A.3. Linear transformation from x to X. The window in x is bounded by x_a, x_b, the viewport in X by X_a, X_b.

A.2.2.2 The Chain of Transformations C3→W3→W2→D

We have just seen that the transformation from C3 coordinates x to W3 coordinates X is

$$X = X_a + (x - x_a)\frac{X_b - X_a}{x_b - x_a}$$

and therefore is completely defined once the window (x_a, x_b) and the viewport (X_a, X_b) are given. Of course, completely analogous formulae hold for the transformations $y \rightarrow Y$ and $z \rightarrow Z$.

The transformation from W3 coordinates (X, Y, Z) to W2 coordinates (ξ, η) is a parallel projection and is given by

$$\xi = -X \sin \varphi + Y \cos \varphi \quad ,$$

$$\eta = -X \cos \varphi \cos \vartheta - Y \sin \varphi \cos \vartheta + Z \sin \vartheta \quad .$$

It is completely determined by the polar angle ϑ and the azimuthal angle φ of the direction $\hat{n}$ from the origin of W3 space to the observer.

Finally we define a window

$$\xi_a \leq \xi \leq \xi_b \, , \, \eta_a \leq \eta \leq \eta_b$$

in W2 space and a viewport in D space

$$u_a \leq u \leq u_b \, , \, v_a \leq v \leq v_b$$

and project that part of W2 space which falls inside the window onto the viewport in D space (i.e., the sensitive plotting device surface).

A.2.2.3 Defining the Window and Viewport in W3 and C3 Coordinates: The Commands XX, YY, ZZ

The viewport (X_a, X_b) in W3 and the window (x_a, x_b) in C3 coordinates are defined in the descriptor by the parameter group XX which has the general form

XX X_a X_b x_a x_b

Inspection of the descriptor in Table A.1 and the corresponding plot in Fig. A.1 shows that the viewport in X spans the range $-3 \leq X \leq 3$ whereas the window in x extends from -1 to 1. You may change that parameter group by the method described in Sect. A.1.2.

The *default value for the C3 window* is the size of the W3 viewport. That means you have to define only the W3 window if you want the C3 viewport to be of the same size. Therefore

    XX -1 1 0 0

is equivalent to

    XX -1 1 -1 1

There is also a *default value for the W3 window*. If you just give

    XX

which is the same as

    XX 0 0 0 0

IQ takes it to be

    XX 0 1 0 1

that is, the default for (X_a, X_b) is (0,1).

Commands of the form

YY Y_a Y_b y_a y_b
ZZ Z_a Z_b z_a z_b

set viewport (window) for the Y (y) and Z (z) coordinate, respectively. You should note one difference between the commands XX, YY on the one hand and ZZ on the other. Plots of the type of Fig. A.1 are illustrations of the function

$$z = f(x, y) \quad ,$$

in which x and y are varied over the ranges $x_a \leq x \leq x_b$ and $y_a \leq y \leq y_b$ of the windows x and y, respectively. The function $z = f(x, y)$ is drawn irrespective of the window (z_a, z_b). Viewport (Z_a, Z_b) and window (z_a, z_b) are given to establish the relation between C3 and W3 coordinates. Their use is convenient to *magnify* or *reduce* the plot in the Z direction. If, for example, instead of the default

ZZ 0 1 0 1

you use

ZZ 0 1 0 0.5

or the equivalent form

ZZ 0 2 0 1

the plot is magnified by a factor of 2 the Z direction.

A.2.2.4 Projecting from W3 to W2: The Command PJ

The polar angle ϑ and the azimuthal angle φ, defining the position of the observer with respect to the origin of the W3 coordinate system, are set by the command

PJ ϑ φ

(**projection**), where ϑ and φ are given in degrees.

A.2.2.5 Defining the Window in W2 Coordinates: The Command SI

A window is defined in W2 space by the command

SI X_t Y_t Z_t W

(**size** of window in W2). Here (X_t, Y_t, Z_t) are the coordinates of a *target point* in W3 coordinates. The projection (ξ_t, η_t) of this point is taken as the center of the rectangular window in W2. The window has the *width* W, i.e., in ξ it extends over the range

$$\xi_t - W/2 \leq \xi \leq \xi_t + W/2$$

The *height* of the window in η is chosen such that the window has the same width-to-height ratio (*aspect ratio*) as the viewport in D coordinates; see next section.

The SI command allows you to do *zooming* and *panning*. You zoom by changing the variable W, e.g., by typing

```
SI (4) .5
```

(followed, of course, by the PL command for plotting). To zoom in you decrease W, to zoom out you increase W. Changing the target point (X_t, Y_t, Z_t) you may move about your window in the W2 plane. (This technique is called "panning".) The default values chosen by **IQ** are $X_t = Y_t = Z_t = 0$, $W = 5$.

A.2.2.6 Defining the Viewport in D Coordinates: The Command FO

To determine the viewport

$$u_a \leq u \leq u_b \quad , \quad v_a \leq v \leq v_b$$

in D coordinates you use the first two variables of the FO command. It has the general form

```
FO u_b v_b
```

(format), and produces a viewport

$$0 \leq u \leq u_b \quad , \quad 0 \leq v \leq v_b \quad .$$

The numerical values of u_b, v_b correspond to centimeters on plotting paper. Thus

```
FO 10 5
```

will result in a plot 10 cm wide and 5 cm high.

If either the first or the second variable are given as zero the FO command chooses a DIN format.

```
FO 5 0
```

will result in DIN A5 (long) format

$$0 \leq u \leq 21.0\,\mathrm{cm} \quad , \quad 0 \leq v \leq 14.8\,\mathrm{cm} \quad ,$$

whereas

```
FO 0 5
```

gives rise to a plot in DIN A5 (high)

$$0 \leq u \leq 14.8\,\mathrm{cm} \quad , \quad 0 \leq v \leq 21.0\,\mathrm{cm} \quad .$$

The *default format* is DIN A5 (long). It will be produced by

```
FO 0 0
```

The format given in the way described so far is to be taken literally only for plotters which are able to produce plots of the required size. If the FO command asks for a size larger than can be accommodated on the paper the plot is shrunk so that it will fit on the paper without being distorted, i.e., the width-to-height ratio is preserved. The plot will be rotated through 90 degrees if that makes better use of the plotting paper. Finally the plot is centered on the plotting paper.

Plots on a display screen are never rotated but they always make optimal use of the screen size, since there is no reason why you would want a small plot on the screen.

A.3 The Different Types of Plot

A.3.1 Choosing a Plot Type: The Command CH

IQ provides for 5 quite different plot types:

type 0 plots: 3D plots based on a grid in Cartesian coordinates

type 1 plots: 3D plots based on a grid in polar coordinates

type 2 plots: 2D plots

type 3 plots: 3D column plots

type 10 plots: special 3D plots

With the command

   CH p_1 p_2 p_3 p_4

(**ch**oose plot type) you choose the type of plot you want. Here p_1 is the plot-type number. The variables p_2, p_3, p_4 are used to specify the particular problem for which you want the answer given as a plot of type p_1.

A.3.2 Cartesian 3D Plots (Type 0 Plots)

The plot of Fig. A.1 is of type 0. The function

$$z = f(x, y)$$

is plotted as a surface embedded in 3D space. In fact, only two sets of lines

$$z = f(x_i, y) \quad , \quad i = 1, \ldots, n_x \quad ,$$
$$z = f(x, y_i) \quad , \quad i = 1, \ldots, n_y$$

are drawn, where the x_i or y_i are constants for a given line. The two sets of lines correspond to a *Cartesian grid* in the xy plane. They are placed evenly in the xy window in C3 coordinates which corresponds to the XY viewport in W3 coordinates.

You may choose the number of lines in each set with the command

   NL n_x, n_y

(**number of lines**).

Within one line only a certain number of points $z = f(x, y)$ are computed. They are connected by straight lines. You can alter the number of points with the command

   AC ΔX

(**accuracy**). Here ΔX is the distance in the W3 coordinate X between two adjacent points. The corresponding quantity ΔY is set equal to ΔX. If you do not specify ΔX it is assumed to be $(X_b - X_a)/100$, i.e., one hundredth of the width of the viewport in W3. The values of ΔX and ΔY are still somewhat modified by **IQ** to match the grid defined by the XY viewport and the NL command.

Besides the surface which illustrates the function $z = f(x, y)$ several other items are contained in Fig. A.1 such as scales, arrows, letters and numbers. They make up the *background* of the plot. The commands controlling the background are discussed in Sect. A.4.

A.3.3 Polar 3D Plots (Type 1 Plots)

So far we have used Cartesian coordinates only. We now introduce polar coordinates r, φ in the xy plane

$$x = r \cos \varphi \quad , \quad y = r \sin \varphi$$

and construct two sets of lines in the xy plane

rays $\varphi = $ const
arcs $r = $ const

We call these two sets of lines a *polar grid*. The function

$$z = f(x, y) = g(r, \varphi)$$

is then illustrated by two sets of lines

$$z = g(r_i, \varphi) \quad , \quad i = 1, \ldots, n_r \quad ,$$

$$z = g(r, \varphi_i) \quad , \quad i = 1, \ldots, n_\varphi \quad .$$

We then again have a surface in 3D space on which two sets of lines are drawn. The projection of these lines onto the xy plane, however, forms a polar rather than a Cartesian grid as in type 0 plots. An example is given in Fig. A.4. Of course, the polar grid does not extend over the whole xy plane.

With the command

RP $R_1, R_2, \varphi_1, \varphi_2$

(**r** and **phi**) you specify the region in W3 coordinates

$$R_1 \leq R \leq R_2 \quad , \quad \varphi_1 \leq \varphi \leq \varphi_2$$

over which you want to extend the polar grid. (The angles are given in degrees.) Reasonable default values are used, i.e.,

RP 0 0 0 0

is equivalent to

RP 0 1 0 360

With

NL n_r, n_φ

(**number of lines**) you specify the number of arcs and rays, respectively.

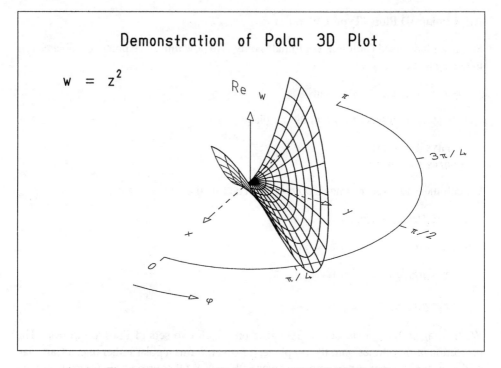

Fig. A.4. Plot corresponding to descriptor 22 on file IQ000.DES

The command

 AC $\Delta R, \Delta \varphi$

(**accuracy**) specifies the spacing of computed points on the rays and arcs, respectively. The *default values* are

$$\Delta R = (R_2 - R_1)/100 \quad , \quad \Delta \varphi = 1°$$

Again you can put *background* items in type 1 plots. See Sect. A.4 for commands controlling them.

A.3.4 2D Plots (Type 2 Plots)

The simplest illustration of a function

$$y = f(x)$$

of one variable is a graph in Cartesian xy coordinates. Such graphs are made by **IQ** in type 2 plots. An example is given in Fig. A.5. The function is drawn for the x region within the C3 window. The number of points computed is determined by the variable ΔX defined in the **AC** command, see Sect. A.3.2. It is possible to show a set of n curves in a type 2 plot. This is done with the command

 NL n

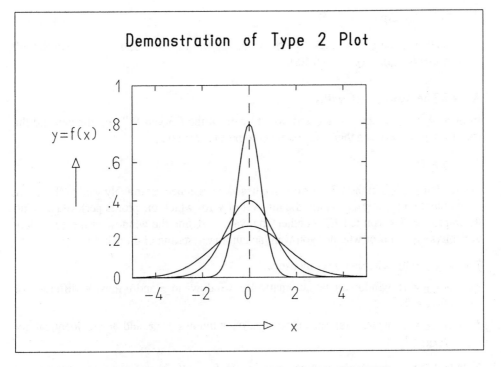

Fig. A.5. Plot corresponding to descriptor 23 on file IQ000.DES

(**number of lines**). In our example we have $n = 3$. The default value is $n = 1$.

Note the close similarity between type 2 plots and type 0 plots. In particular the background of both plot types is controlled in the same way, see Sect. A.4.

IQ offers you more possibilities for producing 2D plots. To use them you specify more parameters in the NL command

$$\text{NL } n \ f_{\text{VAR}} \ f_{\text{AUTOSCALE}} \ f_{\text{POLYMARKER}}$$

A.3.4.1 Choice of Independent Variable

The flag f_{VAR} specifies the independent variable:

$f_{\text{VAR}} = 0$: x is the independent variable. The graph plotted illustrates the function $y = f(x)$

$f_{\text{VAR}} = 1$: y is the independent variable. The graph plotted illustrates the function $x = f(y)$

$f_{\text{VAR}} = 2$: Neither x nor y is the independent variable. Instead, a parameter p serves as independent variable. The graph plotted illustrates the function $x = x(p)$, $y = y(p)$. The range of the parameter in this case is given by the RP command

$$\text{RP } p_1 \ p_2$$

(**range of parameter**) and the step width by the AC command

$$\text{AC}\ \ \Delta p$$

so that the parameter p takes on the values $p_1, p_1 + \Delta p, p_1 + 2\Delta p, \ldots, p2$. The default value for Δp is $(p_2 - p_1)/100$.

A.3.4.2 The Autoscale Facility

In Sect. A.2.2.3 we have discussed how to control the C3 and W3 coordinates and the transformation between them. If you want to plot the function

$$y = f(x)$$

the *window* (y_a, y_b) in the C3 coordinate y has to be matched reasonably well to the *range* of the function (y_{min}, y_{max}) within the interval of x for which the plot is performed. With the *autoscale flag* you tell **IQ** whether you want to define the window in the dependent variable(s) yourself or whether you want an automatic setting of the window(s).

$f_{\text{AUTOSCALE}} = 0$: window(s) set by user

$f_{\text{AUTOSCALE}} = 1$: window(s) set automatically, windows in x and y may be different for
$\quad\quad f_{\text{VAR}} = 2$

$f_{\text{AUTOSCALE}} = 2$: window(s) set automatically, windows in x and y are identical for
$\quad\quad f_{\text{VAR}} = 2$

Note that the window is set automatically in y if $f_{\text{VAR}} = 0$, in x if $f_{\text{VAR}} = 1$ and in x and y if $f_{\text{VAR}} = 2$. The autoscale facility is restricted to windows of "reasonable" width, i.e., $0.001 \leq \Delta y \leq 200000$. Automatic scales are particularly useful for multiple plots (see Sect. A.5.2) where the range of the dependent variable may vary between individual plots within a multiple plot, see Fig. 9.3.

A.3.4.3 The Polymarker Facility

In some cases it is useful to have marks placed on the graph at intervals which are equidistant in the independent variable. In particular in the case $f_{\text{VAR}} = 2$ this allows you to indicate the variation of the parameter p. In the jargon of computer graphics these marks are called *polymarkers*. Their use in type 2 plots is steered by the *polymarker flag*:

$f_{\text{POLYMARKER}} = 0$: *no* polymarkers are drawn

$f_{\text{POLYMARKER}} = 1$: polymarkers are drawn (as circles)

$f_{\text{POLYMARKER}} = 3$: polymarkers are drawn (as triangles)

$f_{\text{POLYMARKER}} = 4$: polymarkers are drawn (as squares)

With the AC command you specify further details

$$\text{AC}\ \ \Delta p\ \ \Delta n\ \ n_0\ \ R$$

where

Δp: step width of independent variable between two computed points of function. The computed points are numbered by $n = 0, 1, 2, \ldots$

Δn: difference in n between two consecutive points on which polymarkers are placed. (Default value: $\Delta n = 10$.)

n_0: first point on which polymarker is placed

R: half diameter (in W3 coordinates) of polymarker. (Default value: 2% of W3 viewport in X.)

For an example plot see Fig. 4.5. Note that the polymarkers are drawn as plot item 5, i.e., you can change their color or line width by changing C2(1), see Sect. A.5.1.2.

A.3.5 3D Column Plots (Type 3 Plots)

With the plot types discussed so far we can illustrate functions of one or two *continuous* variables $y = f(x)$ or $z = f(x, y)$. For functions of two variables which can take only *discrete* values we use type 3 plots. An example is given in Fig. A.6. It illustrates the

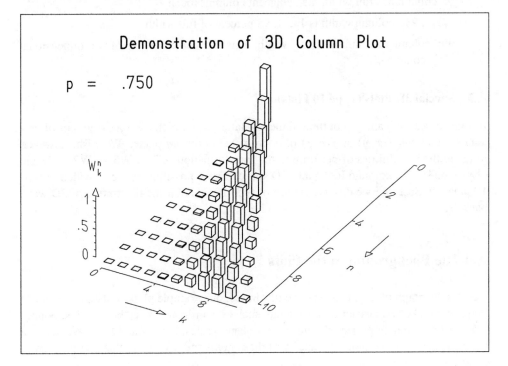

Fig. A.6. Plot corresponding to descriptor 24 on file IQ000.DES

binomial distribution W_k^n where both variables n and k can take only integer values. The plot is composed of vertical columns of height W_k^n placed at the positions $x = k$, $y = n$, in the xy plane.

The number of columns is determined by the NL command which, for type 3 plots, has the form

NL n_x n_y

with

n_x number of columns in the x direction
n_y number of columns in the y direction

Note that not all columns are actually drawn, because W_k^n is defined only for $k \leq n$.

With the AC command you can define the width of the columns. For type 3 plots the command takes the form

AC 0 0 w_X w_Y

where

w_X, w_Y indicate the widths of each column in X and Y direction, respectively. The numerical values of w_X and w_Y are interpreted as follows:

0: column has full width, i.e., adjacent columns touch

1, ..., 98: column width is 1, ..., 98 percent of full width

99: column has "infinitesimal" width. This allows the drawing of two-dimensional columns.

A.3.6 Special 3D Plots (Type 10 Plots)

In some cases we want to plot three-dimensional objects which are not the graphs of uni-valued functions $f(x, y)$ or $f(r, \varphi)$ of coordinates in the xy plane. We wish, however, to retain the possibilities of performing the transformations C3 → W3 → W2 → D and of placing the background items into 3D space. These possibilities are provided in type 10 plots. In Sect. 6.5 we discuss the use of type 10 plots for the illustration of 3D wave packets.

A.4 The Background in the Plots

Besides the graph of the function to be illustrated, our example plots contain a number of items that make them easier to understand, such as coordinate axes, "boxes", i.e., bound-aries of the graph as projected onto the xy plane, scales, arrows and text. We say that the *background* of our plots is made up of these items and discuss, in this section, how to control them.

A.4.1 Boxes and Coordinate Axes: The Command BO

It is often useful to draw a *box* indicating the range in the XY plane for which a function is drawn. You may also wish to see the coordinate axes, i.e., the x axis and the y axis of the C3 coordinate system. Boxes and coordinate axes are controlled by the command

BO b_1 b_2 a_1 a_2

Here b_1, b_2, a_1 and a_2 have the following meaning:

$b_1 = 0$: *no box* will be drawn

$b_1 = 1$: box will be composed of *continuous lines*

$b_1 = 2$: box will be composed of *dashed lines*

$b_2 = 0$: there will be *no ticks* on the lines forming the box

$b_2 = 1$: there will be *ticks* pointing into the box

$b_2 = 2$: there will be ticks and *numbers* on the two sides of the box which are best visible to the observer. (Details about ticks and numbers are controlled by the commands X1, X2, Y1, Y2; see Sect. A.4.2)

$a_1 = 0$: *no coordinate axes* will be drawn

$a_1 = 1$: axes will be drawn within the W3 viewport as *continuous lines*

$a_1 = 2$: axes will be drawn as *dashed lines*

$a_2 = 0$: there will be *no ticks* on the axes

$a_2 = 1$: there will be *ticks*

$a_2 = 2$: there will be ticks and *numbers*

Details of ticks and numbers are controlled by the commands on x and y scales; see next section. Note that for type 1 plots the "box" is not rectangular, since the range for which the function is drawn is limited by rays and arcs. In the example of Fig. A.4 it is a semi circle. For type 2 plots graphs of functions are *clipped* (if necessary) at the upper and lower edges of the box, if you ask for the box to be drawn.

A.4.2 Scales

A.4.2.1 The Scale in x: The Commands X1 and X2

The scale in x is defined by the two commands:

```
X1  f  d  l_d  e
X2  x_0  δx  n_i  l_t
```

with

$f = 0$: *no scale*

$f = 1$: scale as a *continuous line*

$f = 2$: scale as a *dashed line*

d: *distance* of scale from box (in W3 coordinates)

l_d: *length of dashes* (in W3 coordinates). If l_d is not specified, a reasonable default value is used. Note that not only the dash length of the scales is determined by l_d but also the dash length used for the box and the coordinate axes

$e = 0$: numbers are written in decimal notation, e.g., 0.012

$e = 1$: numbers are written in exponential notation, e.g., 1.2×10^{-2}

x_0: a value of x (in C3 coordinates) at which you want a tick on the scale accompanied by a number

δx: a positive number (in C3 coordinates) defining the distance between two ticks with numbers. You will get ticks with numbers at $\dots, x_0 - \delta x, x_0, x_0 + \delta x, \dots$ (If δx is set equal to zero, a reasonable default value is taken.)

n_i: the *number of intervals* between two numbered ticks which are marked off by additional ticks. Default is $n_i = 1$, i.e., no additional tick

l_t: *length of ticks* (in W3 coordinates). If l_t is set larger than the box size it is restricted to the box size. If l_t is zero, a reasonable default value is taken

Note that the X2 command also controls the ticks and numbers on the x coordinate axis and on the edges of the box parallel to that axis.

A.4.2.2 The Scale in y: The Commands Y1 and Y2

The scale in y is defined by the two commands Y1 and Y2 which are completely analogous to the X1 and X2 commands of Sect. A.4.2.1.

A.4.2.3 The Scale in z: The Commands Z1 and Z2

The scale in the z direction is defined by the two commands

$$\text{Z1} \ f \ X_z \ Y_z \ l_d$$
$$\text{Z2} \ z_0 \ \delta z \ n_i \ l_t$$

with

$f = 0$: no scale

$f = 1$: scale as continuous line, numbers in decimal notation, e.g., 0.012

$f = 2$: scale as dashed line, numbers in decimal notation

$f = 3$: scale as continuous line, numbers in exponential notation, e.g., 1.2×10^{-2}

$f = 4$: scale as dashed line, numbers in exponential notation

$l_d, z_0, \delta z, n_i, l_t$: analogous to corresponding variables in X1 and X2 command

X_z, Y_z: W3 coordinates of the point in which the scale intersects with the XY plane

The length of the scale is that of the viewport in W3 coordinates as defined in the ZZ command just as the lengths of the scales in x and y are given by the viewport size in the XY plane.

A.4.2.4 Scale in Phi: The Commands P1 and P2

In type 1 plots you may want a scale for the polar angle φ. It is provided by the commands

$$\text{P1} \ f_1 \ d \ l_d \ f_2$$
$$\text{P2} \ \varphi_0 \ \delta\varphi \ n_i \ l_t$$

with

$f_1 = 0$: *no scale*

$f_1 = 1$: scale as a *continuous line* at radius $R_2 + d$

$f_1 = 2$: scale as a *dashed line* at radius $R_2 + d$

$f_1 = -1$: scale as a *continuous line* at radius $R_1 - d$

$f_1 = -2$: scale as a *dashed line* at radius $R_1 - d$

d: *radial distance* (in W3 coordinates) of scale from box, i.e., the scale is an arc of radius $R_2 + d$ or $R_1 - d$, respectively. For R_1 and R_2 see the RP command in Sect. A.3.3

l_d: *dash length* analogous to the command in X1 or Y1 (in degrees)

$f_2 = 0$: *no ticks* on the circle-segment

$f_2 = 1$: ticks pointing outwards

$f_2 = 2$: with ticks pointing outwards and numbers in decimal notation

$f_2 = 3$: with ticks pointing outwards and numbers in exponential notation

$f_2 = -1$: ticks pointing inwards

$f_2 = -2$: with ticks pointing inwards and numbers in decimal notation

$f_2 = -3$: with ticks pointing inwards and numbers in exponential notation

$\varphi_0, \delta\varphi, n_i, l_t$: analogous to the corresponding arguments in the X2 command. For simplicity you enter the numerical values of φ_0 and $\delta\varphi$ in *degrees*, although the scale plotted will be numbered in radians. If $\delta\varphi$ is set to zero the default value 45° is taken

A.4.3 Arrows

A.4.3.1 The Arrows in the xy Plane: The Commands A1 and A2

You may place two arrows in the xy plane using the command

A1 f_1 f_2 l d

(arrow 1) or the completely analogous A2 command. The parameters of A1 (f_1, f_2, l, d) determine the position, orientation and length of the arrow. A *side* of the box (or rather XY viewport) is chosen by specifying f_2 and the arrow is placed on that side of the box. Details of the location and orientation are fixed through f_1, see Fig.A.7.

$f_1 = 0$: *no arrow*

$f_1 = 1$: arrow *perpendicular* to box side in the *middle* of the side

$f_1 = 2$: arrow *perpendicular* to box side at the *lower* (C3) *end* of the side

$f_1 = 3$: arrow *perpendicular* to box side at the *upper* (C3) *end* of the side

$f_1 = 4$: arrow *along* (C3) *coordinate axis*

$f_1 = 5$: arrow *parallel* to box side in the *middle* of the side

$f_1 = -5$: arrow *antiparallel* to box side in the *middle* of the side

$f_1 = 6$: arrow *parallel* to box side at the *lower* (C3) *end* of the side

$f_1 = -6$: arrow *antiparallel* to box side at the *lower* (C3) *end* of the side

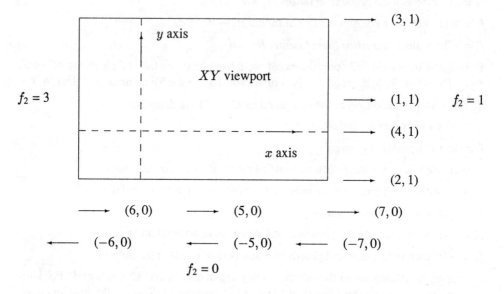

Fig. A.7. Different arrows characterized by the pairs of variables (f_1, f_2)

$f_1 = 7$: arrow *parallel* to box side at the *upper* (C3) *end* of the side

$f_1 = -7$: arrow *antiparallel* to box side at the *upper* (C3) *end* of the side

$f_2 \leq 0$: chosen box side is *bottom* (low y)

$f_2 = 1$: chosen box side is *right* (high x)

$f_2 = 2$: chosen box side is *top* (high y)

$f_2 \geq 3$: chosen box side is *left* (low x)

l: length of arrow (in W3 units)

d: distance of arrow from box side (in W3 units)

A.4.3.2 An Arrow in the z Direction: The Command AF

An arrow perpendicular to the xy plane is produced by the command

   AF l X_b Y_b Z_b

(arrow in function direction, i.e., z direction). The parameters are

$l = 0$: no arrow drawn. Base of arrow serves to position text

$l > 0$: *length* of arrow. Arrow points in $+z$ direction

$l < 0$: arrow points in $-z$ direction and has the length $|l|$

X_b, Y_b, Z_b: W3 coordinates of the arrow base

A.4.3.3 An Arrow in the Azimuthal Direction: The Command AP

In type 1 plots which employ polar coordinates, it can be useful to have an arrow in the xy plane pointing in the azimuthal direction. Graphically, the arrow is a section of a circle in the xy plane centered at the origin of the W3 coordinate system. It is provided by the command

AP φ_l R_b φ_b

(arrow in phi). The parameters signify

$\varphi_l = 0$: *no arrow*

$\varphi_l > 0$: *length* of arrow in degrees. Arrow points in $+\varphi$ direction

$\varphi_l < 0$: arrow points in $-\varphi$ direction and has length $|\varphi|$

R_b, φ_b: polar coordinates of *arrow base* in the xy plane (R_b in W3 coordinates, φ_b in degrees)

A.4.4 Text and Numbers

A.4.4.1 The Different Text Strings

Each descriptor can contain up to 11 text strings with up to 75 characters each which serve different purposes:

Title TI: a line of text, serving as title for the descriptor, but which is not shown in the plot

Caption CA: a line of text, shown in the plot, and which may serve as a short figure caption

Additional text TX: a further line of text shown in the plot

Text at arrow 1 in the xy plane (T1)

Text at arrow 2 in the xy plane (T2)

Text at arrow in the z direction (TF)

Text at arrow in the φ direction (TP)

Four additional text strings which may be used in special applications (F1, F2, F3, F4)

In a descriptor listing the *string names*, TI, CA, TX, T1, T2, TF, TP, F1, F2, F3, F4 are shown followed by the text strings themselves. A string is entered by giving the string name as command. If, for example, you type

TI

(title). **IQ** uses the string name to answer with the *text input prompt*

TI>

asking you to type any alphanumeric text of up to 75 characters (followed by <RET>). After you have provided that input **IQ** will answer with the normal command prompt IQ>.

A.4.4.2 Text Size: The Command TS

The sizes in which the different texts appear in the plot are controlled by the command

$$\text{TS} \quad s_{CA} \quad s_{AR} \quad s_{NB} \quad s_{TX}$$

with

s_{CA}: size of letters in caption

s_{AR}: size of text at arrows

s_{NB}: size of numbers at scales

s_{TX}: size of additional text

The unit in which the sizes (the vertical extensions of capital letters) are given is *percent of the plot diagonal*. This ensures that the text size grows or shrinks in a reasonable way with the plot size. *Default values* are

$$s_{CA} = s_{AR} = s_{NB} = s_{TX} = 2 \quad .$$

A.4.4.3 Text Location: The Command TL

The positions within the plot for text at arrows, or numbers at scales, is determined through the commands defining the scales or arrows. Thus you have to give only the position of the caption and the additional text. These texts are always written horizontally on the plot. It is therefore sufficient to specify the position of the beginning (actually the lower-left point of the first letter) of a text string. This is done with the command

$$\text{TL} \quad u_{CA} \quad v_{CA} \quad u_{TX} \quad v_{TX}$$

(**text** location) where

u_{CA}: horizontal location of caption

v_{CA}: vertical location of caption

u_{TX}: horizontal location of additional text

v_{TX}: vertical location of additional text

The numbers are given in *percent of the horizontal or vertical plot sizes*, respectively. For $u_{CA} = 0$ or $u_{TX} = 0$ the corresponding strings are *centered* horizontally. The default values are

$$u_{CA} = 0 \quad , \quad v_{CA} = 90 \quad , \quad u_{TX} = 0 \quad , \quad v_{TX} = 80 \quad .$$

In many cases texts are provided automatically. If you do not want to display them you may push them outside the plot by giving a vertical position larger than 100.

A.4.5 Mathematical Symbols and Formulae

Text appearing in the plot may contain mathematical symbols and simple formulae. To achieve this the characters in a text string are interpreted by **IQ** before they are plotted. You have *three different alphabets* at your disposal: *Roman*, *Greek* and *Math*, see Table A.2. You choose the alphabet with special characters called *selection flags*:

@ for Roman,

& for Greek,

% for Math.

The selection flag closest to, and on the left of, a character, determines the alphabet for that character. The default selection flag is @, i.e., characters not preceded by any selection flag are interpreted as Roman.

Besides selection flags there are *position flags*:

^ for superscript,

_ for subscript,

for normal line level,

" for backspace.

The default position flag is #, i.e., characters appear on the normal line level if not preceded by any position flag. You can use up to two consecutive steps of subscript or superscript leading you away from the normal line, e.g., A_{α_β}, A_{α^β}, A^{α^β}, A^{α_β}. The backspace flag (") causes the character following it to overwrite the preceding character rather than to be placed to the right of it. It allows the writing of expressions like W_n^k instead of $W_n{}^k$. For examples, see the plots in Sects. A.1 and A.3.

A.5 Further Commands

A.5.1 Line Styles

A.5.1.1 Hidden Lines

In our example plots of 3D objects, i.e., type 0, type 1 and type 3 plots, a *hidden-line technique* is used. This means that only those lines of the 3D object are shown which are visible to the observer if the object is considered solid and nontransparent. In the usual case of type 0 and type 1 plots the 3D object is just a surface. Of this surface one usually sees part of the *top side* and part of the *bottom side*. (You would only see the top side if you were watching with a polar angle $\vartheta = 0$ and only the bottom side for $\vartheta = 180°$.) The hidden-line technique allows you to treat the top side and the bottom side differently. Lines hidden by the object itself are not drawn. If ever you do not want to hide these lines you have to use the hidden-line flag f_{HL} in the PJ command

$$PJ\ \vartheta\ \varphi\ f_{HL}\ n_{PD}$$

with

Table A.2. The three alphabets available for plotting text

	Select Flags				Select Flags		
	Roman	Greek	Math		Roman	Greek	Math
Input	@	&	%	Input	@	&	%
A	A	A(ALPHA)	Ä	a	a	α(alpha)	ä
B	B	B(BETA)	B	b	b	β(beta)	b
C	C	X(CHI)	⌐	c	c	χ(chi)	c
D	D	Δ(DELTA)	Δ	d	d	δ(delta)	d
E	E	E(EPSILON)	E	e	e	ϵ(epsilon)	e
F	F	Φ(PHI)	F	f	f	φ(phi)	f
G	G	Γ(GAMMA)	$\neq$	g	g	γ(gamma)	g
H	H	H(ETA)	H	h	h	η(eta)	h
I	I	I(IOTA)	$\int$	i	i	ι(iota)	i
J	J	I(IOTA)	J	j	j	ι(iota)	j
K	K	K(KAPPA)	K	k	k	κ(kappa)	k
L	L	Λ(LAMBDA)			l	l	λ(lambda)	l
M	M	M(MU)	$\pm$	m	m	μ(mu)	m
N	N	N(NU)	N	n	n	ν(nu)	n
O	O	Ω(OMEGA)	P	o	o	ω(omega)	ö
P	P	Π(PI)	Ö	p	p	π(pi)	p
Q	Q	Θ(THETA)	Q	q	q	ϑ(theta)	q
R	R	R(RHO)	o	r	r	ρ(rho)	r
S	S	Σ(SIGMA)	ß	s	s	σ(sigma)	s
T	T	T(TAU)	¦	t	t	τ(tau)	t
U	U	O(OMICRON)	Ü	u	u	o(omicron)	ü
V	V		Ü	v	v		v
W	W	Ψ(PSI)	$\surd$	w	w	ψ(psi)	w
X	X	Ξ(XI)	X	x	x	ξ(xi)	x
Y	Y	Υ(UPSILON)	Å	y	y	υ(upsilon)	y
Z	Z	Z(ZETA)	Z	z	z	ζ(zeta)	z
~	~	~	~	—	—	—	—
!	!	!	!	=	=	=	$\equiv$
$	$	$	$	{	{	{	{
*	*	#	$\times$	}	}	}	}
(	(	↑	←	|	|	|	|
)	)	↓	→	[	[	&	[
+	+	+	+	]	]	@	]
'	'	'	'	\			
1	1	1	1	:	:	:	:
2	2	2	2	;	;	;	;
3	3	3	3	,	,	'	,
4	4	4	4	<	<	$\subset$	$\leq$
5	5	5	5	>	>	$\supset$	$\geq$
6	6	6	6	?	?	§	~
7	7	7	7	.	.	.	.
8	8	8	8	.	.	.	.
9	9	9	9	/	/	\	%
0	0	0	0				

ϑ, φ: as in Sect. A.2.2

$f_{HL} = 0$: hidden-line technique used, top and bottom side shown in same color

$f_{HL} = 1$: hidden-line technique not used,

$f_{HL} = 2$: hidden-line technique used, only top side shown

$f_{HL} = 3$: hidden-line technique used, top and bottom side shown in different colors

n_{PD}: number of plotting device for which changes of color index are to be performed, see next section.

Usually both f_{HL} and n_{PD} will be zero, so you need not give these arguments in the PJ command.

A.5.1.2 Colors and Line Widths

With many plotting devices it is possible to draw lines in different colors or widths. **IQ** makes use of this possibility of plotting different items in a plot in different colors or line widths. There can be up to 8 different items in a plot which are distinguished by their *item number*, see Table A.3. It is possible to draw each item in a different color or line width. Of course, the choice of color or line width depends on the plotting device you use. For the choice of plotting devices see Sect. A.5.4. Each device is identified by its *device number*. A device may be able to draw different colors or line widths which are specified by a *color index*. As an example, a plotter may have 6 pens which are addressed by a color index which can take the values 1,2,...,6.

Table A.3. The different plot items and their numbers

Item number	Item Name
1	function (3D surface) top side only for $f_{HL} = 3$, top side and bottom side for $f_{HL} = 0$
2	box, scales, numbers, arrows with text
3	caption
4	additional text
5	item 5
6	item 6
7	item 7
8	function (3D surface) bottom side for $f_{HL} = 0$

IQ uses a *table of default color indices* (see Appendix D) to find for each plot item the color index which is to be used to draw that item on a given device. You may change that color index with the commands

$$\text{C1} \quad c_1 \quad c_2 \quad c_3 \quad c_4$$

(colors part **1**) and

$$\text{C2} \quad c_5 \quad c_6 \quad c_7 \quad c_8$$

(colors part **2**). Here

$$c_i \ (i = 1, \ldots, 8)$$

is the color index to be used for plot item i.

If a c_i is zero, the color index from the table of default color indices is taken instead. The color indices in the C1 and C2 commands apply only to one particular plotting device (see Appendix D for details). The default for this device usually is your display screen. If you want to change the indices for another device you specify it as the argument n_{PD} in the PJ command; see the preceding section.

A.5.2 Multiple Plots

For some applications one might like to see series of plots in which one or two parameters are varied. With the command

$$\text{NG} \quad h_1 \quad v_1 \quad h_2 \quad v_2$$

(**n**umber of **g**raphs) you will get such a series of plots. The arguments h_1, h_2, v_1, v_2 are integer numbers, such that the individual plots in the series can be characterized by a pair of indices (h, v) where h and v take the values

$$
\begin{array}{llll}
(h_1, v_1) & , & (h_1 + 1, v_1) & , \quad \ldots \quad (h_2, v_1) & , \\
(h_1, v_1 + 1) & , & (h_1 + 1, v_1 + 1) & , \quad \ldots \quad (h_2, v_1 + 1) & , \\
\quad \vdots & & & \\
(h_1, v_2) & , & (h_1 + 1, v_2) & , \quad \ldots \quad (h_2, v_2) & .
\end{array}
$$

Parameters in an application can be functions of h and v so that they change from plot to plot.

You may get the series of plots as separate outputs, i.e., on separate sheets of paper if you use a plotter, or one after the other on the display screen. It is also possible to have all plots in the series arranged like a matrix on the same output page on which h runs horizontally and v runs vertically, see for example Fig. 9.1. You can choose either mode with the command

$$\text{FO} \quad u_b \quad v_b \quad f_m \quad f_s$$

with

u_b, v_b: as in Sect. A.2.1.4

$f_m = 0$: *multiple* plots as separate outputs

$f_m = 1$: *multiple* plots as single output

$f_s = 0$: individual plots on single output *not separated* by lines

$f_s = 1$: individual plots *separated* by lines

A.5.3 Combined Plots

The multiple plots of the preceding section are based on a single descriptor. All individual plots within one multiple plot are therefore necessarily very similar in their structure, in particular they are of the same plot type. It is sometimes useful to combine plots corresponding to different descriptors. Such a *combined plot* is defined by a *mother descriptor* which in turn *quotes* individual descriptors.

A descriptor is identified as a mother descriptor for a combined plot through the command

        CH -1

The number of graphs within the combined plot is given by the command

        NG h_1 h_2 h_3 h_4

which was discussed in the preceding section. It defines a matrix of $(h_2 - h_1 + 1) \times (v_2 - v_1 + 1)$ plots. Each plot is defined by a descriptor which has to be on the same file as the mother descriptor. Up to 40 descriptors may be quoted. Their numbers $d_1, d_2, \ldots, d_{40}$ are given by the commands

        VO d_1 d_2 d_3 d_4
        V1 d_5 d_6 d_7 d_8
        $\vdots$
        V9 d_{37} d_{38} d_{39} d_{40}

The descriptor d_1 gives rise to the leftmost plot in the top row, d_2 to the second plot in the top row and so on row-wise until the whole matrix of plots is filled. Additional descriptor numbers are ignored. If a descriptor number is zero the corresponding plot is left empty.

The FO command in the mother descriptor

        FO u_b v_b f_m f_s

as discussed in the preceding section also applies to combined plots. In particular it defines the format of all individual plots within a combined plot.

Nesting of multiple plots within combined plots is not possible. Therefore do not quote a descriptor defining a multiple plot in the sense of the preceding section in a mother descriptor. For an example of a combined plot see Fig. 4.5.

A.5.4 Using Different Plotting Devices

With the command

        SP

(show **p**lotting devices) you will get a list similar to that in Table A.4 on your display.

The table contains one line per device. Column 1 contains the *device number* by which you refer to the device. In column 2 you find a flag indicating whether the device is *activated*(1) or *not activated*(0). The PL command yields plots simultaneously on all activated devices. In column 3 you find the *device name* and in column 4 the device type. Devices

Table A.4. Example of a device table

```
NUMBER OF DEVICES IN TABLE:  5
NUMBER OF ACTIVE DEVICES  :  1
```

```
ACTIVE DEVICES ARE:  1
DEV DEVICE DEVICE              DEV. DEVICE    SURFACE SIZE    SURFACE SIZE
NB  ACTIVE NAME               TYPE INSTALLED     (CM)        (INT.UNITS)
 1    1    vga                 0     1        .00    .00    640    480
 2    0    8514/A              0     0        .00    .00   1024    760
 3    0    HERCULES            0     0      -4.00  -3.00    720    348
 4    0    HP-GL PLOTTER       0     1      27.25  19.12  10900   7650
 5    0    Metafile            0     1      28.99  20.11   6850   4750
```

```
YOU CAN ACTIVATE (DEACTIVATE) A PLOTTING DEVICE BY THE CP-COMMAND
FORMAT:CP    device nb.    device active status
e.g.  :CP    2 0    will turn device 2 off
```

can be of type 0 or type 1. There may be several type 0 devices active at the same time but only one type 1 device. The flag in column 5 is 1 if the device is installed and 0 otherwise. You can activate only installed devices (see Appendix D for the installation procedure). The remaining columns give the size of the active surface of the device in centimeters (column 6 and 7) and in internal units. (These numbers need not concern you unless you want to install the device.)

You can activate or deactivate a plotting device by using the command

$$CP \quad n_{PD} \quad s_{PD}$$

(choose **p**lotting device) with

n_{PD}: *device number*

s_{PD}: *device status*, i.e.,

$\quad s_{PD} = 0$: not active

$\quad s_{PD} = 1$: active

You may have a maximum of 3 devices active. **IQ** will warn you if you want to activate a 4th device. Since there cannot be more than one active type 1 device, **IQ** will deactivate any active type 1 device automatically if you want to activate another type 1 device.

A.5.5 The Different Running Modes

A.5.5.1 Interactive Mode: The Command IM

When you start **IQ** you are in *interactive mode*, i.e., the program gets its commands from the keyboard as input unit. **IQ** is designed primarily for interactive mode, hence its name

INTERQUANTA. You may, however, also work in an *automatic mode* in which a prerecorded list of commands is read from a file. You may switch back from automatic mode with the command

    IM

(interactive mode). To serve a nontrivial purpose, of course, the command IM has to be given while **IQ** is in semiautomatic mode, Sect. A.5.5.3, or in automatic mode. In the latter case the command IM has to be the last of series of commands in a command input file.

A.5.5.2 Automatic Mode: The Command AM

You switch from interactive to automatic mode with the command

    AM

(automatic mode). **IQ** asks you with the prompt

    ENTER 3 CHARACTERS TO COMPLETE INPUT FILE NAME IQ???.INP>

to enter 3 characters followed by <RET>, e.g. D12 <RET>. You can abbreviate this procedure by providing the 3 characters directly with the AM command, e.g.,

    AM D12

If a file with the name you have completed, e.g., IQD12.INP, does not exist **IQ** will tell you and you can repeat the procedure using the AM command again. **IQ** will then read its next command from the specified input file, display it on the screen and carry it out, read the next command from this file, etc. until it encounters either the command IM which switches back to interactive mode or the command ST which stops **IQ** altogether.

A.5.5.3 Semiautomatic Mode: The Command SM

Semiautomatic mode invoked by the command

    SA

(semiautomatic mode) differs from the automatic mode only in so far as **IQ** expects you to press <RET> after the display of every plot and every *comment* (see Sect. A.5.5.5) and will only then proceed. In this way you can study plots and comments for as long as you wish. You may switch from semiautomatic mode to interactive mode by typing IM and terminate **IQ** with ST.

A.5.5.4 Automatic Mode for Plot Production

You may want to use the fully automatic mode to produce a number of plots when you do not want to be present at your computer. The following could be an example of your input file which you prepare with the editor you have available on your machine:

```
CD
010
CP 5,1
PL 3
PL 9,11
PL 5
ST
```

The first line invokes the CD command with which you choose the descriptor file. With the second line you complete the descriptor file name to read IQ010.DES. With the next command you choose the plotting device number 5. Then plots are made corresponding to descriptors 3, 9, 10, 11, 5, and finally **IQ** is stopped.

A.5.5.5 Automatic Mode for Demonstrations

Automatic mode is also useful for demonstration. These may be shown in a lecture on quantum mechanics or as an introduction to an interactive session a student group may have with **IQ**. For demonstration the input file can contain *comments* which are displayed on the screen. A comment begins with a line containing only the character ' [' followed by any number of text lines and is concluded by a line containing just the character ']'. If you want to study an example of a command input file print one of the files listed in Appendix C.4.

In a demonstration one wants to give the spectator time to analyze the graphics, the comments or other items shown on the screen. This can be done either in the semiautomatic mode (which gives the user complete control over the waiting times after plots and comments) or in automatic mode. In this mode **IQ** waits for predetermined times which are measured in *units of sleeping time*. **IQ** will wait

- for 20 units after displaying a command
- for 100 units after displaying a comment
- for 100 units after displaying a plot

You may adjust the time interval of one unit to your needs using the command

```
CS n
```

(choose sleeping time). One unit of sleeping time is the time your machine needs to compute $100n$ times the sine of some number.

You may also put the command

```
WT n
```

(wait) at any position of your input file. It will cause **IQ** to wait for n units of sleeping time before proceeding.

The command input file for a demonstration may end in three different ways. Normally the last command will be

```
ST
```

so that **IQ** terminates. If the last command is

    IM

you can directly continue with an interactive session. Finally the last command can be

    RI

(**rewind input file**). In this case the input file is read again from the beginning so that you have an "endless" demonstration.

A.5.6 Introducing Physical Variables: The Commands V0 to V9

The commands V0, V1,...,V9 are used to define physical parameters. Their meaning depends on the physics problem at hand, see Chapters 2 to 9.

A.5.7 Reserved Commands

The commands R1 and R2 appearing at the end of the numerical part of each descriptor (see Table A.1) are reserved for future use.

Appendix B: How to Install IQ

B.1 Hardware Requirements

IQ will run on any IBM PC/XT or AT personal computer with at least 640 kbyte memory, on any IBM PS/2 or on any compatible computer. For the graphics devices which are supported by **IQ** see Appendix D. A *mathematical coprocessor* is strongly recommended although **IQ** will run without one, however, very, very slowly. Also strongly recommended is a *hard disk*. If you have a diskette drive only you have to assemble all files necessary for your session on a diskette. (You may omit the HELP files but then you must not try to use the HELP facility.)

B.2 Operating-System Requirements

You *must* work under the operating system PC DOS or MS DOS version 3.3 or higher.

B.3 Diskette Format

The program is distributed on a high-density (1.2 MByte) $5\frac{1}{4}''$ diskette. We are aware of the fact that you may not have a drive for such diskettes. But for economic reasons we had to write in high density and at the time of writing $5\frac{1}{4}''$ diskette drives are still more common than $3\frac{1}{2}''$ ones. If on your computer you have only a drive with a different format, find a friend or colleague with a PC and a $5\frac{1}{4}''$ high-density drive and follow the procedures in Sects. B.4 and B.5.

B.4 Installation

Make sure that you work in the root directory of your hard disk. We assume that the name of your $5\frac{1}{4}''$ diskette drive is **A**. If it is not, replace **A** by the appropriate letter in the following command. Insert the diskette labeled "INTERQUANTA" into the diskette drive and type

```
A:INSTALL
```

After a while you will get the message

```
Installation of IQ completed.
```

In the installation procedure a new directory \IQ was created and all files as described in Appendix C were copied to this directory. You can now start **IQ** by just typing

```
IQ
```

(see Sect. 2.2 or Appendix A).

B.5 Reformatting IQ for Different Types of Diskette

Once you have **IQ** installed on a hard disk you may write the complete content of the directory \IQ onto a diskette of any format, provided you have the appropriate drive. We assume that this drive is labeled B, otherwise replace B by the appropriate letter in the following commands. Now insert a formatted diskette into drive B and type

```
BACKUP *.* B: /F
```

You will be prompted for the insertion of more diskettes until the backup procedure is complete. Do not forget to label the diskettes in consecutive order.

You can take these diskettes to a computer with a corresponding diskette drive, create a directory \IQ and in this directory give the command

```
RESTORE B:
```

You will be prompted for the insertion of all diskettes and after completion of the restore procedure you will again find all necessary files in the directory \IQ.

Appendix C: Lists of All Provided Files

C.1 Command Files

File Name	Contents
IQ.BAT	DOS commands to start **IQ**
INSTALL.BAT	DOS commands for installation of **IQ**

Command files are formatted sequential files which can be printed or listed using an operating system command. The file INSTALL.BAT will not be copied to the hard disk during the installation procedure.

C.2 Program File

File Name	Contents
IQEXE.EXE	**IQ** program

The program file is unformatted. It cannot be printed or listed.

C.3 Descriptor Files for Examples and Exercises

File Name	Number of Descriptors	Topic
IQ012.DES	11	Free particle motion in one dimension (Ch.2)
IQ010.DES	19	Bound states in one dimension (Ch.3)
IQ011.DES	43	Scattering in one dimension (Ch.4)
IQ040.DES	10	Two-particle system (Ch.5)
IQ032.DES	6	Free particle motion in three dimensions (Ch.6)
IQ030.DES	13	Bound states in three dimensions (Ch.7)
IQ031.DES	19	Scattering in three dimensions (Ch.8)
IQ000.DES	24	Mathematical functions (Ch.9, Appendix A)

Descriptor files are unformatted direct-access files which can be printed or changed only from **IQ**.

C.4 Command Input Files and Associated Descriptor Files for Demonstrations

Input File Name	Descriptor File Name	Topic
IQD12.INP	IQD12.DES	Free particle motion in one dimension (Ch.2)
IQD10.INP	IQD10.DES	Bound states in one dimension (Ch.3)
IQD11.INP	IQD11.DES	Scattering in one dimension (Ch.4)
IQD40.INP	IQD40.DES	Two-particle system (Ch.5)
IQD32.INP	IQD32.DES	Free particle motion in three dimensions (Ch.6)
IQD30.INP	IQD30.DES	Bound states in three dimensions (Ch.7)
IQD31.INP	IQD31.DES	Scattering in three dimensions (Ch.8)
IQD00.INP	IQD00.DES	Mathematical functions (Ch.9, Appendix A)

Input files are formatted sequential files which can be printed or listed using an operating system command and changed using an editor.

C.5 Data File Specifying Graphics Devices

File Name	Contents
IQINI.DAT	File used by **IQ**

The data file is a formatted sequential file which can be printed or listed using an operating system command and changed using an editor.

C.6 Help Files

File Name	Contents
HELPIQ.HLP	Information on general **IQ** commands
PHYSIQ.HLP	Information on physics-related **IQ** commands

Help files are formatted sequential files which can be printed or listed using an operating system command.

Appendix D: Graphics Devices and Metafiles

While working directly with **IQ** you may get your graphics on a *display screen*, on a *mechanical plotter* using the HPGL command language or you can write the plot information on disk as a POSTSCRIPT file for later printing. You may use a maximum number of 3 *devices* at a time: one display screen, one plotter and one metafile.

Table D.1 is a listing of the file IQINI.DAT which provides **IQ** with the information about all allowed plotting devices. The first line contains 3 integers which are the *number*

Table D.1. Listing of the File IQINI.DAT

```
5,1,1
1 vga           0 112    640    480    0.00000    0.00000 14 15  3 12 12 11  2 10
2 8514/A         0 1    1024    768    0.00000    0.00000 14 15  3 12 12 11  2 10
3 HERCULES       0 1     720    348   -4.00000   -3.00000  1  1  1  1  1  1  1  1
4 HP-GL PLOTTER  0 1   10900   7650   27.25000   19.12500  1  2  3  4  5  6  7  8
5 Metafile       0 1    6850   4750   28.99000   20.11000  2  1  4  3  3  2  3  1
```

of devices in the table, the device number of the *default device*, i.e., of the device that is activated when you start **IQ** (usually your display screen) and the device number of the device for which *color changes* are made through the commands C1 and C2 unless you specify another device by setting n_{PD} see Sect. A.5.1.

The following lines (one per device) contain the *device number*, the *device name*, the *device type*, the *device installation status*, the *video mode* (hexadecimal) for **vga**-type devices, the *device surface size* in internal device units and in centimeters (for plotters), and *default color indices*. Most of these quantities are explained in Appendix A.5.4. The remaining ones (video mode and default color indices) will be discussed below as properties of the individual devices.

Before starting to work with **IQ** you may want to change the contents of the file IQINI.DAT using the editor of your computer. Since you will have only one display screen on your computer you should set the device installation status of the other two displays in the list to zero. (This is not really necessary, but if you choose a nonexisting device which has installation status one, **IQ** will be aborted.) Similarly, if you have no HPGL plotter, you should set its installation status to zero. If you have a display with **vga** graphics the corresponding line in IQINI.DAT is set up properly. For **ega** and **cga** you have to change the video mode and the default color table, see below. Finally, for each device you may change the default color indices, i.e., the last eight numbers in the line for that device. They are the color codes as used in the graphics language of that device which **IQ** uses normally for item 1 through item 8 of Table A.3. For a change of colors from these default values see Appendix A.5.1.

We conclude this appendix by giving a few details for the different graphics devices you may use.

vga Display:

As already stated, line 2 of the file IQINI.DAT is set up properly for vga graphics. This is, for instance, the standard graphics mode for the IBM PS/2 models 50 and upwards. Table D.2 is the default color palette for vga, ega and 8514/A. The default color indices in IQINI.DAT are 14 for item 1, 15 for item 2, etc. Thus item 1 is shown in yellow, item 2 in white, etc. You can change these colors in an individual plot with the commands C1 and C2, which replace the default color indices by input values that you can choose from the palette. You can also change the default color indices in IQINI.DAT and thus affect all plots. (Technical remark: if you have the hardware for vga graphics with higher resolution, replace the video mode 12 by the proper value and replace the two integer numbers 640 and 480 by the number of pixels in the horizontal and vertical direction, respectively.)

Table D.2. Default Color Palette of vga, ega and 8514/A Displays

Color Index	Color (Grey Level)	Color Index	Color (Grey Level)
0	black (0)	8	gray (14)
1	blue (5)	9	light blue (24)
2	green (17)	10	light green (45)
3	cyan (28)	11	light cyan (50)
4	red (8)	12	light red (32)
5	magenta (11)	13	light magenta (36)
6	brown (20)	14	yellow (56)
7	white (40)	15	intensive white (63)
The gray levels refer to a monochrome screen and a scale between 0 and 63			

ega Display

If you have an ega display replace line 2 of IQINI.DAT by

```
1 ega             0 110    640    350   -1.37000  -1.00000 14 15  3 12 12 11  2 10
```

The two numbers with minus signs signal the nonquadratic pixel size.

cga Display

Replace line 2 of IQINI.DAT by

```
1 cga             0 1 6   640    200   -2.40000  -1.00000  1  1  1  1  1  1  1  1
```

You have only two color indices: 0 (black) and 1 (white)

mcga Display (IBM PS/2, model 30)

Replace line 2 of `IQINI.DAT` by

```
1  mcga          0 111    640    480    0.00000    0.00000  1  1  1  1  1  1  1  1
```

You have only two color indices: 0 (black) and 1 (white)

8514/A Display
(High-Resolution Graphics Adapter on the IBM PS/2)

Line 3 of `IQINI.DAT` is set up properly for this device.

HERCULES Graphics Display

Line 4 of `IQINI.DAT` is set up properly for this device. You have only two color indices: 0 (black) and 1 (white)

HPGL Plotter

Line 5 of `IQINI.DAT` is set up properly for a mechanical plotter using the HPGL language. Note, however, that such plotters exist with different accuracy and for different paper formats. If you are not satisfied with plots obtained by the settings in Table D.1 change the numbers 10900 and 7670 to the maximal length and width of a plot in the internal units of your plotter and the numbers 27.25000 and 19.12500 to the corresponding lengths in centimeters. A color index is simply the number of a pen in the pen magazine of the plotter.

Metafile

Rather than use the graphics information generated by **IQ** to steer a graphical output device directly, you can also write it into a data *metafile* for later use. **IQ** can write a metafile in POSTSCRIPT which is a device independent graphics language. You activate POSTSCRIPT just as any other device using the CP command, see Appendix A.5.4. Together with the activation, a file `IQ000.MET` is opened. If such a file already exists, you are warned and can decide whether you want to overwrite that file. You may at any time open a file with a different name by typing

    CM

(choose metafile) and you are then prompted for the file name. You may display the current metafile name by

    SM

(show metafile).

Line 5 of the file `IQINI.DAT` is set up properly for POSTSCRIPT on a printer with A4 paper. For a printer with larger paper adjust the plot size in internal units (1/300 of an inch) and in centimeters accordingly. The color indices are interpreted as line widths. For color index 0 no line is drawn. An index $n > 0$ signifies a line width of approximately $n \times 25.4/300$ mm.

After termination of **IQ** you can produce the graphical output corresponding to the information in the metafile(s) by printing the file(s) `IQ???.MET` on a printer (which of course has to be able to understand POSTSCRIPT) following the instructions provided with that printer.

Index of IQ Commands

Subject Index